Wireless Sensor Networks

Wireless Sensor Networks

Edited by **Kevin Merriman**

Clanrye International
New Jersey

Published by Clanrye International,
55 Van Reypen Street,
Jersey City, NJ 07306, USA
www.clanryeinternational.com

Wireless Sensor Networks
Edited by Kevin Merriman

International Standard Book Number: 978-1-63240-551-7 (Hardback)

Printed in China.

Contents

Preface

This book has been an outcome of determined endeavour from a group of educationists in the field. The primary objective was to involve a broad spectrum of professionals from diverse cultural background involved in the field for developing new researches. The book not only targets students but also scholars pursuing higher research for further enhancement of the theoretical and practical applications of the subject.

This book provides comprehensive insights into the field of wireless sensor networks. It traces the progress of this field and highlights some of its key concepts and applications. Wireless sensor networks refer to the sensors which are used to monitor and collect data in order to save it in a central location. They have varied applications such as measuring environmental conditions like humidity, temperature, wind speed, or detecting air pollution, landslides, natural disasters, water quality, etc. This book will discuss in detail the various applications of WSNs and will provide in-depth knowledge of different techniques used in this area. The various studies that are constantly contributing towards advancing technologies and evolution of this field are examined in detail in this text. Students, scientists, engineers, researchers, etc. will benefit alike from this text.

It was an honour to edit such a profound book and also a challenging task to compile and examine all the relevant data for accuracy and originality. I wish to acknowledge the efforts of the contributors for submitting such brilliant and diverse chapters in the field and for endlessly working for the completion of the book. Last, but not the least; I thank my family for being a constant source of support in all my research endeavours.

Editor

1

A Distributed Trust Based Secure Communication Framework for Wireless Sensor Network

Geetha V.[1], K. Chandrasekaran[2]

[1]Department of Information Technology, National Institute of Technology, Karnataka, Surathkal, Karnataka, India
[2]Department of Computer Science and Engineering, National Institute of Technology, Karnataka, Surathkal, Karnataka, India
Email: geethav@nitk.edu.in, kch@nitk.ac.in

Abstract

The wireless sensor network is an emerging technology, which is used to sense and monitor the environment. As the nodes are deployed in an open environment, the security is one of the essential factors. The cryptography techniques can ensure confidentiality, integrity and authentication. However, wireless sensor network also needs to deal with inside and outside attackers. To deal with outside attackers, attacks by compromised or malicious nodes, trust management system is suggested by many researchers in the area of wireless sensor network. Trust management system can be implemented in various applications for security management such as secure data aggregation, secure cluster head selection, trusted routing, access control, etc. Many researchers provide different kind of solutions for these secure applications based on trust management. However, to incorporate, all such applications on a single sensor node in the network, it is essential to design and develop a trust management system, which considers various aspects and applications of wireless sensor network. As a result, in this paper, we would like to propose a parameter and trust factor based secure communication framework and design a trust management system for wireless sensor networks. Our main contribution is to identify various parameters and trust factors which influences on trust in wireless sensor network and developing a framework for a trust management system based on various parameters and trust factors. The working of the proposed model is shown by simulation experiments conducted in MATLAB for the application of secure communication, data aggregation and intrusion detection in wireless sensor networks.

Keywords

Wireless Sensor Network, Trust Management System, Black Hole, Sinkhole, DOS, Trust Factors

1. Introduction

The Wireless Sensor Network (WSN) is an emerging technology where a number of tiny nodes are deployed in an open environment to sense various phenomena [1]. As the sensor nodes are resource constraint and are deployed in an open environment, the nodes are more prone to inside and outside attacks. Cryptography technique can ensure authentication, confidentiality and integrity. However, to deal with outside attacks such as black holes, sink hole, Denial of Service (DOS) attacks, researchers propose a trust based system. Trust management has proven good results in other network areas such as social networks, ad hoc networks and P2P networks. The trust models and methods which are applicable to other networks are not directly applicable to wireless sensor network, as WSN is a resource constraint network.

Trust in WSN can be viewed as communication trust and data trust or node trust, path trust and service trust. The following definition of trust provides general factors needs to be addressed in wireless sensor network.

"*Trust is a subjective opinion on the reliability of the other entities or functions, including veracity of the other data, connectivity path, processing capability of the node and availability of service etc.*"

The trust is subjective, non-transitive, reflexive and asymmetric in characteristics. The trust between node increases if the node performs every action according to the specific rules of networking. If a node violates the rules of the network, then the nodes must be identified as malicious nodes and further eliminate these nodes from further communication in the network. The node can build the trust based on its direct observation and recommendation. The recommendation, trust ensures convergence of trust values faster based on neighbor recommendations. However, these recommendation systems are more helpful, when the topology of nodes changes dynamically. The exchange of recommendation information in secure way increases communication cost. As a result, it is better to use direct observation based trust for calculating the trust of a node in the network.

Belief is the probability of a node that decides the level of trust. For example, 0 indicates complete distrust and 1 indicates complete trust. The expected probability of belief is trust and actual probability is said to be trustworthy. Miscalculation of trust and trustworthiness difference will leave scope for poor risk estimation over vulnerabilities. In some cases, trust alone is not sufficient in all operations. However, risk, quality of service and trust need to be dealt separately before they are included in the trust computation.

Many researchers propose trust model and trust management system for one particular layer of the network protocol stack. But, the trust management system must deal with various kinds of application in each layer of the protocol stack. Most of the time, the trust management system considers very few parameters and trust factors for evaluation of trust. A trust management framework must consider various parameters and trust factors associated with building trust among nodes in wireless sensor network.

In trust management system, each sensor node must observe its neighbor based on various parameters such as packets forwarded, broadcast packets, etc. Based on these observed parameters the trust factors are evaluated. The combined value of these trust factors ensures the trustworthiness of a neighbor node. The framework proposed in this paper considers these aspects in developing a trust management system.

The main questions to be addressed are: How best does the trust model detect malicious nodes? How best does it take action on detection of malicious nodes? What is the rate of false positive and false negative? What is its impact on wireless sensor network? etc. In this paper, we propose a framework for trust based secure communication in wireless sensor network. The rest of the paper is organized as follows: Section 2 discusses about related work in the area of trust based framework proposed for various kinds of network. Section 3 presents our proposed trust based secure communication framework for wireless sensor network. Section 4 discusses about simulation results and discussion, followed by conclusions and reference.

2. Related Work

The researchers proposed various frameworks for trust based secure communication for different kinds of networks such as a social network, ad hoc network, p2p network and wireless sensor networks. Farruh Ishmanov *et al.*, [2] provides a detailed insight on the trust management system in wireless sensor networks. They have discussed about the importance of trust management in wireless sensor network as well as compared various kinds of trust models. They have also listed various open research issues such as monitoring and learning, trust evaluation, trust propagation, attack resistance and performance comparison of trust management systems. Since, wireless sensor network itself is an emerging area, the trust system for wireless sensor network is still new which needs further improvement in various aspects. As a result, the trust management system essentially has a major role in providing secure communication. Fenye *et al.*, [3] proposes a highly scalable cluster-based hierarchical trust management protocol for wireless sensor networks to effectively deal with selfish or malicious nodes. The trust is evaluated based on multiple attributes. Ganeriwal *et al.* [4] Proposed reputation based framework for data integrity in wireless sensor networks. The proposed reputation system takes information collected by each node using a watchdog mechanism to detect invalid data and uncooperative nodes. Yao *et al.* [5] proposed a parameterized and localized trust management scheme for WSN security, particularly for secure routing, where each node maintains a highly abstracted parameter to evaluate its neighbors. Junqi Duan *et al.* [6] provides a framework for trust aware secure routing. They have given a framework for considering a trust model for routing protocol. Javier Lopez *et al.*, [7] provides a list of best practices for developing a good trust management system for wireless sensor network. Chen *et al.*, [8] provides an event based trust management framework model. The trust is calculated based on event occurs and confidence. Most of the trust model focus on one type of application of trust management systems. It is essential to provide a framework, where the trust management system is considered across various layers of the protocol stack and which is applicable for detecting various kinds of attacks.

3. Proposed Trust Based Secure Communication Framework for Wireless Sensor Network

Secure communication is one of the essential factors in wireless sensor network as the nodes are deployed in an open environment. Each node in the network, must be capable of identifying malicious nodes, securely aggregate the data, identify trustworthy nodes in its neighbor, and cooperate for all the activities of the network with its group. We propose, a parameter and trust factor based trust management system for wireless sensor network. The system design of the trust management system is shown in **Figure 1**.

The trust management system is not a single layer in the protocol stack of wireless sensor network. We propose that the trust management system must be designed across the layers, as the security has to be ensured in all layers of wireless sensor network. As a result, the protocol stack, Application, Transport, Network, Data link and Physical layer, interacts with the trust management system as shown in **Figure 1**.

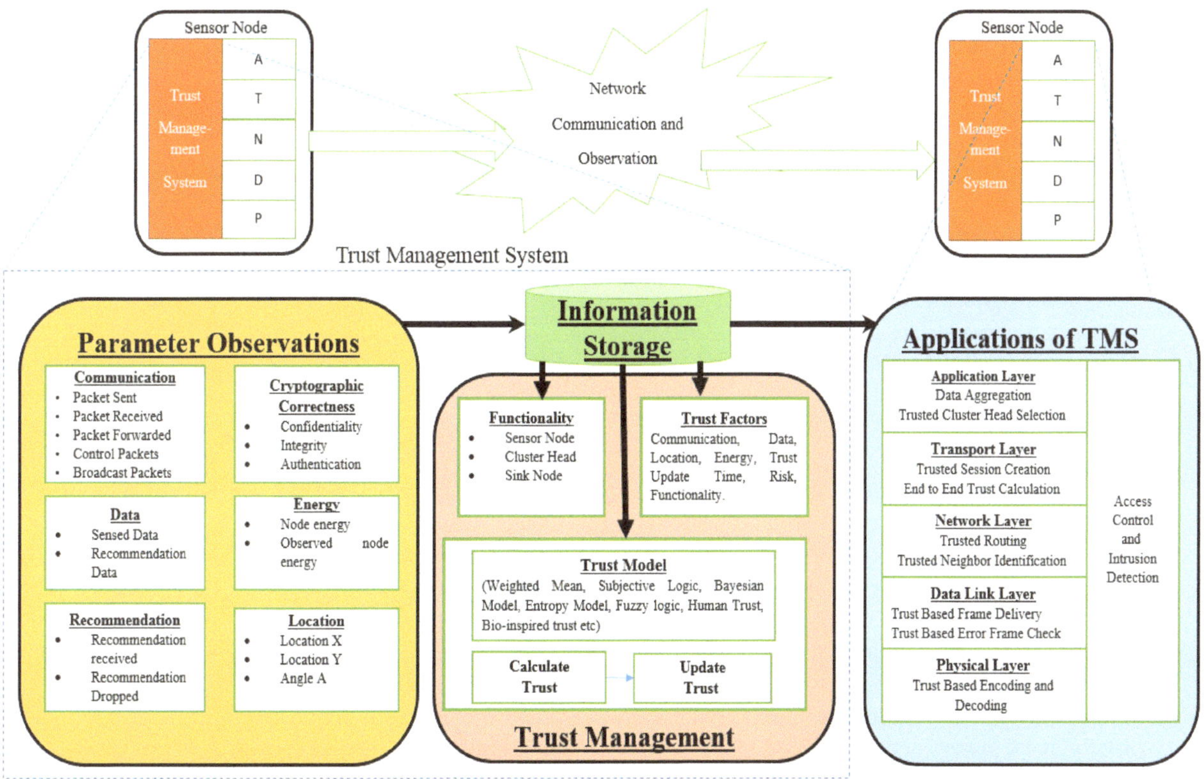

Figure 1. Proposed trust management system for wireless sensor networks.

Node A can communicate to Node B over the wireless media. Each event is observed in the network for developing trust on neighbor nodes. The proposed trust management system, mainly consider following components.

- Parameter Observation
- Trust Management
- Information Storage and
- Trust Applications

Each component is discussed in detail in following subsections.

3.1. Parameter Observation

To calculate trust between two nodes, a node *i* has to observe its neighbor *j* for its interactions with node *i*. Every event *e* is observed on the network. We have identified total six groups of data to be observed on the network. They are Communication, Data, Recommendation, Location, Energy, and Cryptographic Correctness.

- Communication: The node *i* observes node *j* for all the events related to communication. The parameters associated with *communication* are as follows: Number of packets sent (PS), Number of packets received (PR), Number of packets forwarded (PF), number of control packets sent (PCS), number of control packets received (PCR) and number of broadcast packets received (PB), number of data packets sent (PDS), number of data packets received (PDR).
- Data: The data of sensor node contain two parts. Either a node has to store its own sensed data or the data which it has forward towards sink node. The sensor node can sense static data or multimedia data. The information about data has to be stored for further processing. If the trust management uses indirect trust, then, the recommendation sent by neighbor nodes can also be treated as data.
- Recommendation: The number of recommendation packets sent and received also helps to monitor, the neighbor node with respect to the node behavior in sending recommendations.
- Location: In most of the routing protocols, the protocol considers geographical location based routing. A node may lie on its location information. To monitor such events, a node can request location information of node *j*, either from node *j* itself or node *i* can find out based on node *j*'s received signal strength.
- Energy: Energy is one of the essential aspect of wireless sensor network, as the nodes are resource constraint.
- Cryptographic correctness: This parameter is used to check, whether a node in the network is behaving properly according to cryptographic rules.

3.2. Trust Factors

The trust of a neighbor node is calculated based on evaluation of trust factors. Each trust factor is evaluated based on observed parameters. We have identified seven trust factors which mainly influence on trust of a node. Each trust factor is a function of a set of parameters.

- **Communication Trust:** This communication trust factor is a function of parameters observed for communication behavior. The communication trust factor is evaluated based on parameters as follows:

  ***Communication Trust* (*PS*, *PR*, *PF*, *PCS*, *PCR*, *PB*, *PDS*, *PDR*);**

  The communication trust factor may consider weighted average technique to evaluate all parameter values to communication trust value.
- **Data Trust:** The trust factor data is used to calculate trustworthiness with respect to data in the network. Attacks such as the stealthy attack can effect on the data aggregation. Similarly, in case of recommendation based trust calculation, the verification of recommendation data is also essential. As a result, the trust factor data contains two subcomponents: *Sensed data* and *Recommendation data*. The trust factor *Data Trust* is evaluated based on these two parameters as follows:

  ***Data Trust* (*Sensed Data*, *Recommendation Data*);**

  The data trust can provide certain weights for *sensed data* and *recommendation data*. If recommendation system is not used, then only *sensed data* information can be used to calculate Data Trust.
- Functionality Trust: Based on topology the wireless sensor network can be classified as flat based network or hierarchical based network. For flat based network topology, the nodes are of two types. 1) Sensor node which senses the data and routes it to sink node and 2) Sink node which collects data from sensor nodes. In case of hierarchical based network topology, the nodes are of three types: 1) sensor node which senses the

data and routes it to cluster head; 2) Sink node which collects data from sensor nodes and 3) cluster heads, which collects data from sensor nodes in its region, aggregate data and forwards it to sink node. Based on their functionality, the trust of each node can be calculated. As a result, the functionality trust is a trust factor which contributes to trust of a sensor node. The parameters of functionality are derived from other observed parameters such as communication parameters and data parameters, etc. As a result, in **Figure 1**, the functionality is shown with separate block, where it is derived from the observed information stored in Information storage. The functionality trust is evaluated based on observed functional parameters as follows:

Functionality Trust (Sensor Node, Cluster Head, Sink);

If a node is not able to perform its functions in the network properly, those nodes a can be eliminated fro network based on functionality trust.

- Location Trust: The node trust has to be evaluated based on location information, if the routing algorithm is based on geographical location. The location trust is one of the factors for evaluation of trust of a neighbor node. Based on the current location information obtained, the node has to calculate its trust. The location trust is calculated based on parameters of LocX, LocY and Angle observed or explicitly obtained by the neighbor node as follows:

Location Trust (LocX, LocY, Angle);

If the location informed by node *j* is same as location information calculated at node i, then node can be considered as trustworthy.

- Energy: Energy is one of the major factors in trust calculation. A neighbor node may have enough energy and may not cooperate for network functions which clearly indicates the node is a malicious node. A node may not be functioning properly as it is not having sufficient energy for communication. Care must be taken for not to take actions of the malicious node detection, in case of node is dead in the network.

Energy Trust (Current Energy);

- Trust Update Time: This trust factor affects overall performance of a trust management system. If *trust update time* is less, most of the time a node may be busy in trust management rather than packet transfer. If the *trust update time* is large then the malicious nodes may get advantage of this and its activity may down the entire network. We consider that instead of having static *trust update time*, dynamically changing trust update time is better. The interval of trust update time can be considered based on the rate at which the events are occurring in the network.

Trust Update Time (Event occurrence Time);

- Risk: Risk is the factor which is related to every other trust factor. If a certain amount of information is not available about the neighbor, then the factor of *Risk* has to be considered for evaluation of trust factors. The input for *Risk* function contain all other six trust factors as follows:

Risk (Communication Trust, Data Trust, Functionality Trust, Location Trust, Energy Trust, Trut Update-Time).

3.3. Trust Model

Trust models are used to evaluate trust factors based on various theoretical concepts. The most widely used trust models are weighted mean, Bayesian model, subjective logic, entropy based model, fuzzy logic based model, game theoretic based model, human trust model, bio inspired models etc. The trust model mainly contains two major functions. 1) Calculate trust: The trust value is calculated based on trust factors and trust model 2) Update trust: the trust value is updated in the information storage.

We have considered only one type of trust model in this paper where other models can also be applied to the proposed trust management systems. The trust is defined as a belief level that one node can put on another node for specific action according to previous direct or indirect information of observations on behaviors. The belief level is the extent that one node believes that another node is willing to and able to obey the protocol and act normally. Let us consider an example, as a Bayesian based trust model. It is assumed that the subject node believes the object node behaves normally with probability θ, which can also be described as $p(\text{Belief})$. Here *Belief* denotes the trust in a node to perform normal behavior. Also, we use *Observation* to represent the observations one node obtains on another node. Then similar to [4], the formula for the standard Bayesian approach to be used for trust management can be provided as follows:

$$p(\text{Belief}/\text{Observation}) = \left[p(\text{Observation}/\text{Belief}) \cdot p(\text{Belief})\right] / \text{Normalizing constant} \quad (1)$$

where *p*(Belief) is the prior probability, *p*(observation/Belief) is the posterior probability, *p*(Observation/Belief) is the likelihood function, and *p*(Belief/Observation) is the posterior probability.

Based on Ganeriwal *et al.*, [4] trust model, it can be identified that, the probability of succession can be obtained by Bayesian inference, by observing on two parameters α and β. The expected value can be obtained as $\alpha/(\alpha+\beta)$ according to Baye's and $(\alpha+1)/(\alpha+\beta+2)$ according to Laplace Law, which considers that at least one "success" and one "failure" were observed before observing n trials where $n = (\alpha+\beta)$.

A node will observe a neighboring node's behavior and build a trust for that node based on the observed information. The neighboring node's transactions are direct observations referred as first-hand information. For each observation, the node i maintains two parameters α and β which indicates the number of "successful" and "unsuccessful" operation by a neighbor node j.

$$T_{ij} = (\alpha+1)/(\alpha+\beta+2) \tag{2}$$

The communication trust denoted as T_{ij}, is initialized to 0.5 based on Laplace Law. The Trust is calculated as shown in Equation (2) where α and β represents the number of "successful" and "unsuccessful" cooperation by node j to node i respectively.

As the sensor nodes are resource constraint, maintaining the history of all observed trials is resource consuming. To solve this issue, the α and β are updated periodically, based on r and s where s indicates number of "successes" and r indicates number of "unsuccessful" cooperation in a time window t.

$$\alpha_j = \alpha_j + r \quad \text{and} \quad \beta_j = \beta_j + r \tag{3}$$

Then α_j and β_j can be updated as shown in Equation (3).

$$\alpha_j = W_{age} \cdot \alpha_j + r \quad \text{and} \quad \beta_j = W_{age} \cdot \beta_j + r \quad \text{where} \quad 0 <= W_{age} <= 1. \tag{4}$$

As the data becomes old, the oldest information has to be discarded to provide a higher preference for latest information. Ganeriwal *et al.* [4] provides the concept of aging factor, W_{age} to update α_j and β_j as shown in Equation (3). Since it provides high weightage for past interactions, a node can perform ONOFF attack very easily. A ONOFF attack is one where a node behaves benevolent until it obtains a high trust value with good history of records, and then starts dropping packets (ON) and forwarding packets (OFF) periodically. As a result, the nodes can launch attacks even while maintaining its trustworthiness. To overcome such attacks, we have proposed exponential decrease Bayesian Trust model to detect ONOFF attack in [9].

3.4. Applications for Trust Management in Wireless Sensor Networks

The trust management system can be used for various applications in wireless sensor networks. The application of trust management related to each layer of the protocol stack is shown in **Figure 1**.

- Application Layer: The data aggregation and cluster head selection are two major applications of the trust management system in the application layer. In case of hierarchical networks, the cluster head collects the sensor data from all sensor nodes, which are joined to the particular cluster head. The secure data aggregation needs to check a data stealthy attack, and aggregate data of only trusted nodes. Many of the hierarchical protocols for WSN selects a new set of cluster head in each round. The trust management system helps to identify trusted nodes for cluster head selection, thereby increases the security of the network.
- Transport Layer: Even though for simple sensed static data based wireless sensor network, UDP is sufficient. In case of multimedia wireless sensor networks, it needs TCP for streaming. As a result, it needs trust for end-to-end communication and trusted session operations which can be provided by the trust management system.
- Network layer: Each node in the network, sends or forwards the packet to sink node. The trust management in network layer basically has two roles. 1) Trusted neighbor identification: To identify a trusted neighbor for one hop communication; 2) Trusted routing path selection: The routing path must contain trusted path for communication in the network. Trust management can help to identify trusted nodes and trusted routing path in the network.
- Data link layer: The sensor node may try to access the network continuously in case of guaranteed services in MAC protocol of network. The DoS attack and unfairness access to channel can be dealt with trust management systems.

- Physical layer: The nodes in physical layer are prone to various kinds of attacks. The intrustion detection and reliable packet transfer at physical layer are main applications which needs trust management systems.

The complexity of the proposed framework, lies in observing neighbors based on specified parameters. As the nodes are resource constraint, watchdog techniques are costlier in terms of energy dissipation. The trust factors an be calculated based on observed trust parameters. Since, we assume nodes in network are static, it's enough to calculate trust based on direct observation. To converge the trust values faster, or if nodes are mobile, then it needs to consider indirect trust or recommendation trust, which further increases communication cost, as nodes should share their recommendations among its neighbours.

4. Simulation Results Discussion

The proposed trust management system is implemented in MATLAB for data aggregation and trusted routing applications along with intrusion detection. We have used the first order radio model for energy calculation. The sensor nodes are assumed to be deployed in an open environment. We use a tree based topology for routing in sensor network. The sink initiates process of routing. Every node observes each interaction with its neighbors and calculates trust during certain intervals of time. The malicious nodes are eliminated for further interactions in the network. The simulation details are explained in detail in following subsections.

4.1. Topology Creation

The sink node sends a beacon node which can reach up to radius R, where R is the communication range of a sensor node. When a sensor node, receives a beacon node, it initiates neighbor discovery based on HELLO packets and updates it's neighbor information table. We consider all nodes which are at one hop distance to sink node are first level parent nodes or cluster heads. A node sends a JOINREQUEST packet for requesting for join operations. The parent replies with JOINREPLY along with TDMA schedule or CDMA code for further communication. Once the first level parent node joins to Sink node, then it broadcasts the information along with number hops to sink as one of the parameters. When a node receives these broadcast packets, then based on the nearest distance from sink node, a second level node joins to its parent and further continues the process of sending broadcast messages. This process continues until all nodes in the network joins to the network. **Figure 2** shows the topology created for 100 nodes deployed in the area of 100 × 100 m area. The sink node is located at the center, and JOIN operation of each node is shown with dotted lines between two nodes. The node towards sink node are parent nodes for children nodes in the network.

4.2. Data Aggregation and Data Transfer

After a node joins to its parent node towards the sink, the node starts sensing the channel and sending data. The one level immediate parent node aggregates the data from its child and sends it to sink node. Each node checks for a stealthy attack before aggregating the data. We use standard deviation and Bayesian based estimation for trust calculation. If a node's data are within the range of valid data range calculated by standard deviation, then the operation is considered as "successful" operation else it is considered as "unsuccessful" operation. By taking it as α and β respectively, the trust model explained in Section 3.3 is used for calculating a node's trust with respect to data aggregation.

4.3. Attacker Model

Simulation is conducted for three kinds of attacks related to communication: Blackhole attack, Selective forward attack, and ONOFF attack, and one type of attacks related to data: Stealthy attack. In case of blackhole attack, a node drops all packets received from its neighbor. A selective forward attack, forwards the received packet to sink with a probability p. The ONOFF attack is one where a node initially forwards all the packets until it obtains a high trust value among neighbors. Later, periodically it forwards data (attacker is OFF) and drops the packets (attacker is ON). Data stealthy attack is a kind of attack where a node sends either very low or very high value to the cluster head instead of actually sensed value. This kind of attacks effects data aggregation value at cluster head.

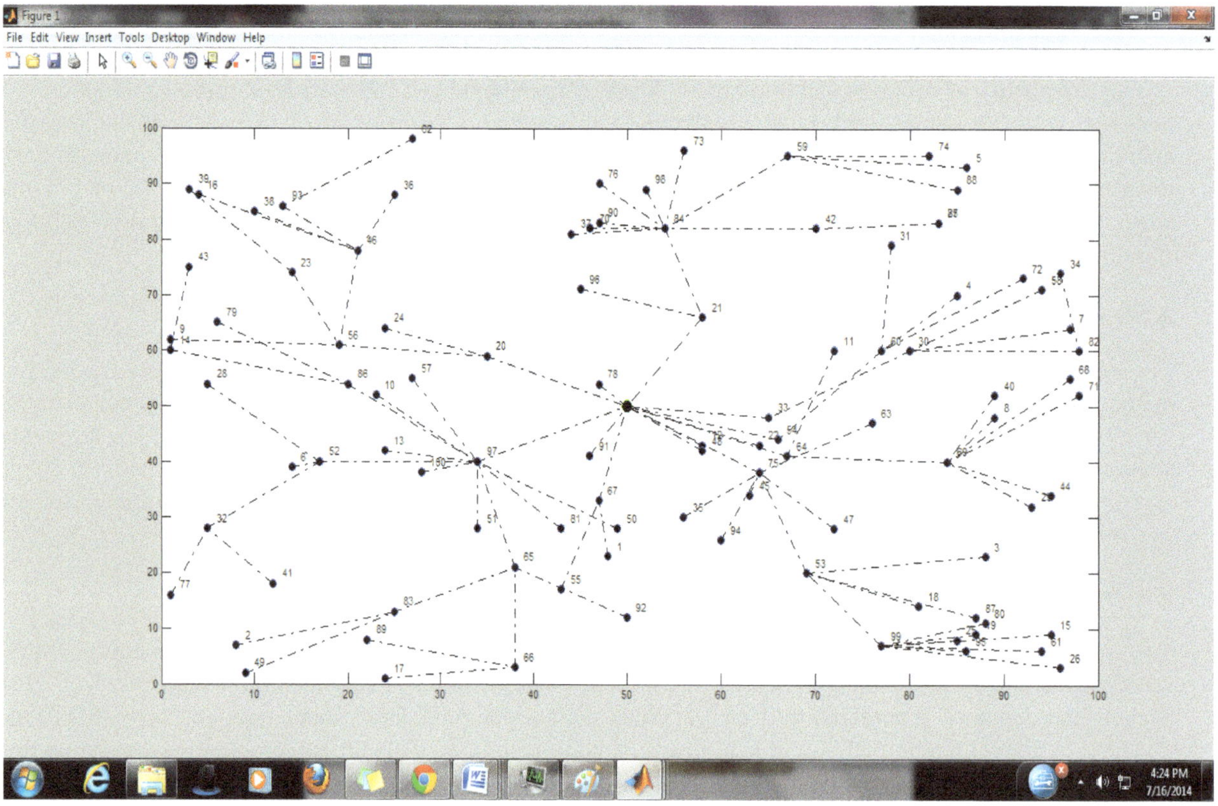

Figure 2. Topology of wireless sensor network for simulation.

4.4. Results and Discussions

The simulation is carried out for trust based secure communication and data aggregation along with intrusion detection for black holes, selective forward, ONOFF and stealthy attacks. The results are analyzed for 10%, 20% and 30% attacker nodes. The metrics used for comparison of results is based on True Positive (TP), True Negative (TN), False Positive (FP) and False Negative (FN). The results are compared to standard Bayesian trust model (STM) with exponential decrease Bayesian trust model (ETM).

Table 1 shows the results for Black hole attack. In case of black hole attack, since all the packets are dropped by an attacker, the standard Bayesian trust model and proposed Exponential decrease Bayesian Trust model are able to identify the black hole attack 100% for scenarios with 10%, 20% and 30% attackers.

Table 2 shows the results for selective forward attack. The experiment is conducted for selective forward attack, where a node forwards the packet with a probability of 0.6. The results show that the exponential decrease Bayesian trust model is having 100% attacker detection compared to the standard Bayesian trust model.

Table 3 shows the results for ONOFF attack. ONOFF attack is an attack, where a node initially tries to establish high trust value in its neighbors and then starts dropping packet (ON) and forwarding packet (OFF) for a certain period of a time in the fashion of cycles. We can observe that, the standard Bayesian trust model, does not detect ONOFF attack, where the exponential decrease Bayesian model detects ONOFF attack with 100%. **Figure 3** shows percentage of True Positive (TP) for various kinds of attacks, with Standard and Exponential decrease Bayesian trust models. The results show that Exponential decrease trust model, performs better than Standard Trust model.

Table 4 shows the results of simulation with data stealthy attack. A stealth attack is one where a node simply sends either low or very high value as sensed data. This results in variation of aggregation value. The results are analyzed for 10%, 20% and 30% data stealthy attackers. The sensed value for the data stealthy attacker is analyzed for two different ranges. (i) Data with range (45 - 75) and (45 - 95). We can observe that the TP decreases as the number of attackers increases. As the number of nodes become stealthy attackers, the detection level decreases.

Table 1. Results of simulation with black hole attack.

Trust model	10% attacker nodes				20% attacker nodes				30% attacker nodes			
	TP	TN	FP	FN	TP	TN	FP	FN	TP	TN	FP	FN
Number of nodes in case of Standard Bayesian Trust Model	10	90	0	0	20	80	0	0	30	70	0	0
Number of nodes in case of Exponential Decrease Bayesian Trust Model	10	90	0	0	20	80	0	0	30	70	0	0

Table 2. Results of simulation with selective forward attack.

Trust model	10% attacker nodes				20% attacker nodes				30% attacker nodes			
	TP	TN	FP	FN	TP	TN	FP	FN	TP	TN	FP	FN
Number of nodes in case of Standard Bayesian Trust Model	9	90	0	1	18	80	0	2	27	70	0	3
Number of nodes in case of Exponential Decrease Bayesian Trust Model	10	90	0	0	20	80	0	0	30	70	0	0

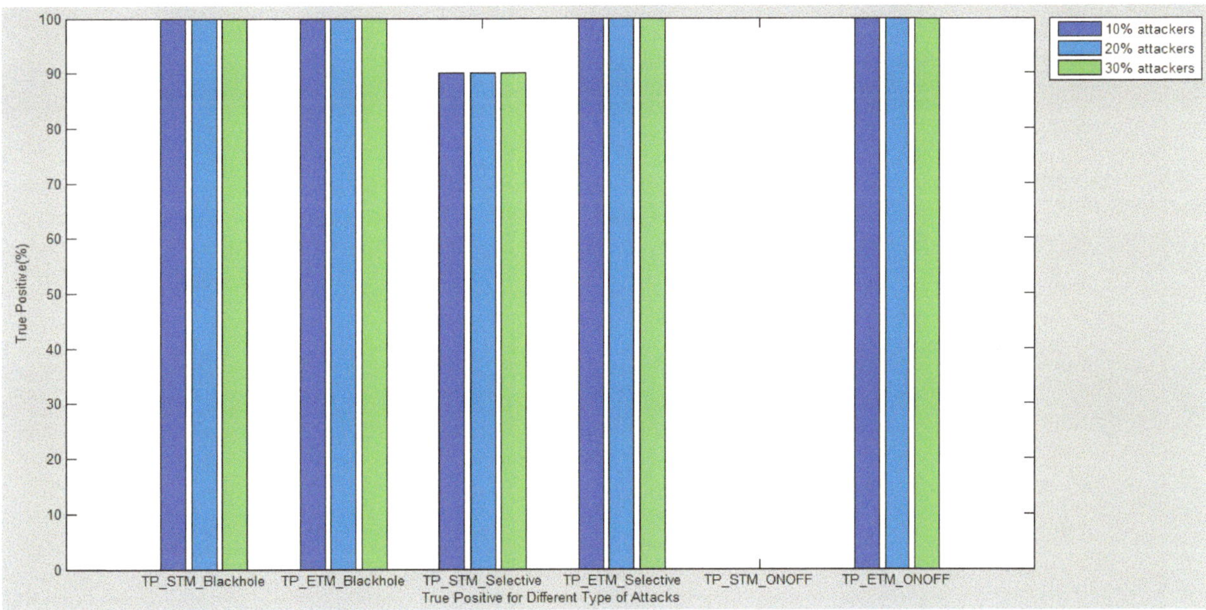

Figure 3. Analysis of True positive percentage for standard Bayesian trust model and Exponential decrease based trust model for various types of attacks.

Table 3. Results of simulation with ONOFF attack.

Trust model	10% attacker nodes				20% attacker nodes				30% attacker nodes			
	TP	TN	FP	FN	TP	TN	FP	FN	TP	TN	FP	FN
Number of nodes in case of Standard Bayesian Trust Model	0	90	0	10	0	80	0	20	0	70	0	30
Number of nodes in case of Exponential Decrease Bayesian Trust Model	10	90	0	0	20	80	0	0	30	70	0	0

The trust value of a benevolent node and a stealthy attacker is shown in **Figure 4** with respect data aggregation rounds. The result is taken for making a node as stealthy attacker after 20th round. The curve shows that the trust model decreases the thrust of the attacker, for each observation of stealthy attack hosted by a malicious node.

Trust management system, detects communication attacks such as blackhole attack (100%), selective forward attack (100%), ONOFF attack (100%). Data stealthy attack gets detected 100% if the number of attackers in the network are small. As the number of attacker nodes increases, the detection rate decreases (0%). It needs more sophisticated and robust technique for identifying data stealthy attacks. Our simulation has focused on attacker detection based on trust management system. Theoretically, detection of attacker and eliminating those nodes for further activities in the network, improves the performance, as the packets get transferred to sink node

Table 4. Results of simulation with stealthy attack.

Trust Model	10% attacker nodes				20% attacker nodes				30% attacker nodes			
	TP	TN	FP	FN	TP	TN	FP	FN	TP	TN	FP	FN
Number of nodes in case of Standard Bayesian Trust Model (data range 45 - 95)	10	90	0	0	19	80	0	1	15	70	0	15
Number of nodes in case of Standard Bayesian Trust Model (data range 75 - 95)	10	90	0	0	20	80	0	0	0	70	0	30
Number of nodes in case of Standard Bayesian Trust Model (data range 45 - 95)	10	90	0	0	20	80	0	0	30	70	0	0
Number of nodes in case of Standard Bayesian Trust Model (data range 75 - 95)	10	90	0	0	20	80	0	0	0	70	0	30

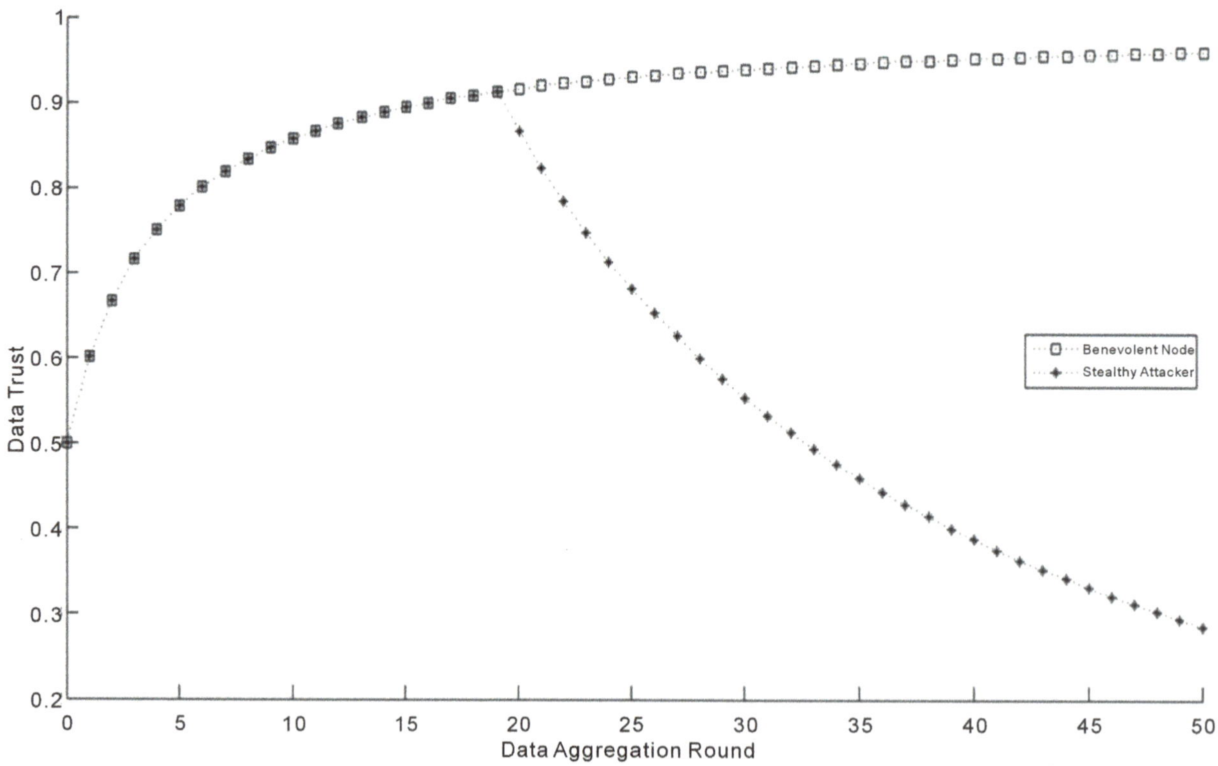

Figure 4. The trust value changes in benevolent and stealthy attacker for data aggregation rounds.

through trusted nodes in the network.

In summary, we have proposed a trust management system. The idea of simulation experiment is to show that the trust proposed trust management system can be used for various applications such as secure communication data aggregation, intrusion detection, etc. Even though we have analyzed very few applications here, we ensure that the trust management system along with proposed framework can be extended fro any kind of applications for secure communication in wireless sensor networks.

5. Conclusion

Wireless sensor network is an emerging technology for monitoring environment. The resource constraint sensor nodes are more vulnerable to attacks in wireless sensor networks as the nodes are deployed in open environment. A trust management system is essential for WSN, where cryptography techniques fail to address some issues. The frame work for secure communication and trust management systems are proposed in this paper. The simulation results show that the proposed model works for secure communication, data aggregation and intrusion. However, the proposed trust management system can be further extended for various other applications. This is an ongoing work, and as a future work we would like extend this work with more sophisticated trust models for various applications.

References

[1] Akyildiz, I.F., Su, W., Sankarasubramaniam, Y. and Cayirci, E. (2002) Wireless Sensor Networks: A Survey. *Computer Networks*, **38**, 393-422.

[2] Ishmanov, F., Malik, A.S., Kim, S.W. and Begalov, B. (2013) Trust Management System in Wireless Sensor Networks: Design Considerations and Research Challenges. *Transactions on Emerging Telecommunications Technologies*.

[3] Bao, F.Y., Chen, I.-R., Chang, M.J. and Cho, J.-H. (2012) Hierarchical Trust Management for Wireless Sensor Networks and Its Applications to Trust-Based Routing and Intrustion Detection. *IEEE Transactions on Network and Service Management*, **9**, 169-183.

[4] Ganeriwal, S., Balzano, L.K. and Srivastava, M.B. (2008) Reputation-Based Framework for High Integrity Sensor Networks. *ACM Transactions on Sensor Networks*, **4**, 1-37.

[5] Yao, Z., Kim, D. and Doh, Y. (2006) PLUS: Parameterized and Localized Trust Management Scheme for Sensor Networks Security. 2006 *IEEE International Conference on Mobile Adhoc and Sensor Systems* (*MASS*), Vancouver, October 2006, 437-446.

[6] Duan, J.Q., Yang, D., Zhu, H.Q., Zhang, S.D. and Zhao, J. (2014) TSRF: A Trust-Aware Secure Routing Framework in Wireless Sensor Networks. *International Journal of Distributed Sensor Networks*, Article ID: 209436. http://dx.doi.org/10.1155/2014/209436

[7] Lopez, J., Roman, R., Agudo, I. and Fernandez-Gago, C. (2010) Trust Management Systems for Wireless Sensor Networks: Best Practices. *Computer Communications*, **33**, 1086-1093.

[8] Chen, H.G., Wu, H.F., Hu, J.C. and Gao. C.S. (2008) Event-Based Trust Framework Model in Wireless Sensor Networks. *Proceedings of International Conference on Neworking, Architecture, and Storage*, 359-364.

[9] Geetha, V. and Chandrasekaran, K. (2013) Enhanced Beta Trust Model for Identifying Insider Attacks in Wireless Sensor Networks. *International Journal of Computer Science and Network Security* (*IJCSNS*), **13**, 14-19.

2

Detection of Insider Selective Forwarding Attack Based on Monitor Node and Trust Mechanism in WSN

Yu Hu, Yuanming Wu, Hongshuai Wang

School of Optoelectronic Information, University of Electronic Science and Technology, Chengdu, China
Email: 903821402@qq.com, ymwu@uestc.edu.cn, 478284495@qq.com

Abstract

The security problems of wireless sensor networks (WSN) have attracted people's wide attention. In this paper, after we have summarized the existing security problems and solutions in WSN, we find that the insider attack to WSN is hard to solve. Insider attack is different from outsider attack, because it can't be solved by the traditional encryption and message authentication. Therefore, a reliable secure routing protocol should be proposed in order to defense the insider attack. In this paper, we focus on insider selective forwarding attack. The existing detection mechanisms, such as watchdog, multipath retreat, neighbor-based monitoring and so on, have both advantages and disadvantages. According to their characteristics, we proposed a secure routing protocol based on monitor node and trust mechanism. The reputation value is made up with packet forwarding rate and node's residual energy. So this detection and routing mechanism is universal because it can take account of both the safety and lifetime of network. Finally, we use OPNET simulation to verify the performance of our algorithm.

Keywords

Insider Attack, Selective Forwarding, Trust Mechanism, Monitor Node, Secure Routing Protocol

1. Introduction

Wireless sensor network (WSN) is composed of a set of sensor nodes in the network. It has the features of self-organization, multiple-hop and limited energy resources, etc. In the WSN which is based on the center of data, because of jamming, collision and bad channel environment, the data packets which are far from the base station (BS) may lose or make mistakes in the route of being forwarded hop by hop. When the attacker intrudes the network, the forwarded data packets may be stolen, tampered, destroyed and discarded so that it makes more serious damage to the network. The attack types to the network include insider attack and outsider attack, but insider attack [1] is hard to defense effectively.

The classic encryption [2] and authentication mechanism can only resist outsider attackers who steal the data or tamper the information and route. However, to the insider attack, after the attackers capture the normal nodes, they acquire the identity and encryption information and put the malicious nodes to destroy the network. The malicious nodes work in the network with the legal identity so that the method of encryption and authentication can't defense the insider attack. Therefore, an Intrusion Detection System (IDS) [3] should be injected in WSN as the second line of defense. Such a IDS works monitors the nodes which behave abnormally, then decides whether these nodes are malicious nodes according to their behaviors, and finally takes some measures to deal with these nodes.

Selective forwarding attack [4] is such a form of attack which is hardest to resist among various kinds of insider attacks. The malicious node drops all the forwarded packets to launch a black hole attack. This attack can be detected easily. However, the malicious node can launch the selective forwarding attack more intelligently and covertly, such as taking part of packet loss, periodic packet loss and intermittent packet loss in order to elude various detection mechanisms. If malicious nodes are more difficult to be detected or take us more time to find them, they will do more damage to the network. Thus, we have to find a more effective detection mechanism to remit the damage of the selective forwarding attack. This kind of security mechanism should have the following features:

1) During the lifetime of the network, it can detect the existence of the malicious nodes accurately and quickly as well as it can determine the type of attack. It should also indicate the location and ID of the malicious nodes.

2) After the security mechanism has inspected the malicious nodes, it should provide the feedback information to the network and update the routing, and reconsider the forwarding nodes and forward path.

3) In consideration of limited conditions of sensor nodes, the complexity of the design scheme should be lightweight. The scheme should try to reduce the communication cost and the node energy consumption in different working mode.

4) The proposed mechanism should try to solve the means of attack as more as possible not only a single type of attack. Otherwise, it is lack of value.

For selective forwarding attack, most of the existing solutions to them are based on monitoring mechanism and reputation value estimation. Different designing scheme has different idea and performance. The traditional watchdog [5] mechanism overhears the packets forwarded from node N to decide whether N is a malicious node. The system calculates the forwarding rate as reputation value to compare with an appropriate threshold. Ultimately, it can estimate whether node N has launched selective forwarding attack or not through the result of the comparison. Although the computation complexity of watchdog mechanism is ultra-lightweight and it meets the characteristics of WSN, it can resist very limited situations of attack. Only when a single malicious node intrudes in the network and the node launches a simple selective forwarding, watchdog can detect the malicious node successfully. However, if multiple nodes make collusion attack or one single node does other forms of attack, simple watchdog mechanism cannot defense effectively. Other methods are improved on the basis of the watchdog which can resist more forms of attack and all have their characteristics and advantages, but they also have some drawbacks.

In this paper, a scheme based on the monitor node managing the network and the watchdog mechanism is put forward referring to the characteristics of different defense mechanisms. Moreover, unlike general arithmetic based on static threshold, our algorithm is contained of dynamic threshold. The threshold is adjusted in real time according to the overall packet loss situation of the network so that the loss detection rate and false alarm rate can be decreased.

The paper's other paragraphs are as follows. Section II summarizes related work, elaborates and analyzes the advantages and disadvantages of some popular schemes against the attack. Our secure mechanism and network design is proposed in Section III. Section IV explains the ways of attack which can be solved by our secure scheme. Section V presents the simulation performance and interpretation of results. At last, we conclude the paper in Section VI.

2. Related Work

In this section, the related works against the insider selective forwarding attack are summarized specifically. Nowadays, to defense and reduce this kind of attack, the main solution includes watchdog, trust mechanism, neighbor-based monitoring, multipath routing avoidance, multiple data flow scheme, etc. Such kinds of schemes show a good defending effect through continuous improvement. However, they are also hard to avoid exposing the defects of their own. So we should find a trade-off between their advantages and disadvantages in order to propose a more optimal scheme. We expound the characteristics of some mechanisms as following.

2.1. Watchdog

Watchdog monitoring mechanism was introduced by Marti *et al.* [5] to identify misbehaving nodes in wireless ad hoc networks. This scheme can be applied in WSN and it was the earliest trust mechanism which is the basis of many defense methods. In their approach, each sensor node has its own watchdog that monitors and records the behaviors of its one hop neighbors. The watchdog system of each node stores the routing table which is about the performance good or bad of the neighbor nodes. When the node M sends a packet to its neighbor node N, the watchdog of M verifies whether N forwards the packet toward the BS or not by using the sensor's over-hearing ability.

The advantage of this kind of security mechanism is that the principle is simple, complexity is low and it adapts to the WSN completely. But this scheme can only solve limited conditions of malicious nodes packet dropping. It can just detect the behavior that a single node drops packet in the simple form, but it can't defend that more than two nodes collude to launch packet loss attack.

2.2. Trust Mechanism

On the basis of Watchdog scheme, Yu *et al.* [1] describe several representative approaches to build a trust model. In their paper, Bayesian approach, Entropy approach, Game-theoretic approach, and Fuzzy approach are proposed. Through the different algorithms, we convert the information which comes from the watchdog to statistical data in order to weigh the behavior of monitored node whether good or bad. The design of trust measurement is aimed at optimizing the condition of determining a forwarding node. We expect to choose the node whose all kinds of attributes are comprehensively perfect as the next-hop node.

Using different trust model, the reputation value of a node will be different. When the trust model is determined, we should set a reasonable threshold. If a node's reputation value below the threshold, it will be recorded as a malicious node. Moreover, depending on the WSN's trust mechanism, the detection of such malicious node will or will not be broadcast to the rest of the nodes and BS in the WSN.

2.3. Other Kinds of Defense Mechanism

1) Neighbor-based monitoring

Malicious node may be the intersection of several path, that is to say this kind of node may forward more than one source node's data. It can drop the data package of one or some node among those source nodes so that the estimation to this malicious node from each source node is different. Therefore, this kind of malicious node cannot be ruled out in time. Thus, a detection mechanism based on neighbor nodes monitoring was proposed. By listening to each other, each node takes statistics of the number of forwarded packets by neighbor nodes. Then, each node calculates the reputation values about its neighbor nodes. The information of all reputation values will be put together in order to compute each node's weighted credibility.

Several works [6] [7] that used neighbor-based approach have been introduced in order to mitigate selective forwarding attacks. Lu *et al.* [8] have proposed a neighbor-based monitoring mechanism. In their design, each node computes the trust value of its 1-hop neighbors based on their multiple behavior evaluation and builds a trust management so that it can confirm whether a node is a malicious node or not.

The benefit of this method is that we can find the captured node more accurately and faster through multiple nodes cooperation. However, there are some restrictions in these approaches. First, they do not describe how their approaches can counter a node's selective packet drops against source node. Secondly, because every node must overhear each other and calculate every neighbor node's trust value, it will need more communication overhead and computational cost. Thirdly, the conflict may take place when information is exchanged between each node.

2) Multipath-routing avoidance

No matter how optimizing the detection mechanism is, it always establishes on the situation that malicious nodes have already discarded a mass of data packets. Thus, if we want to make sure that the packets are forwarded to the base station perfectly, we should let the packets avoid the malicious node. Karlof and Wagner [9] mentioned k disjoint multipath routing avoidance can completely defend against selective forwarding involving at most k compromised nodes. Several works illustrating that multipath routing defends against packet dropping attack can be found in [10] [11]. Although using this mechanism can make the information transmitted to BS more effectively, the communication cost is very high. Obviously, if there is k path, it will need k times communication overhead compared with a single path.

3) Multiple data flow

Similarly, Sun *et al.* [12] introduced multiple data flow scheme using multiple disjoint topologies. In this approach, a node sends its packets through one or more randomly chosen topologies among the pre-established multiple topologies to mitigate insider packet dropping. The theory of this scheme is similar to multipath routing avoidance, but there is some trade-off between communication consumption and the intension of avoidance function.

3. Network Design

In this section, combining with the advantages of watchdog algorithm and node monitoring mechanism in related work, we propose a new network scheme. We have designed a kind of local network security routing algorithm based on trust mechanism and monitor node. The monitor node overhears the forwarding rate and surplus energy of ordinary nodes, and weights the two factors as the composite reputation value of each node in order to determine the routing node.

3.1. Network Topology

1) In a certain range, a number of ordinary nodes which are used for gathering around data and forwarding information are sprinkled random and uniformly. 2) Then monitor nodes are sprinkled random and uniformly which are a certain percentage of ordinary nodes. For example, we set 10 monitor nodes for 100 ordinary nodes. That is to say, each monitoring node can regulate 10 common nodes. 3) Monitor nodes broadcast information around in turn with a settled communication radius, to attract the ordinary nodes nearby. Each monitor node owns its monitor area and member nodes. 4) According to the location to the sink node, monitor nodes acquire ID and allocate ID for each ordinary node in their area.

3.2. Monitoring Mechanism

In this mechanism, the head of forwarding packet should include the ID of source node, ID of forward node and ID of next-hop. When the packet is forwarded, the monitor node overhears the packet and just takes out the head information to achieve the ID information as the determination whether the monitored node's behavior is normal or not.

We mainly care about the design of local safety monitoring in network, and focus on the research about how to choose the node of the highest reputation value, how to detect and exclude the malicious node, how to set of threshold and other problems. The schematic diagram of local monitoring is in **Figure 1**.This picture shows the workflow of monitoring nodes which is like that, it overhears forwarding packets of ordinary nodes, then calculate the reputation value of all the monitored nodes in this area, and find the node with the highest reputation value. When the routing node is confirmed, if there is no collecting data task in this area, other nodes can move to hibernation mode in order to save some energy. The process of monitoring mechanism is in **Figure 2**. Monitoring nodes only read the data according to the structure of different packets, that is to say, they only read the information of packet's head. So it can reduce the resource consumption of monitoring nodes. And some security detection function can be carried out through testing the header information.

3.3. Setting of Reputation Value

The reputation value of each node is as follows:

$$\mathrm{Val}\left[\mathrm{node}_{\mathrm{id}}\right] = a * \mathrm{Pr}_{\mathrm{id}} + b * \mathrm{Power}_{\mathrm{id}} \tag{1}$$

Where $0 < a < 1$, $0 < b < 1$, $a + b = 1$.

The $\mathrm{Pr}_{\mathrm{id}}$ in Equation (1) is the trust value of this node. The trust value of a node may be different when we

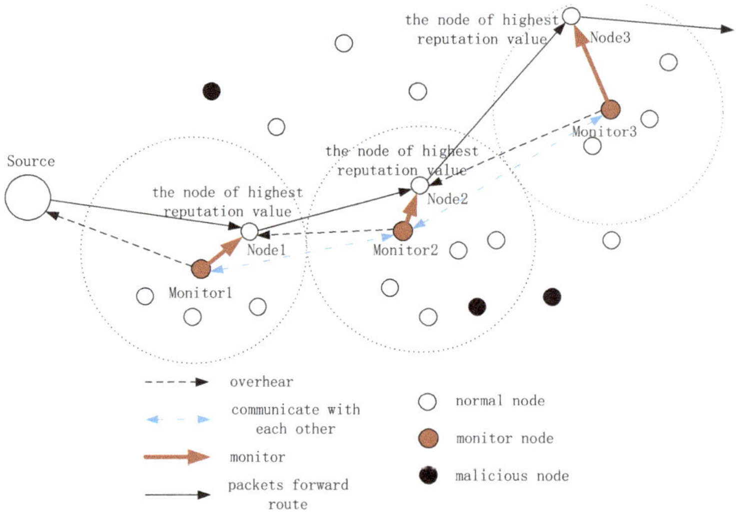

Figure 1. The local process of secure routing.

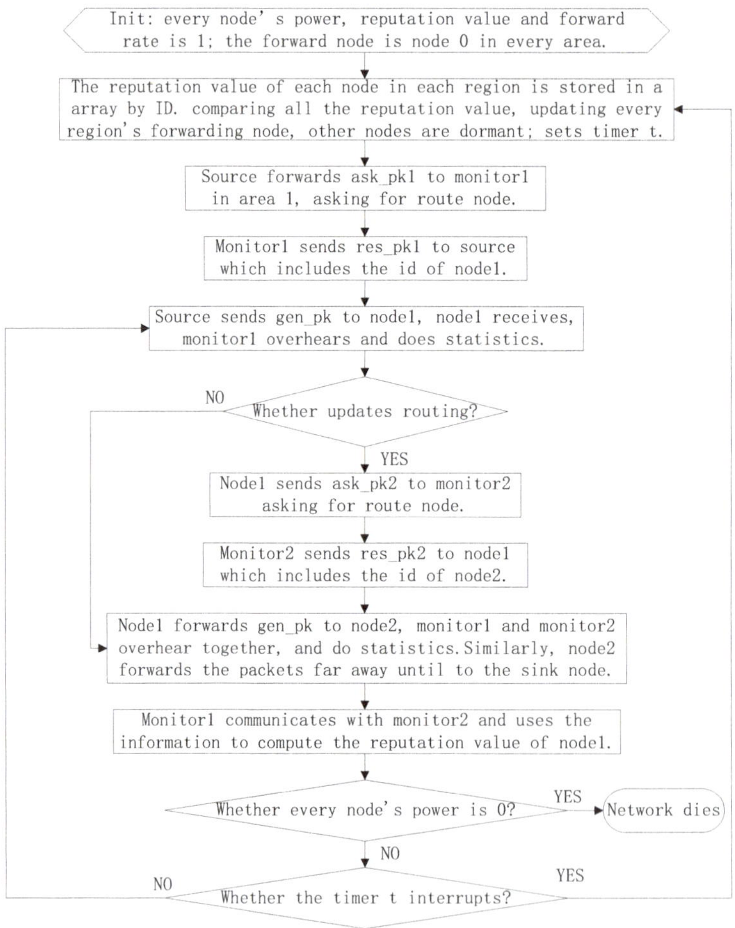

Figure 2. The process of monitoring mechanism.

use different trust models [12]. When a node is observed to forward the packet s times and drops the packet f times, the beta trust model will assign trust value Tr (0 < Tr < 1) to this node using the following formula:

$$\mathrm{Tr} = (s+1)/(s+f+2) \quad (2)$$

Entropy trust model uses entropy function $H(p)$, whose input p is the trust value that can be obtained from beta trust model, to determine the trust value Tr. The entropy function:

$$H(p) = -p\log_2 p - (1-p)\log_2(1-p) \tag{3}$$

The trust value Tr ($-1 < \text{Tr} < 1$) is defined by

$$\text{Tr} = \begin{cases} 1 - H(p), \text{for } 0.5 \le p \le 1; \\ H(p) - 1, \text{for } 0 \le p < 0.5. \end{cases} \tag{4}$$

In our simulation, we use the beta trust model. In Equation (1), $\text{Pr}_{\text{id}} = \text{Forward\_pks}/\text{Receive\_pks}$ is the forward rate of the node; Power_{id} is the node's residual energy; $\text{Val}[\text{node}_{\text{id}}]$ is composite reputation value of the node. Each monitor node calculates the reputation value of all nodes within its administering area and stores this information in an array. Then, the node of the highest value is selected as the routing node of this area. In order to ensure that monitoring node can listen to the general node, the radius of monitoring area should less than the quarter of normal node's forwarding radius.

- Routing node is updated periodically in order to prolong the network lifetime. Because the residual energy of nodes is taken into consideration, the node of high forward rate won't be overused. If some nodes of high credibility are overused, they will die prematurely so that the connectivity, cover degree and lifetime of network will decrease.

4. Function of Our Monitor Mechanism

According to the description of network in section Ⅲ, our secure routing protocol can not only accomplish the general sensor network' task about data collection and transmission, but also detect a variety of malicious nodes' internal selective forwarding in network. The secure mechanism can solve these ways of attack which are as the following conclusion.

4.1. Black Hole of One Node

When a normal node M is captured and becomes the malicious node, it can launch black hole attack. On the basis of Equation (1), due to the high node energy of black hole, M can achieve a high value, so that it will becomes the forwarding node. However, because M doesn't transmit packets, its forward rate will decline soon. After its forward rate is below the threshold T ($0 < T < 1$), it will be regarded as a malicious node by the monitor node. And the Val of M will be set at 0 so that M is excluded out of network.

4.2. Selective Forwarding Attack of One Node

Similar to the black hole attack, when the node N drops the received packets selectively, its general forward rate will decrease. If N's Val is below the threshold T ($0 < T < 1$), it will be regarded as a malicious node by the monitor node. This way of attack will need more time to be detected compared with black hole.

4.3. Colluding Attack of Multiple Nodes

As shown in **Figure 3**, the malicious node 1 forwards packets to malicious node 3 instead of normal node 2, and node 3 drops the packets. Because node 1 received m packets and sends m packets successfully, it attempts to deceive monitor node. However, according to the request of this routing algorithm, the packets must include the information about source ID, forward node ID and ID of next-hop. Whether node 2 receives packets is based on the head information from node 1's packet. If malicious node 1 changes the next-hop ID to node 3's ID, it will be caught by monitor node and be regarded as tampering the route.

This way of attack is similar to wormhole. With the control of monitor node, even though there is a high energy node attracting the packet stream, we don't change the route and follow the original route. Otherwise, if some node tampers the head information by itself, it will be regarded as malicious node.

4.4. Malicious Nodes Control the Transmitting Distance

As shown in **Figure 4**, malicious node 1 forward the received packets with small power, in order to control the

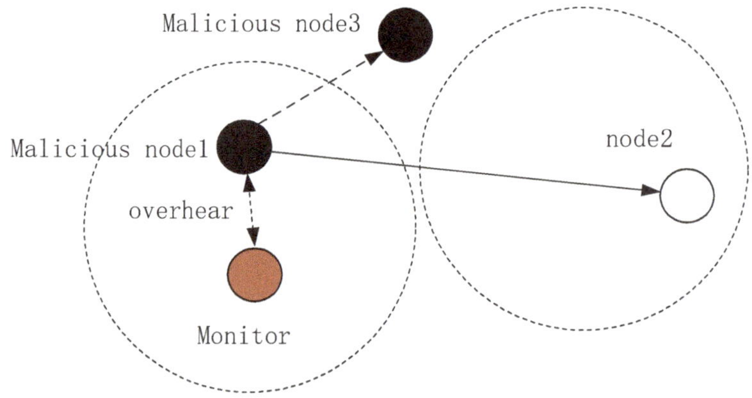

Figure 3. Colluding attack of multiple nodes.

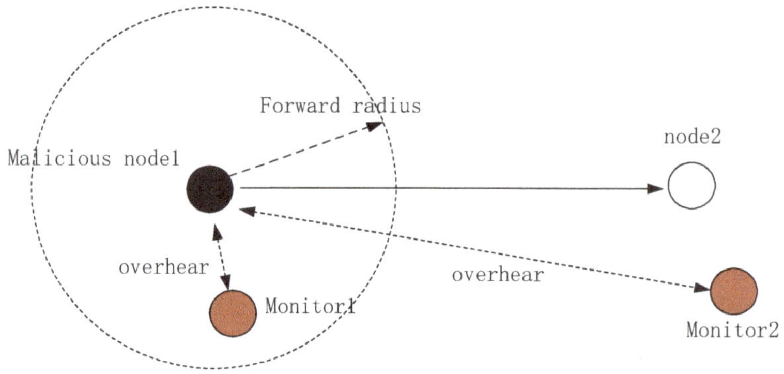

Figure 4. Malicious nodes control the transmitting distance to drop packets.

transmitting distance so that node 2 can't receive the packets. On the other hand, monitor 1 can hear the packets, so node 1 is aimed at cheat monitor 1 to achieve a high reputation value. However, both node 2 and monitor 2 can't receive the packets. After a period of time, monitor 2 will change some information with monitor 1. Monitor 1 hears m packets from node 1 and monitor 2 hears n packets from node 1. If the value of n/m is below a threshold t ($0 < t < 1$), node 1 will be regarded as a malicious node.

4.5. Trade-off of Energy Balance and High Forward Rate

In Equation (1) we can adjust the value a and b, in order to control the condition of becoming a high reputation node. So we can control the usage time of every node. That is to say, the characteristic of the network will be an energy balance one or the one of high forward rate according to the physical truth.

5. Simulation and Analysis

In this section, we use OPNET simulation to evaluate the performance of our trust monitor model and routing scheme. In our simulation, 100 sensor nodes and 10 monitor nodes are randomly distributed over a 100 m × 100 m area.

We only do the simulation of local network. A monitor node administrates 10 common nodes in each region. Firstly, the nodes will be numbered to meet the design requirements of the software. The size of node's ID is 1 byte (8 bits). The ID of source node is 00000000. The monitor node's ID in first region is 00000001. The ID of 10 common nodes within first region is from 00010000 to 00011001 (0001 indicates in region 1; 0000-1001 indicates 0-9). Therefore, the decimal ID of nodes in region 1 is from 16 to 25. Similarly, the ID of 10 nodes in region 2 is from 32 to 41. So every node in the network is numbered reasonably.

For each ordinary node in region 1, we set their forward rate respectively. The forward rate of node 16 and 24 is 70% - 90%; 20 and 25 is 80% - 90%; 19 and 21 is 80% - 100%; 22 is 65% - 85%. The node 18's init forward rate is 70% - 100%. After a certain number of turns, it will become a malicious node to launch selective forwarding attack with the forward rate of 30% - 50%.

We assume that each node's initial energy is 1, and suppose that receiving a packet's energy consumption is 0.0001 and that sending a packet's energy consumption is 0.0003. It is 50 seconds each turn. That is to say, after requesting routing node, the source node sends packets at a speed of 2 packets per second for 20 seconds. At the same time, monitor node listens and forward node receives the packets. In the next 20 seconds, the forward node forwards the 40 packets random based on its forward rate; at the same time, the next-hop node receives the packets and monitor nodes listen and do some statistics and computation. There is 10s time for monitor node to calculate the reputation value of nodes and update the route node. Above is a turn of local data forwarding process and we cycle the process like that.

Suppose that node 18 turned to malicious node in 2000 s, its forward rate became 30% - 50%. As shown in **Figure 5(a)**, monitor node identified the malicious node, and showed the finding time and malicious node's ID. It's not hard to think of that if the normal node is captured later, it will cost more time to detect it out, because the node has performed well for a long time and established a relatively stable credibility. **Figure 5(b)** shows that the different finding time of node 18 which is captured in turn 50 (2500 s), turn 100 (5000 s), turn 150 (7500 s), turn 200 (10,000 s) and turn 250 (12,500 s). Therefore, the detection mechanism should be improved. If a node's forward rate is below the threshold three times continuously, it will be regarded as a malicious node.

In Equation (1), if value a is bigger enough than b, the nodes of high forward rate will always be selected

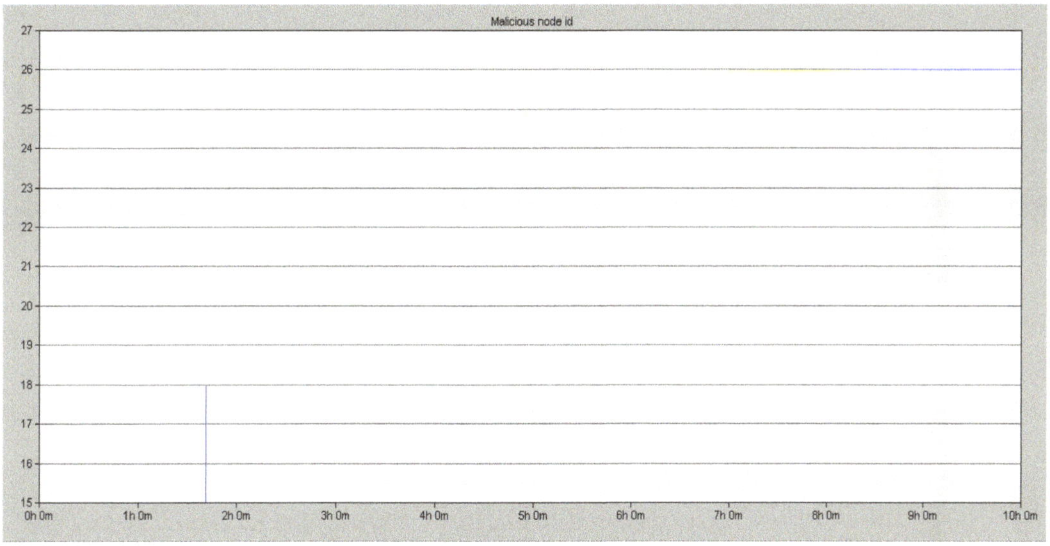

(a) Node 18 turned to malicious node in time 2000 s.

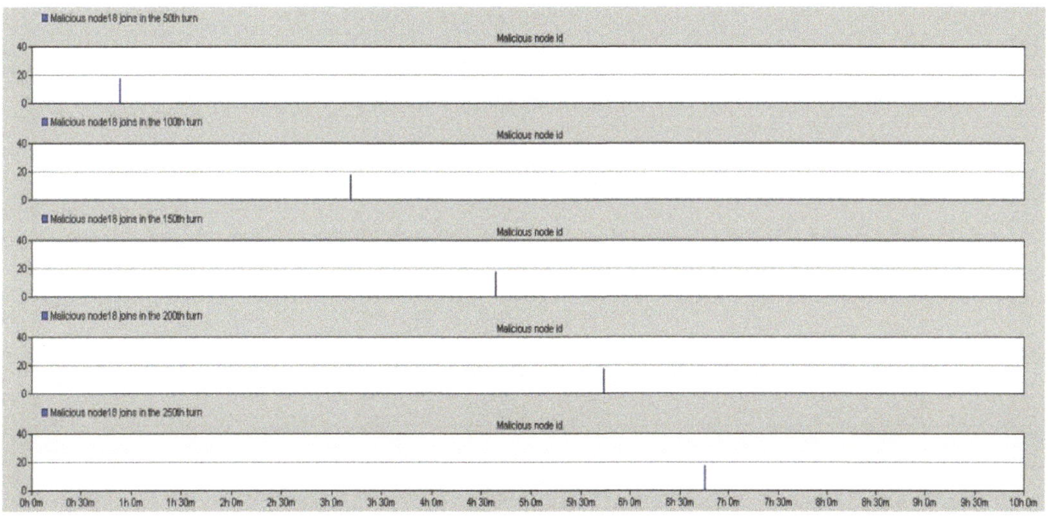

(b) Node 18 turned to malicious node in different time.

Figure 5. The time of detecting the malicious node 18.

as the forwarding nodes so that it will die soon. On the other hand, if a is bigger enough than b, the network will turn to an energy-balanced network. Every node will be used equally so that data forward rate of network is not stable. Thus, we should set a and b reasonably according to physical truth.

We don't put malicious nodes in the simulation. We get **Figure 6(a)** after the simulation of 6000 seconds

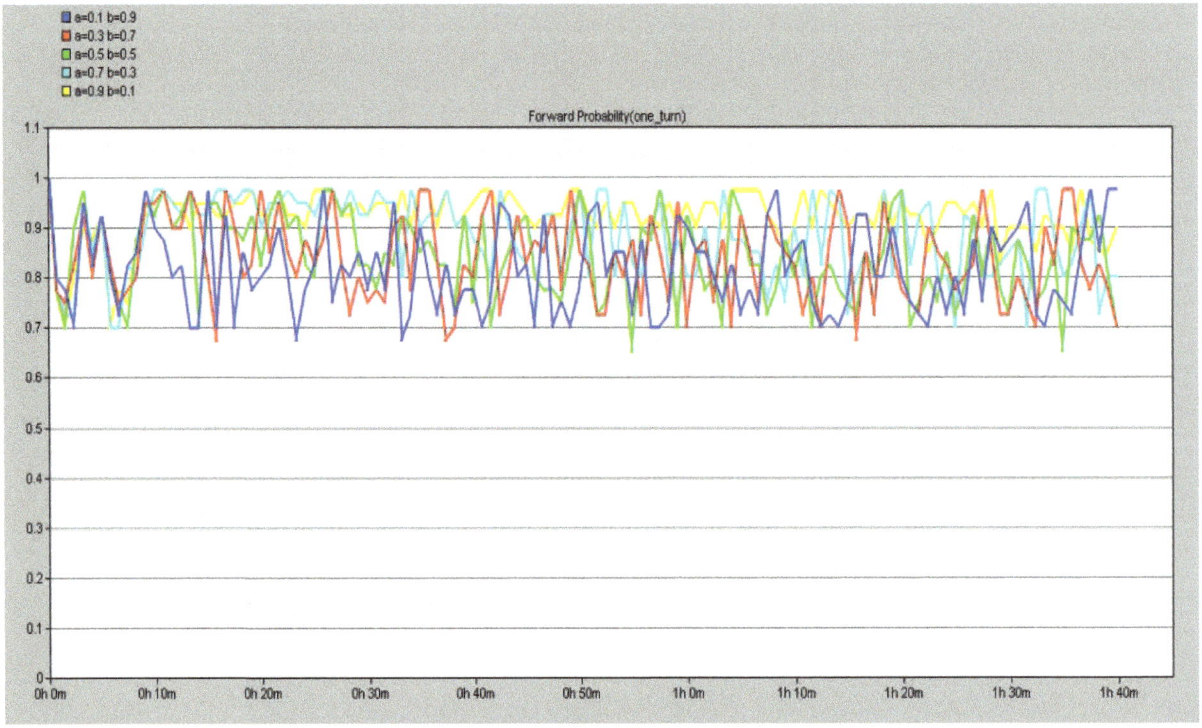

(a) Simulation with different value of a and b.

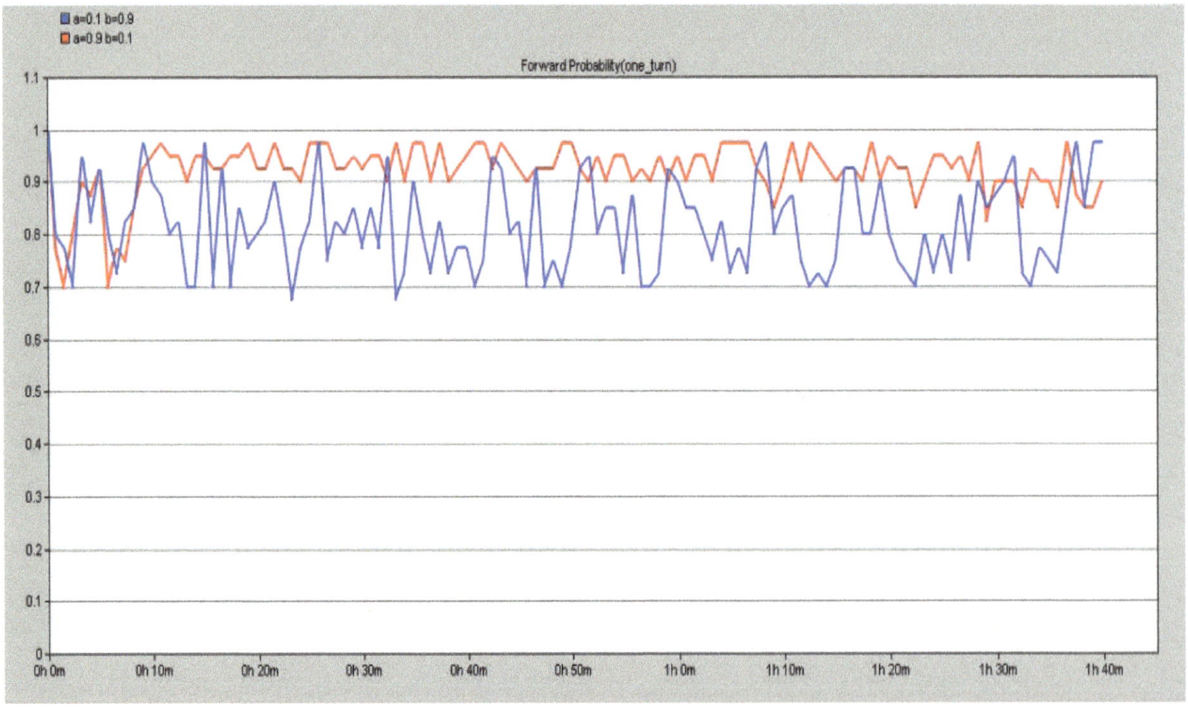

(b) Simulation when a = 0.1, b = 0.9 and a = 0.9, b = 0.1.

Figure 6. The changing of forward rate (one turn) over time.

(120 turns). It shows that if a is big, nodes of high forward rate will be used more. So the whole forward rate of the network will be higher. When a = 0.1, b = 0.9 and a = 0.9, b = 0.1, the network's forward rate is as **Figure 6(b)**.

On the other hand, the nodes' use time is different because of different value of a and b. As shown in **Figure 7(a)** and **Figure 7(b)**, when a = 0.1, b = 0.9, the nodes were used equally. But when a = 0.9 and b = 0.1, the nodes were used not equally.

We define that the first node's death time is the network's lifetime. That is to say, although the early stage of forward rate in the network is very high, if the nodes of high forward rate are used excessively, the network's lifetime will decrease. We do a 40000 s simulation, until all the nodes die. In **Figure 8(a)** and **Figure 8(b)**, we can see when a = 0.9 and b = 0.1, the network has a high forward rate at first. But because of using the good nodes excessively, the best node dies so early. If we don't add new nodes instead of died nodes, the connectivity and

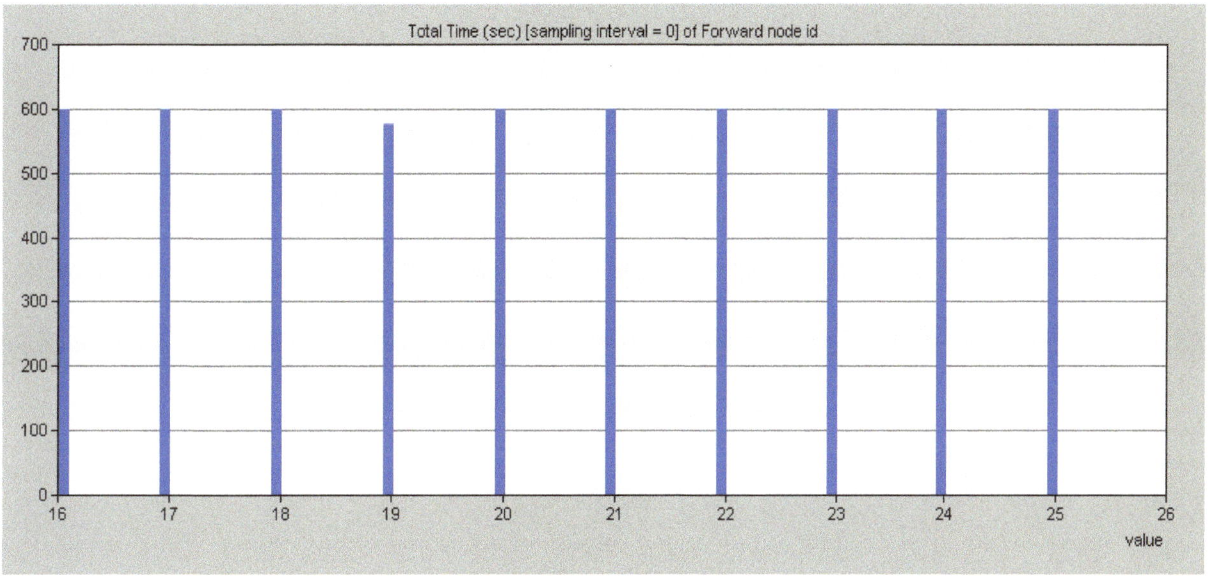

(a) When a = 0.1, b = 0.9.

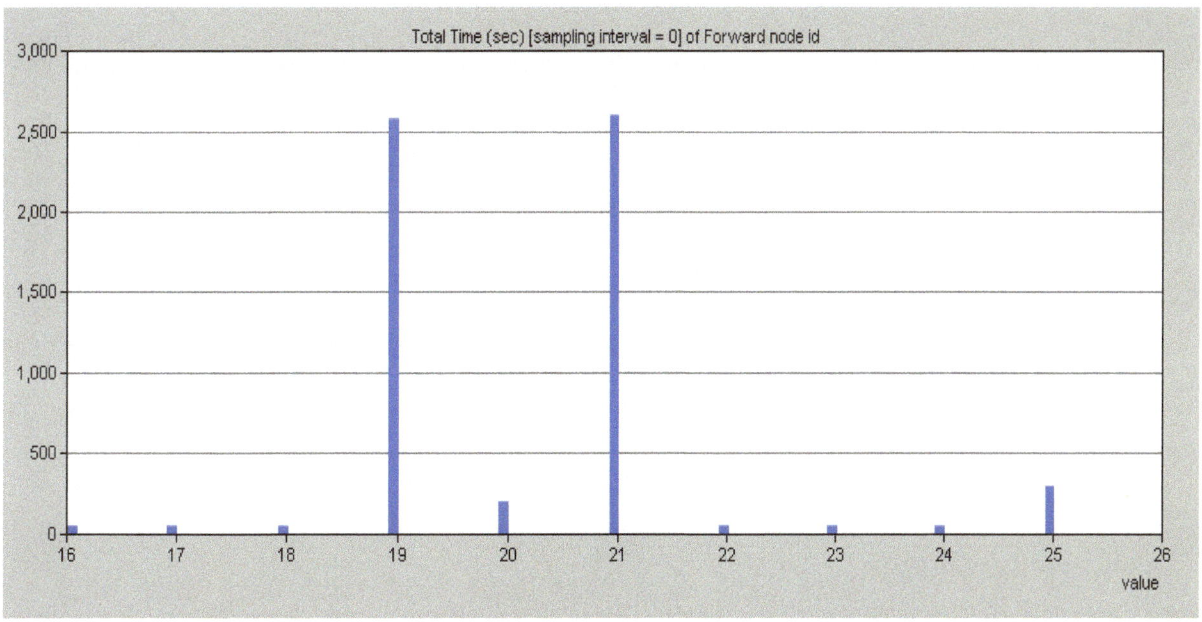

(b) When a = 0.9, b = 0.1.

Figure 7. Every node's using time.

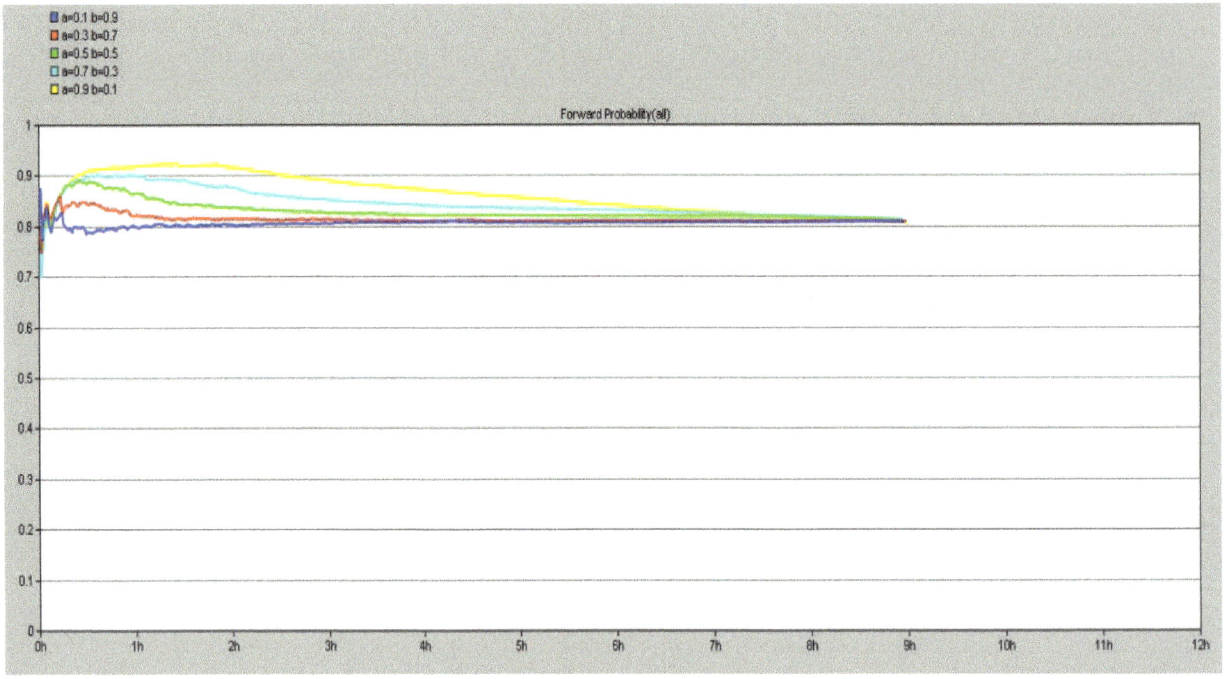

(a) The forward rate (average) of the network.

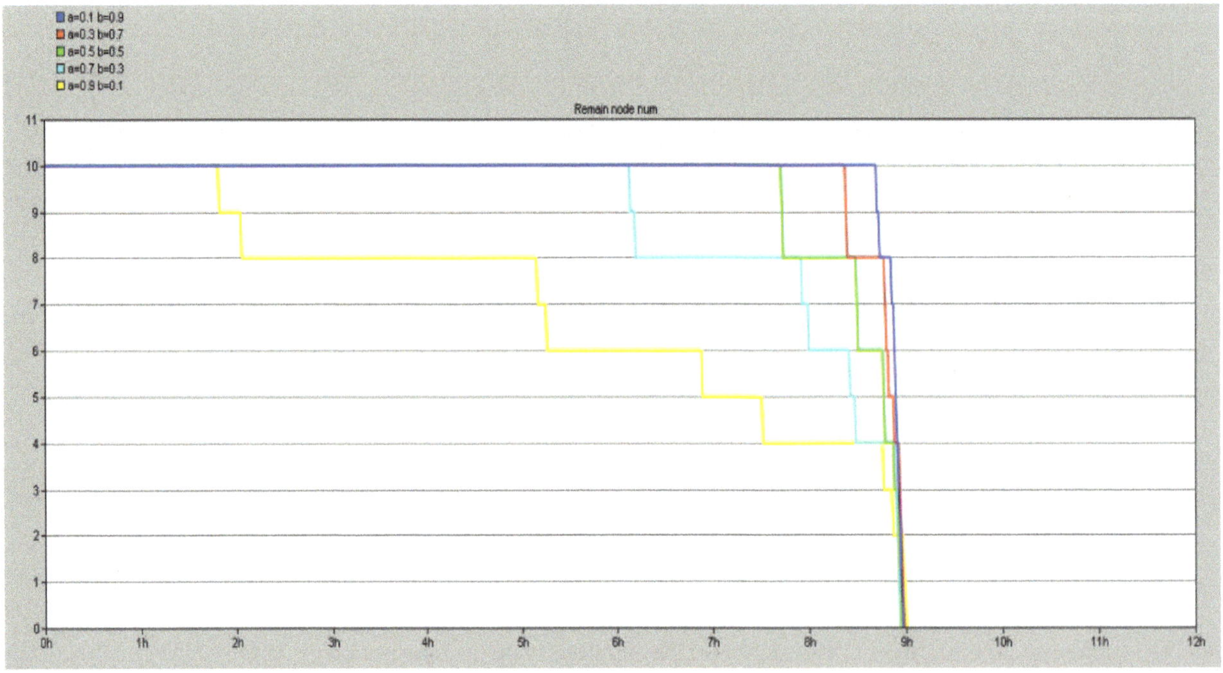

(b) The remainder node of the local network.

Figure 8. The simulation until all the nodes' power using up with different value a and b.

cover degree of network will be destroyed.

Acknowledgements

The authors wish to acknowledge the anonymous reviewers for their valuable comments and suggestions. We also thank OPNET Technologies, Inc., for providing us with OPNET Wireless Modeler to validate our approaches.

References

[1] Yu, Y.L., Li, K.Q., Zhou, W.L. and Li, P. (2011) Trust Mechanisms in Wireless Sensor Networks: Attack Analysis and Countermeasures. *Journal of Network and Computer Applications*, **35**, 867-880.

[2] He, B., Joshi, S., Agrawal, D.P. and Sun, D. (2010) An Efficient Authenticated Key Establishment Scheme for Wireless Mesh Networks. 2010 *IEEE Global Telecommunications Conference*, Miami, 6-10 December 2010, 6-10.

[3] Butun, I., Morgera, S.D. and Sankar, R. (2013) A Survey of Intrusion Detection Systems in Wireless Sensor Networks. *IEEE Communications Surveys & Tutorials*, **16**, 266-282.

[4] Cho, Y. and Qu, G. (2013) Detection and Prevention of Selective Forwarding-Based Denial-of-Service Attacks in WSNs. *International Journal of Distributed Sensor Networks*, **2013**, Article ID: 205920, 16 p. http://dx.doi.org/10.1155/2013/205920

[5] Marti, S., Giuli, T.J., Lai, K. and Baker, M. (2000) Mitigating Routing Misbehavior in Mobile and Ad Hoc Networks. *Proceedings of International Conference on Mobile Computing and Networking*, Boston, 6-11 August 2000, 255-265.

[6] Hai, T.H. and Huh, E.-N. (2008) Detecting Selective Forwarding Attacks in Wireless Sensor Networks Using Two-Hops Neighbor Knowledge. 2013 *IEEE* 12*th International Symposium on Network Computing and Applications*, Cambridge, 10-12 July 2008, 325-331.

[7] Khalil, I., Bagchi, S., Rotaru, C.N. and Shroff, N.B. (2009) UnMask: Utilizing Neighbor Monitoring for Attack Mitigation in Multihop Wireless Sensor Networks. *Ad Hoc Networks*, **8**, 148-164. http://dx.doi.org/10.1016/j.adhoc.2009.06.002

[8] Lu, Y.F., Lin, K. and Li, K.Q. (2012) Trust Evaluation Model against Insider Attack in Wireless Sensor Networks. *International Conference on Cloud and Green Computing*, Xiangtan, 1-3 November 2012, 319-326.

[9] Karlof, C. and Wagner, D. (2003) Secure Routing in Wireless Sensor Networks: Attacks and Countermeasures. *Ad Hoc Networks*, **1**, 293-315. http://dx.doi.org/10.1016/S1570-8705(03)00008-8

[10] Challal, Y., Ouadjaout, A., Lasla, N., Bagaa, M. and Hadjidj, A. (2011) Secure and Efficient Disjoint Multipath Construction for Fault Tolerant Routing in Wireless Sensor Networks. *Journal of Network and Computer Applications*, **34**, 1380-1397. http://dx.doi.org/10.1016/j.jnca.2011.03.022

[11] Shu, T., Liu, S. and Krunz, M. (2009) Secure Data Collection in Wireless Sensor Networks Using Randomized Dispersive Routes. *Proceedings of the IEEE INFOCOM Conference*, Rio de Janeiro, 19-25 April 2009, 2846-2850.

[12] Sun, H.-M., Che, C.-M. and Hsiao, Y.-C. (2007) An Efficient Countermeasure to the Selective Forwarding Attack in Wireless Sensor Network. *IEEE Region* 10 *Conference* (*TENCON*), Taipei, 30 October-2 November 2007, 1-4.

3

Visualization and Management Platform with Augmented Reality for Wireless Sensor Networks

Kenya Sato, Naoya Sakamoto, Hideki Shimada

Department of Information Systems Design, Doshisha University, Kyoto, Japan
Email: ksato@mail.doshisha.ac.jp, naoyas@mail4.doshisha.ac.jp, hideki-s@is.naist.jp

Abstract

Recently a ubiquitous sensor network which collects our environmental information gets increasingly popular, a visualization application is necessary for users to manage complicated wireless networks, however, these applications are developed individually for wireless communication standard or a type of wireless device. Therefore, users are forced to adopt and use the application individually according to the target of the wireless network. In this paper, we propose a visualization platform for wireless network environments using augmented reality technology, and evaluate the effectiveness of the platform. From the result of the evaluation, we have confirmed the proposed platform has availability for visualization and management of wireless networks.

Keywords

Network Management, Visualization, Augmented Reality

1. Introduction

Recently, wireless networks of various standards such as Wi-Fi [1], Bluetooth [2], and ZigBee [3] are proliferating around us. Networks of small-scale wireless sensors, called the Internet of Things [4] or M2M [5], are also being built, and research is being done on technology to sense and automate our surroundings. The number of wireless devices using these wireless networks continues to increase, wireless networks are becoming increasingly complex, and new wireless communication standards are appearing to accommodate user applications.

On the other hand, little emphasis has been placed on overall visualization and management of wireless networks for the Internet of Things so far. There are visualization applications that can be used to visualize and manage complex wireless networks, but there are no such applications that handle various wireless communication standards or types of wireless devices, so applications must be selected and used depending on the conditions. However, as the number of wireless communication standards and types of wireless devices increases, it is becoming difficult for users who do not understand the differences among the wireless communication standards to select a visualization application and manage their wireless networks. Further, with general visualization techniques, which map wireless devices onto a virtual space to visualize the state of the wireless network, the location of the wireless device in virtual space is unrelated to the position of the device in real space, so it is difficult for users to associate the devices in the virtual space with those in real space, so it is difficult to identify them.

In this paper, we propose a platform that uses Augmented Reality (AR) technology [6] for visualization and management of wireless networks to resolve these problems. Visualizing wireless networks using AR technology allows us to build a user interface based on images of real space, so it is easy to identify wireless devices. The proposed platform also handles multiple wireless communication standards, and provides application programming interfaces (APIs) for visualization and management functions so that applications can be built using the platform functionality. This allows applications to handle multiple wireless communications standards and devices easily, and users can deal with faults and configure wireless networks by using a single visualization application.

2. Wireless Network Visualization and Management Issues

Generally, a visualization application is used to determine and manage the state of a wireless network. Visualization applications retrieve information from wireless devices in the wireless network, perform a mapping of each wireless device onto 2D space, and create a visualization showing the information from the wireless devices and the network. To manage the network, wireless devices are selected on the screen of the visualization application, and control operations are performed.

The three most common visualization functions in a visualization application are visualizing links, visualizing traffic, and visualizing wireless device information. In general, most visualization applications only implement a few of these functions, depending on their purpose.

2.1. Link Visualization

When two wireless devices are connected to each other, they are linked. Links are generally visualized by mapping wireless devices to a virtual space and showing the link status with a line joining the two devices. The network topology is shown by displaying all links in the wireless network.

2.2. Traffic Visualization

Traffic refers to the content of data (or packets) or the data volume flowing on a wireless network. Traffic is generally visualized by analyzing packet content and displaying it, by associating traffic volume with the thickness of the line used to display the link, or by using animation to display the packets.

2.3. Wireless Device Information Visualization

Wireless device information includes the device name, its network address, the type of wireless network, its role in the wireless network (base station, relay node, terminal node, etc.), and its hardware state.

The two most common control functions in a visualization application are link switching and power switching.

1) Link switching

Link switching is the function of specifying wireless devices for a particular wireless device to connect to, and performing the required link connections and disconnections. For ordinary Wi-Fi networks, this refers to connecting and disconnecting wireless devices to a wireless router, and for Bluetooth it refers to connecting and disconnecting between wireless devices.

2) Power switching

Power switching refers to selecting a particular wireless device and turning the power on or off, or re-booting the device.

3. Wireless Network Visualization and Management Platform Proposal

3.1. Proposal Objectives

In this paper, we propose a platform that uses AR technology for visualization and management of wireless networks. The proposal has two objectives as described below.

The first is to abstract the various wireless specifications, such as Wi-Fi and Bluetooth, positioning the platform between the specifications and applications as shown in **Figure 1**. Wireless network visualization and management functions are implemented in the platform and provided through APIs, so visualization applications operating on the platform can handle multiple wireless specifications and wireless devices easily. Because of this, users no longer need to select an application for each wireless specification or wireless device.

The second is to integrate the virtual space, into which wireless devices are mapped, with real space using AR technology. To do this, it is necessary to integrate both position information, which is the basis for displaying supplementary information using AR technology, and the corresponding wireless-device-specific information, such as its ID, into the platform. With AR technology, a real-space image is used for the user interface, so applications can build a user interface by retrieving location information for wireless devices from the platform and using AR technology to overlay information from the virtual space onto the real-space image. This resolves any inconsistencies between the virtual space and real space with earlier visualization applications.

3.2. Functional Requirements

We now describe the functions required by the proposed platform in order to realize the objectives described in the previous section.

1) Abstraction of wireless communication specifications

The proposed platform supports multiple wireless communication standards, abstracts the wireless communication standard layer, and provides APIs for visualization and control functions. Control functions must be handled according to the type of each wireless device, but much of the wireless network information needed for the visualization functions, such as links and topology, does not depend on the wireless communication standard, so there is much in common that can be combined in the APIs, even for different wireless communication standards. Applications can retrieve wireless network information for multiple wireless communication standards by just changing parameters in the APIs.

2) Gathering wireless network and device information

To implement a conventional visualization application operating on the platform, the platform must have functions to gather wireless network information such as link information, signal strength, and packet data, and also wireless device information such as network addresses and hardware information.

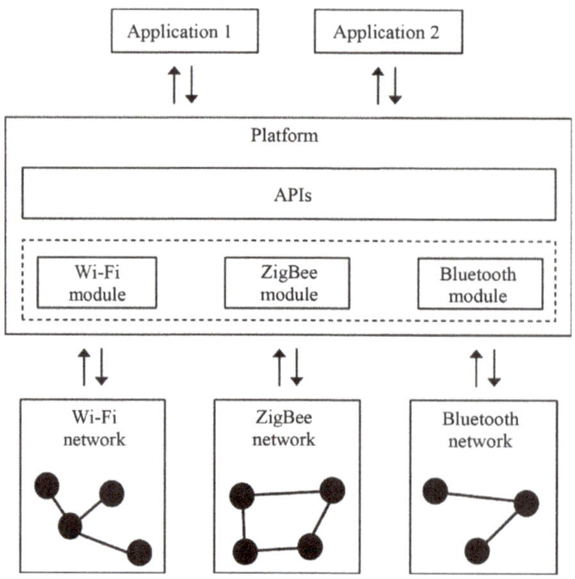

Figure 1. Abstraction of Wireless Network Interface for API.

3) Managing wireless network and device information

Functions for the application to retrieve the gathered wireless network and wireless device information are needed. It is also essential to accumulate logs of wireless network information in cases when the past state of the wireless network is to be reproduced, or when comparing the current state to past state.

4) Wireless device control

To implement a conventional visualization application operating on the platform, the platform must have functions to control the wireless network upon receiving instructions from the application, such as for connecting and disconnecting links, and for controlling the wireless device hardware, such as for turning the power on or off.

5) Location data management

There are several ways to specify location information for objects being handled with AR technology, and each of them has advantages and disadvantages. For this reason, the proposed platform does not limit the method used for positioning objects, but only manages the information needed to implement each such method, and it leaves the object positioning method to be handled by the application.

4. Design

4.1. Architecture and Functions

The architecture of the platform is shown in **Figure 2**, and the functionality of each element is described below.

1) Communication modules: These modules communicate with each wireless network, connect to the wireless network and devices, retrieve packets flowing on the networks and send packets on the networks. There is a communication module for each wireless communication standard. For example, there are Wi-Fi, ZigBee and Bluetooth modules for Wi-Fi, ZigBee, and Bluetooth communication respectively. Communication modules consist of two components: Packet sender and Packet capture.

Packet sender handles packet transmission within the communication module. When a packet needs to be sent (e.g. for an operation on a wireless device, or a packet has arrived for a terminal on the wireless network), this component generates and sends the packet.

Packet capture handles receiving packets within the communication module. It receives packets and sends them to the Network DB manager to register them in the database, according to commands from the Platform manager.

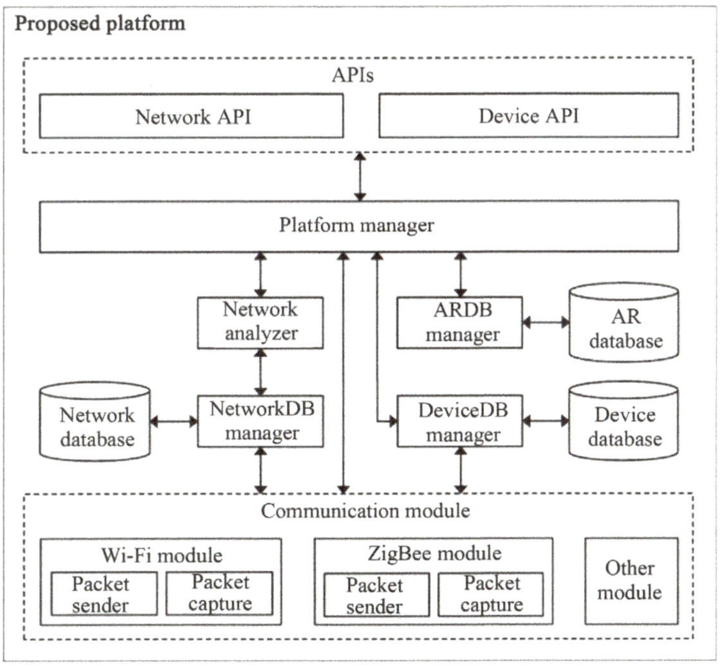

Figure 2. Architecture of proposed platform.

2) Network DB manager: This is the component that manipulates the Network database, which stores information regarding the wireless networks. Wireless network information refers to packet data flowing on a wireless network. The Network DB manager stores, retrieves and deletes packet data in the Network database. The packet data stored in the Network database is obtained from the communication module, and the manager performs queries and sends the results to the Network analyzer when sending stored wireless network information to an application.

3) Network analyzer: This component analyzes and formats wireless network information. When a command to retrieve wireless network information is called using an API, it retrieves the wireless network information from the Network database through the Network DB manager according to the Platform manager's command, and performs operations such as filtering, protocol analysis, and link analysis.

4) Device DB manager: This component manipulates the Device database, which manages information regarding devices on the wireless network. The Device DB manager gets device information from the Communication module and stores it in the Device database. Device information includes device network address, name, type, state, and control messages. It retrieves device information from the Device database upon request from the Platform manager and sends it to the Platform manager.

5) ARDB manager: This component manipulates the AR database, which manages position data for wireless devices. The wireless device position data managed by the AR database is used as the basis for visualizing the wireless network using AR technology. There are two methods for storing and updating position information for wireless devices. The first is for wireless devices equipped with GPS, in which case, the ARDB manager retrieves the wireless device position information from the Communication module and stores or updates it in the AR database. For the second method, the application must call API methods to store or update the location information.

6) Platform manager: Platform is the central processing component and it receives data and passes to the other components. When an application calls an API, the content of the called API is passed to the Platform manager, where it is processed. When doing so, it gets information from each of the DB managers and directly from the wireless networks and devices through the Communication modules.

7) APIs: The set of commands used by an application to use the functionality of the proposed platform. Depending on the command for the API called, the Platform manager gets information from a database, executes a command directly on a Communication module, or reads a command from Device DB and executes it. There are two APIs: the Network API and the Device API.

The Network API is a set of commands to retrieve information from a wireless network and it handles information in the Network database. It does not depend on particular wireless communication standards. An application can call the Network API to retrieve link, topology, packet or statistical data for multiple wireless communication standards.

The Device API is a set of commands to retrieve wireless device information and to manipulate wireless devices. It handles information in the Device database and the AR database. There are differences in retrieving data from and operating different types of wireless device, so different data is managed in the Device database for each device. This data is used to support differences in data retrieval and operation for the different devices.

4.2. Implementation

Wi-Fi and ZigBee Communication modules were implemented based on the design. The implementation environment is shown in **Table 1**. Overall, the proposed platform was built using Java, and MySQL [7] was used for the databases. The APIs are provided to the application through messages using socket communications, and applications receive API results from the platform in XML format.

For communication with Wi-Fi networks, the device in the PC running the proposed platform was used. The Wi-Fi module uses WinPcap [8] for Packet capture, and a Native Wi-Fi API for the Control component.

For communication with ZigBee networks, a Sun SPOT [9] [10] base station connected to the PC was used. The ZigBee module uses a Packet Sniffer in the Sun Spot SDK for Packet capture, and the Sun SPOT API for the Control component.

5. Application

5.1. Functional Definition

We implemented a visualization and management application using AR technology as an example application

Table 1. Implementation modules for proposed platform.

Network	Item	Description
Platform PC	OS	Windows 7 Professional
	CPU	Intel Core i7-2600 3.40 GHz
	RAM	8GByte
	Database	MySQL
	API	Socket
Wi-Fi Module	Device	Realtek Wireless LAN Mini-PCI Express Adapter II
	Wi-Fi API	Native Wi-Fi API
	Packet Capture	WinPcap
ZigBee Module	Device	Sun SPOT Base Station
	ZigBee API	Sun SPOT API
	Packet Capture	Packet Sniffer

using the proposed platform. The visualization and management application defines the following three functions.

1) Display of wireless network information

This function displays the link state between two wireless devices that are communicating. When a wireless device is communicating with another one, a line is shown between the two devices using AR technology. The line disappears when communication stops.

2) Display of wireless device information

This function displays the wireless device network address and hardware state.

3) Control of wireless devices

This function connects and disconnects links between communicating wireless devices. Links shown with lines on the application screen can be manipulated using gestures, such as connecting and disconnecting links.

A wireless network is composed of two wireless devices. The program keeps one wireless device always communicating with another, and communication is started and stopped by connecting and disconnecting links.

5.2. Implementation Environment

The implementation environment and wireless devices used to build the wireless network when implementing the application are shown in **Table 2**. A Web camera is used for visualization using AR technology. A touch display is also used for gesture operations on the application screen. The NyARToolKit for Java [11] is used to display annotation information using AR technology.

Sun SPOTs are used for the wireless devices composing the wireless network. Sun SPOTs are programmable wireless devices, so network control is possible. The ZigBee stack can be used and various sensors are built in. An identifying 2D barcode marker is attached to each wireless device and registered in the platform database beforehand.

6. Implementation Results

A real-space image and the operating screen of the implemented application are shown in **Figure 3** and **Figure 4** respectively. On the real-space image, there are just two wireless devices. On the application screen, wireless network and device information is visualized over the real-space image using AR technology, resolving any inconsistencies between the virtual space and real space. For example, the link between wireless devices in the wireless network is shown as though they were connected by a cable, so it is easy to see the state of links and topology in real space. The network address and hardware state for wireless devices is also shown, so it is easy to know the state of wireless devices. The information for each wireless device can be shown or hidden by touching the device on the operating screen.

With the implemented application, wireless devices can be controlled directly using an interface based on

Table 2. Implementation environment.

Network	Item	Description
Application PC	OS	Windows 7 Professional
	CPU	Intel Core i5-560 M 2.67 CHz
	RAM	4GByte
	Touch Display	I-O DATA LCD-AD221FB-T
	AR Library	NyARToolKit for Java
Wireless Device	Device	Sun SPOT
	CPU	MHz180 T920ARM
	RAM	512KByte
	Wireless Protocol	PHY/MACiZZi802.15.4E
	Sensor	Acceleration, Temperature, Illuminance

Figure 3. Real image for sensor devices.

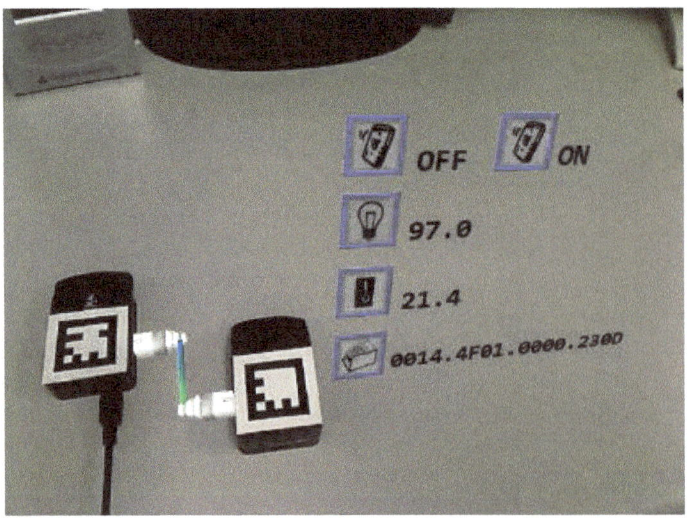

Figure 4. Virtual image (including annotation information) for sensor network.

real-space video, resolving any problems with inconsistencies going from real space to virtual space. On the operating screen, a cable representing the communication between two devices is shown, and the link between the devices can be cut using a cutting (disconnecting) gesture on this cable. If a drag gesture on the operating screen is done between two wireless devices that are not connected by a link, a connection is made between them and communication resumes.

7. Related Research

7.1. Config Free

Yoneyama *et al.* have proposed a wireless network visualization user interface called Config Free [12] [13]. Config Free has three functions: Network diagnostics, network device search, and connecting and disconnecting wireless devices. Config Free visualizes wireless devices arranged on a 2D virtual space. The user's PC is placed in the center of the 2D virtual space, and other wireless devices are arranged on concentric circles around it at distances that are closer as the strength of the signal from them increases. A drag-and-drop operation from the user's PC at the center to one of the devices on a concentric circle creates a connection between the wireless devices. When a connection is created, a line is shown between the PC and the other wireless device. The connection can be cut by clicking on the line when the pointer is shown as a pair of scissors. Config Free supports two types of network: Wi-Fi and Bluetooth.

7.2. 3D-Tcpdump

Saito *et al.* have proposed 3D-tcpdump [14] [15], which is a visualization application that displays network communication state and topology graphically in 3D. 3D-tcpdump has two functions: visualization of traffic on the network and visualization of topology. 3D-tcpdump generates and arranges 3D objects in a virtual 3D space for the user's PC, the default router that is one hop away, and the group of destination PCs. When the network traffic occurs, it can be visualized as spherical objects according to the traffic volume from the 3D object to the destination one.

7.3. uMegane

Imaeda *et al.* have proposed the uMegane system [16], for visualizing sensor information using AR technology. uMegane uses AR technology to abstract and display sensor data from wireless sensor devices as 3D objects, using properties such as color, size, and motion of the objects so that ordinary users can understand them easily. This system can also be applied to detect abnormalities using user-definable filters.

7.4. Extate

Takimoto *et al.* have proposed the Extate system [17], which uses AR technology for visualization of wireless networks. Extate has functions to visualize link states in a wireless network and to visualize wireless device information. A 2D barcode is attached to 2D devices, lines are used to display link state, and text is used to display wireless device network addresses and sensor data.

8. Evaluation

8.1. Items Evaluated

Compared to earlier methods, in which visualization and management applications connected directly to wireless networks, with the proposed method, the platform connects to various wireless networks and provides the results of processing to the application. The platform is able to perform visualization of the wireless networks over real space, but if there is significant delay between retrieving wireless network data when performing visualization, it could cause confusion for users expecting to know the wireless network state in real time.

To implement real-time visualization and management of wireless networks using the proposed platform, any increase in the time taken to complete tasks through the platform is undesirable. Accordingly, we evaluated the performance of the proposed platform implementation.

8.2. Performance Evaluation

Four functions required in the proposed platform are retrieving link state, retrieving traffic information, retrieving device information, and controlling devices (connect or disconnect links). We measured the response time of each of these functions, namely the time from the instant the application calls the platform API until the application obtains the result.

The application can be used in two ways: either with the platform and the application running on the same host, or with the platform and application running on different hosts and the application using a platform on the same Local Area Network (LAN). Thus, performance evaluations were done in both an environment with the components on the same host, and with them on the same LAN. For the environment using the same host, platform API calls were made to the local host. For the same-LAN evaluation environment, the implemented platform was used, and the implemented application called the platform APIs. Response times in each case were measured using Wireshark.

Measurements are shown in **Table 3**. Response times for retrieving information are shorter than for device control functions because the proposed platform is asynchronous and gathers information from the wireless network, so it only needs to return results from its internal database. On the other hand, control commands are only sent to a wireless device after a control API is called, and the result is only sent to the application after the control operation has completed. This adds to the time required to control wireless devices, making response times long compared to information retrieval operations.

9. Discussion

9.1. Discussion of Performance Evaluation

From the performance evaluation results, in both the same host and same LAN cases, the application was able to obtain wireless network state information from the proposed platform with approximately 0.1 to 0.5 seconds of delay. The proposed platform gathers information from the wireless network asynchronously, regardless of whether there is an API call from an application, so delays from the component of the platform gathering information from the wireless network are very small. For wireless networks other than ZigBee, which we used in the evaluation environment, the same asynchronous architecture is used to gather wireless network information so the results will be the same.

For the performance results when controlling wireless devices, the time required to control the wireless device is added, so response times were from 2 to 3 s, but this time depends on the type of device being controlled, so the amount of time added by passing through the proposed platform can be expected to be similar to that required to retrieve wireless network information.

From these results, the delay introduced by the proposed platform is approximately 0.1 to 0.5 s, indicating that the platform will not present problems with delay for real-time visualization of wireless networks using AR technology.

9.2. Comparison with Related Research

We compared the proposed method with related research in terms of functionality, visualization methods, and whether it resolves issues discussed above. The results of this comparison are shown in **Table 4**.

The comparison shows that the application built using the proposed platform is the most effective in terms of

Table 3. Response time for each function.

	Response Time (Same Host) [Sec.]	Response Time (Same LAN) [Sec.]
Retrieving Link State	0.1758	0.4016
Traffic Information	0.1994	0.4381
Devise Information	0.1673	0.3858
Link Connection	2.8449	3.1820
Link Disconnection	2.1426	2.3995

Table 4. Comparison with related works.

	Config Free	3D-Tcpdump	uMegane	Extate	Proposal
Link Visualization	Yes	Yes	No	Yes	Yes
Traffic Visualization	No	Yes	No	No	Yes
Device State Visualization	No	No	Yes	Yes	Yes
Device Control	OK	No	No	No	Yes
Visualization Methods	Virtual 2D space	Virtual 3D space	Real space	Real space	Real space
Location Information	N.A.	N.A.	Image marker	Image marker	Application
Visualization Area	Total area	Total area	Inside camera area	Inside camera area	Application
Multiple Application Availability	No	No	No	No	Yes
Identification of Virtual and Real Space	No	No	?	?	Yes

functionality and resolving the issues we have discussed in this paper.

10. Conclusion

Recently, various devices are being used with wireless networks. Also, with research on the Internet of Things, the number of wireless devices being used by ordinary users is increasing, and we are surrounded by wireless networks based on various standards. With the increasing complexity of wireless networks, an application for visualizing them is essential for understanding and managing their state. However, existing visualization applications are only for particular wireless communication standards or device types, so users must select applications to suit their particular conditions. Also, these visualization applications generally map wireless devices into a virtual space, so there is no relation between the position of devices in the virtual space and their location in real space. Thus, in environments with increasing numbers of wireless devices, it is difficult to identify devices in real space based on information in the virtual space, and to match wireless devices in real space to the corresponding device in the virtual space. In this paper we resolve these issues by developing a wireless network visualization and control platform using AR technology. With the proposed platform, wireless network visualization and management functions are implemented in the platform, and are provided to applications through API functions. This allows applications to support multiple wireless communications standards easily, resolving the problem of having to choose among different applications. Further, visualization and management can be done using AR technology, resolving the difficulty with correspondence between the virtual and real spaces. The results of performance evaluations also demonstrate that the platform does not introduce performance issues. A comparison with related research confirmed that applications implemented using the proposed platform is effective in resolving the problems discussed here.

Acknowledgements

This work was partly supported by JSPS KAKENHI, Grant Number 26540038.

References

[1] Wi-Fi Alliance. http://www.wi-fi.org/

[2] Bluetooth Special Interest Group. http://bluetooth.org/

[3] ZigBee Alliance. http://www.zigbee.org/

[4] Atzoria, L., Ierab, A. and Morabitoc, G. (2010) The Internet of Things: A Survey. *Computer Networks*, **54**, 2787-2805. http://dx.doi.org/10.1016/j.comnet.2010.05.010

[5] Wu, G., Talwar, S., Johnsson, K., Himayat, N. and Johnson, K.D. (2011) M2M: From Mobile to Embedded Internet. *IEEE Communication Magazine*, **49**, 36-43. http://dx.doi.org/10.1109/MCOM.2011.5741144

[6] Azuma, R.T. (1997) A Survey of Augmented Reality, Presence. *Teleoperators and Virtual Environments*, **6**, 355-385.

[7] Oracle Corporation, MySQL. http://www.mysql.com/

[8] Riverbed Technology, WinPcap. http://www.winpcap.org/

[9] Oracle Corporation, Sun SPOT World. http://www.sunspotworld.com/

[10] Smith, R.B. (2007) SPOT World and the Sun SPOT. *Proceedings of the 6th International Conference on Information Processing in Sensor Networks*, Cambridge, 25-27 April 2007, 565-566.

[11] NyARToolKit Project. http://nyatla.jp/nyartoolkit/

[12] Yonenaga, T. and Yoshihara, H. (2004) User Interface for Wireless Networks Featuring Visualization of Information. *Toshiba Review*, **59**, 31-34.

[13] Toshiba Config Free. http://www.toshiba.co.uk/innovation/generic/01_connect_tab_03_EN/z/26/

[14] 3D-Tcpdump. http://www.3d-tcpdump.org/

[15] Ubiquitous Network Browser (3D-Tcpdump) Demo. https://www.youtube.com/watch?v=k7UO45k7Mvw

[16] Imaeda, T., Takashio, K. and Tokuda, H. (2008) uMegane: Visualization System of Sensor Data Using AR Technology. *IPSJ Technical Report*, **66**, 39-44.

[17] Takimoto, T., Namatame, N., Nakazawa, J., Takashio, K. and Tokuda, H. (2010) Extate: Visualizing Wireless Networks by Using AR Technology. *The 4th International Workshop on Ubiquitous Virtual Reality*, Helsinki, 17 May 2010.

4

An IEEE 802.15.4 Based Adaptive Communication Protocol in Wireless Sensor Network: Application to Monitoring the Elderly at Home

Juan Lu[1,2], Adrien Van Den Bossche[1,3], Eric Campo[1,2]

[1]Univ de Toulouse, Toulouse, France
[2]CNRS, LAAS, Toulouse, France
[3]CNRS, IRIT, Toulouse, France
Email: juan.lu@univ-tlse2.fr, vandenbo@univ-tlse2.fr, eric.campo@univ-tlse2.fr

Abstract

Monitoring behaviour of the elderly and the disabled living alone has become a major public health problem in our modern societies. Among the various scientific aspects involved in the home monitoring field, we are interested in the study and the proposal of a solution allowing distributed sensor nodes to communicate with each other in an optimal way adapted to the specific application constraints. More precisely, we want to build a wireless network that consists of several short range sensor nodes exchanging data between them according to a communication protocol at MAC (Medium Access Control) level. This protocol must be able to optimize energy consumption, transmission time and loss of information. To achieve this objective, we have analyzed the advantages and the limitations of WSN (Wireless Sensor Network) technologies and communication protocols currently used in relation to the requirements of our application. Then we proposed a deterministic, adaptive and energy saving medium access method based on the IEEE 802.15.4 physical layer and a mesh topology. It ensures the message delivery time with strongly limited collision risk due to the spatial reuse of medium in the two-hop neighbourhood. This proposal was characterized by modelling and simulation using OPNET network simulator. Finally we implemented the proposed mechanisms on hardware devices and deployed a sensors network in real situation to verify the accuracy of the model and evaluate the proposal according to different test configurations.

Keywords

Wireless Sensor Network, Medium Access Control, Quality of Service, Energy Saving, Mesh

Topology, IEEE 802.15.4, Indoor Monitoring Application

1. Introduction

Nowadays the aging population is constantly increasing and many countries are now facing an urgent requirement to provide appropriate home environment solutions for their citizens [1] [2]. These individuals attach a great importance to the autonomy that allows them to live mostly at home, and in their immediate environment, providing them freedom and a better quality of life. But, in the case of an accident such as a fall, faintness…, that autonomy can quickly turn into dependence. To supply solutions, some people use systems embedded on their body, such as physiological sensors or fall sensors [3]-[5]. These devices are intrusive and limitations become apparent due to the fact that the patient is often unable to use an alert system because either he is not wearing his equipment or, if he suddenly feels unwell, is unable to perform the alert activation gesture.

The solution we consider is to instrument the environment of the person. Indeed, by monitoring the main environmental characteristics of their living space, it seems to be possible to get a lifestyle pattern of the person [6]-[8]. For example, measuring temperature, humidity, luminosity, noise levels, presence..., in many strategic areas at home can provide useful data to interpret a physical activity in space and time. Data processing will determine circadian activity rhythms of the person and so will contribute to detect unusual situations and emergency cases. Generally, the challenge is to propose a suitable sensor network that allows uninterrupted data transmission in a bounded time.

Within this context, the objective of this work is to modelize and implement a complete heterogeneous sensor network allowing the measurement and the transmission of short-range data collected by the environmental sensors. The planned network will be deployed in a house or even building and transmit alert messages caused by a malfunction of environmental parameters via a continuous monitoring. So a limited scale, up to 50 nodes, seems to be sufficient for a home monitoring application. These nodes exchange data between them according to a communication protocol and, obviously, the following performance criteria are crucial to an indoor monitoring application: 1) Different QoS capacities should be provided for message delivery, especially when some time-sensitive messages are zero tolerance for packet loss. 2) Power consumption is a fundamental concern for the battery constrained sensor nodes. A long lifetime network, several months or even years, is eagerly expected. 3) The network should be self-organizing, quickly deployed and robust against link failure or link establishment. For example, the failure of critical nodes can lead to the entire network failure and harm the safety of people monitored by this network.

Our work focuses on MAC layer to address the above constraints. A mesh WSN is considered because it is more flexible and robust than a star or a tree topology. However, how to construct this topology and how to provide the QoS are some issues in a mesh WSN. The avoidance of collisions between 2-hop neighbours is another goal because there is scarcely interference at distance of more than 2 hops [9]. Also, as the largest energy consumption of the nodes is due to the time spent in the receive state [10], so time slot allocation in a mesh topology is an important task to strictly limit non efficient transceiver activities such as idle listening.

This paper aims to present a beacon-enabled MAC over IEEE 802.15.4 PHY which supports mesh topology and enables guaranteed service with low energy consumption. The paper is organized as follows: in Section 2 some related works are presented. In Section 3 the proposed MAC protocol is fully described. Section 4 provides simulation results and Section 5 presents the results obtained through practical measurements in a real environment. The last section concludes the paper.

2. Related Works

Several standards for low-power, low-cost and short-range network have been investigated. Among them, IEEE 802.15.4 is considered as a promising way in terms of energy saving and guaranteed medium access. Many other main standards such as ZigBee, IEEE 802.15.5 and 6LoWPAN are all based on IEEE 802.15.4 PHY/MAC or backwards compatible with this standard. Therefore, we consider IEEE 802.15.4 as a starting point for our work. In part 2.1, IEEE 802.15.4 standard is briefly presented. In part 2.2, previous scientific works including classical WSN MAC protocols and recent works based on IEEE 802.15.4, have been studied and analyzed. The focus is mainly on the metrics: QoS, energy saving and mesh topology.

2.1. IEEE 802.15.4 Standard

IEEE 802.15.4 [11] is a well-known standard for Low-Power and Low-Rate Wireless Personal Area Networks (LP-LR-WLAN). Several physical layers (PHYs) are proposed; the main and widely used 802.15.4 PHY operates in the 2.4 GHz radio band and enables transmissions at 250 kbps in a 30 m indoor range. The protocol supports beacon and non beacon mode. More precisely, in beacon mode, it is possible to achieve variable duty cycles (from 100% down to 0.006%), which is particularly interesting for our application where energy constraint and network lifetime are concerned. In addition, the beacon mode has an attractive feature for time-sensitive applications as QoS properties are available with GTS (Guaranteed Time Slot) mechanism. On the other side, the non beacon mode that has the advantage of a lower complexity and a higher scalability as compared with beacon mode, does not provide any of those features. Thus, we focus our development on the beacon mode of this standard.

IEEE 802.15.4 may operate in two topologies: a star topology or a peer-to-peer topology. Peer-to-peer topology allows a more complex network to be implemented such as cluster-tree topology. However, how to construct a mesh networking topology as required by our application is missing in the standard. In fact, while the current standard supports multi-hop networking using peer-to-peer topology, it restricts its use to a non beacon mode. This contradiction makes the interesting advantages, such as GTS and energy saving, disappear, especially on router nodes.

2.2. MAC Protocols for WSN

Some classical WSN MAC protocols such as S-MAC [12] and ContikiMAC [13] and their optimized works such as T-MAC [14] and X-MAC [15] are based on the sleep-wake cycle to reduce energy consumption of WSN. However, they all access the medium by CSMA (Carrier Sense Multiple Access) mechanism that provides a best-effort service. So the transmission of time-sensitive messages may not be guaranteed.

ZigBee specifications [16] clear the ambiguities of IEEE 802.15.4 in a cluster-tree topology. The centralized PAN (Personal Area Network) coordinator calculates and assigns a beacon transmission offset for each node when it wants to associate the PAN. Therefore, the network scalability and flexibility are both limited. The direct communications between neighbours are not possible.

Anis Koubâa also focuses his work in the field of cluster-tree topology. In TDBS [17] [18], the requirement of different BI (Beacon Interval) and SD (Superframe Duration) for each node is calculated in advance. However, this weakens the flexibility and robustness as well as restricts the scalability of network.

Another example has been proposed in OCARI project [19] [20]. A PAN coordinator is the destination of all association requests and allows a beacon slot for each associated node. The main drawback of this solution is the lack of flexibility, especially regarding a changing topology, for example due to the inconstancy of wireless medium.

P. S. Muthukumaran proposed MeshMAC protocol [21]. This protocol enables mesh networking over the beacon mode through a distributed SDS (Superframe Duration Scheduling) strategy in which each node calculates its schedule to transmit beacons based only on locally available information. The limitations of MeshMAC are: it imposes very low duty cycles for a large scale network so that the end-to-end delay may be extended; the beacon transmission offset is difficult to choose if the topology changes.

B. Carballido Villaverde proposed DBOP MAC protocol [22] [23]. It creates a BOP (Beacon Only Period) where beacons are transmitted at different time slots among neighbours and neighbours' neighbours. However, DBOP introduces an overhead into the IEEE 802.15.4 MAC standard. Another drawback is the inefficient management of BOP length. In addition, our MAC protocol was proposed before this work and we have continued to work on it until a prototype has been totally implemented.

Hence, we consider that all the current standards and protocols cannot satisfy all our application criteria. Many problems such as beacon collision, dynamic timeslot allocation, energy saving on router nodes, determinism medium access in a mesh topology, etc., have not been solved yet. This makes urgent requirement of new MAC protocols.

3. ADCF MAC Protocol

This part presents an original MAC protocol named as ADCF (Adaptive and Distributed Collision Free). The proposed protocol aims to improve robustness and flexibility of IEEE 802.15.4 MAC, which means the capacity of self-organization and auto-reparation. Simultaneously, ADCF should enable energy efficiency and guaranteed slots negotiation [24] [25]. Before showing the details, some assumptions should be highlighted: 1) the main contribution of ADCF concerns the router nodes. We have considered an ad-hoc approach: all the considered nodes have the capacity to be both sensor and router, even if classical sensors (Reduced Function Devices) may be associated to the routers, but this last point is out of the scope of this paper. 2) To ensure full determinism in the build of the topology, nodes addresses must be preliminary set during installation.

3.1. Overview of ADCF

ADCF is based on the IEEE 802.15.4 2.4 GHz DSSS physical layer and classical superframe structure. On the one hand, ADCF proposes a distributed beacon scheduling mechanism in a beacon only period which spatially reuses the timeslots over 2-hop. On the other hand, the contribution of ADCF lies in a data slot allocation mechanism, which makes GTS possible in a mesh topology.

Generally, the nodes can switch between two stages: initialization stage and working stage. In initialization stage, which is not using the superframe structure, the nodes access medium by unslotted CSMA/CA mechanism. In working stage, the time is divided into superframe and each superframe includes three parts: {BOP, Active Period, Inactive Period}.

As shown in **Figure 1**, BOP is organized by CFBS (Collision Free Beacon Slot). Each node has a 2-hop collision-free beacon slot in BOP. Active period is divided into 16 equally sized slots as a classical IEEE 802.15.4 superframe. It starts with the CAP (Contention Access Period) where medium accesses are done by using slotted CSMA/CA mechanism. It ends with CFP (Contention Free Period) where medium accesses are done by using an original mechanism that provides GTS-equivalent for the mesh topology. This original mechanism is based on the CFDS (Collision Free Data Slot). After active period, the optional inactive period allows all the nodes to go to sleep mode for energy saving.

In a mesh network, the node that uses the first CFBS of the superframe is called as initiator. Initiator is not a supernode as each node in the network may be selected as initiator; initiator is not a supernode as it only has a partial knowledge of the 2-hop neighbourhood, like the other nodes; initiator is not a supernode as the network could work properly when initiator fails; initiator is not a supernode as it has not any central role such as a classical IEEE 802.15.4 PAN-Coordinator.

In the rest of this part, CFBS, CFDS mechanisms as well as a smart repair protocol will be presented.

3.2. CFBS Mechanism

The beacon frame is very important as it announces the presence of the node and the presence of its 1-hop neighbours; it maintains network synchronization and is used to request/reply neighbour data slot negotiations, *i.e.* CFDS. The nodes far away more than 2 hops could reuse the same CFBS to enhance channel reutilization. In other words, nodes must determine their CFBSs by taking into account CFBSs used by their 2-hop neighbourhood with a certain priority.

To determine priority, three parameters of the node are employed: 1) ND (Neighbour Density) indicates the number of neighbours within 2 hops, including the node itself. The node with the higher ND is usually in the

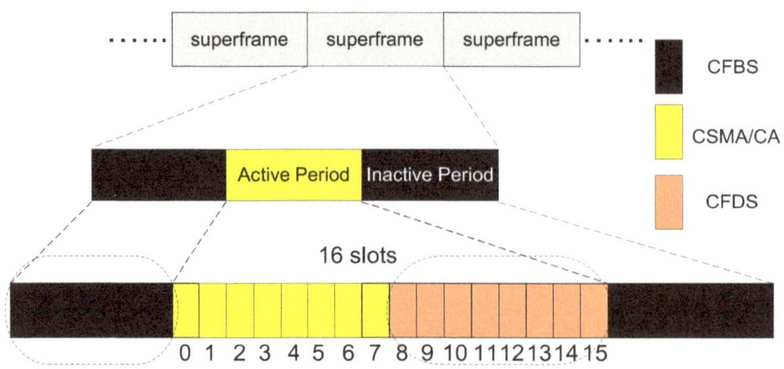

Figure 1. ADCF superframe structure.

center zone of a network and has a faster awareness to the topological changes. 2) NE (Neighbour Energy) indicates the residual energy level of the node and has the profound effect to the network. 3) NA (Neighbour Address) means the 16-bit address of the node and is unique for each node. The node with higher ND, higher NE and smaller NA has higher priority. The comparison order is ND, NE and then NA.

The node with the highest priority in the network is selected as the initiator. It specifies the beginning of BOP and measures the length of BOP (BOPL). The initiator's ND is defined as D_{max} and equals the length of BOP.

When a new node joins the network, it will firstly listen to the channel for a fixed period. Depending on the received beacons during listening, the new node will send its own beacon by different mechanisms. Each node broadcasts its beacon within 1-hop and records direct neighbours' in its NT (Neighbour Table). Therefore, all the 2-hop neighbours' information is obtained by this new node. The interesting information in a beacon includes ND, NE, NA, etc.

The node with the higher priority could choose its beacon slot earlier. Obviously, initiator occupies the first CFBS and its information is broadcasted without the hop limitation. Then other nodes, according to their priorities, can take the slots that are not used by their 2-hop neighbours and store the slot number in their NT. As each node only maintains a 2-hop NT, the nodes may choose CFBS synchronously and therefore quickly. With BOPL, each node can calculate the beginning of a superframe by the time of a received neighbour beacon and its slot number. At this moment, the node synchronizes with its neighbours and can schedule its own superframe. We can say that the node is in working stage now.

More details of CFBS mechanism have been published in [24], including beacon frame format, theoretical analysis and simulation study of protocol cost such as convergence time and the incurred message overhead.

3.3. CFDS Mechanism

For the time-bounded traffic or the traffic of zero-tolerance packet loss, CFDS enables these traffics to be sent in some dedicated slots. Medium access can be done directly, without backoff delays or medium sensing. This aspect is very useful in our application.

Thanks to CFBS, the CFDS negotiations can be achieved between source node and destination node using beacon exchanges. Since the beacons are broadcasted with guarantee through the CFBS, the CFDS allocation process can be time-bounded, unlike the *GTS_request* process of IEEE 802.15.4 that uses CSMA/CA. Similarly, nodes must determine their own CFDSs by taking into account CFDSs used by their 2-hop neighbourhoods.

For the source node, CFDS allocation is triggered by the upper layer, which contains the destination address and the length of requested CFDS. By setting flags of beacons, source node broadcasts this request to all its neighbours. When a source node receives a beacon from a destination node with the valid CFDS slot number and CFDS length, it will update the NT and broadcast the allocated slot in the next beacon. The new time schedule is also calculated in order to achieve this direct collision-free transmission. As there is no extra overhead for the CFDS negotiation, source node can continue requesting until the destination node finds the valid CFDS.

So for destination node, the objective is to search the available CFDS and respond to source node. The slot number must be decided by the receiver (destination) of the traffic to avoid frame collision. Specifically, when a node receives neighbour's beacon and finds its address as the destination, it will check its NT, allocate the first available data slot (e.g. slot 8 in **Figure 1**) to source node and announce this allocation in the next beacon. This available CFDS should not be reused by another node in the 2-hop neighbourhood. When source node requests several CFDSs, the slot number is noted as 8 if the following CFDSs are also available. Else, destination node searches until the end of active period (slot 15). If the available CFDSs are not enough, destination node also returns to the first available CFDS and provides CFDS services as more as possible.

The CFDS deallocation is also invoked by source node of this CFDS. Source node clears up all the flags and slot number and then sends the beacon. Destination node also clears up all the CFDS information with this source node from its NT when it receives the indicated beacon. Therefore, the deallocation is complete and the CFDS is free for other nodes.

CFDS mechanism allows point-to-point bidirectional communications in the mesh topology. When a destination node wishes to reserve CFDS with the corresponding source node, it launches the same procedure as single-direction communication. However, considering the indoor application environment and the possible large overhead (a 3-hop NT), the traffic sending by CFDS has no acknowledgement. This specific point has been successfully studied on our real testbed. In addition, if no CFDS used by the node at all, it may sleep during these data slots for further energy saving.

3.4. Smart Repair Protocol

The objective of this protocol is to minimize the impact of a change of topology as much as possible. The SRP (Smart Repair Protocol) is critical in ADCF as it improves network flexibility and robustness. Generally, topology changes could be classified as two types: node join and node failure. All other complex topological changes such as network separation and network integration are all based on the two basic changes.

In reality, node failure can be caused by energy exhaustion, unexpected damage or even the poor quality of wireless link. So a node considers the failure of its neighbour node only when the neighbour's beacon loss is above a threshold. Similarly, new node is regarded as neighbour only when its beacon loss is below a threshold. In our proposal, beacon loss threshold is a predefined parameter which depends on different wireless environments during the implementation. In other words, we estimate the wireless link quality by beacon loss. Four states are defined for a link or a node connected by the link. 1) Preliminary. 2) Unconfirmed. 3) Confirmed. 4) Deleted. Therefore, node join or node failure here means the presence or disappearance of one node in its neighbours' NT. Both unconfirmed nodes and confirmed nodes are stored as 1-hop neighbours in NT. But only confirmed nodes are chosen as 1-hop neighbours when constructing beacon frame.

Asymmetric links are possible and acceptable. In ADCF, both CFBS and CFDS mechanisms are over a 2-hop NT. If two neighbours connected with an asymmetric link choose the different slot to transmit the beacon or data frame, the collision could be avoided and that's what we expect. If two neighbours connected with an asymmetric link choose the same slot, they will transmit the beacon or data frame at approximate the same time. In this situation, this asymmetric link will be abandoned and other high quality link may be utilized. The redundant mesh link of ADCF also improves the communication reliability.

Based on the above definitions, the following two parts specify different mechanisms depending on different topology change cases.

1) Node join and BOP augmentation. Except full mesh network, multiple-hop network may reuse the medium so that there are always free slots in BOP. A new node may choose its CFBS directly after the listening period and join the network immediately. So ADCF has the great advantage to adapt the topology change. For full mesh network, if a new node wants to join the network and finds BOPL is not sufficient (D_{max} > BOPL), the new node may send its beacon by CSMA until a new initiator is designed with updated D_{max}. In this process, the nodes will return to initialization stage.

2) Node failure and BOP reduction. If link failure is detected by beacon loss mechanism, neighbours will simply delete this failing node from NT. If the initiator fails, others re-select an initiator but keep their BOP with the original slots. Therefore, the network will still work without disruption. In one case (D_{max} > BOPL/2), the nodes will also return to initialization stage for an optimal BOP.

In conclusion, our application requires an adaptive communication protocol providing QoS-guaranteed service with reasonable energy consumption in a mesh network. With ADCF, we proposed two mechanisms called CFBS and CFDS to enable efficient deterministic medium access for beacon and data. The Smart Repair Protocol may further improve the network flexibility and robustness. In the following parts, we will present the performance results obtained by the simulation and the prototyping work.

4. Simulation Study

To study the scope of our contribution, we use OPNET to develop a simulation model that implements the ADCF MAC. Many scenarios and metrics were simulated. For example, we investigated the protocol cost of initialization stage, the network performance with different topology change cases, and the network performance in large scale and high-density situations, etc. [25]. In this paper, we choose to present the comparison of ADCF with IEEE 802.15.4 (ZigBee version implementation [26]). An ideal IEEE 802.15.4 physical layer, indicating the channel without random packet loss, is applied for both protocols. A more practical energy model [27] is used in the simulation. A static routing mechanism achieved by prior manually adding routes is above ADCF in order to simulate application traffics over the network. Buffers sizes are consistent with the prototype. In addition, the topology is a random mesh configuration. Other basic parameters are shown in the **Table 1**.

According to our energy model, the actual energy consumption in reception mode is larger than the energy consumption in transmission mode. Therefore, even though there is no application traffic, the active period always keeps in receive mode, which consumes more energy. So only beacons are delivered in order to compare the protocols cost. As shown in **Figure 2**, it can be seen that ADCF consumes less energy, about 37%, compared to IEEE 802.15.4 as time goes by. This is because the active period is the same for all the ADCF nodes, while IEEE 802.15.4 router nodes have two active periods. One is for the communication with parents and the other is for the communication with children. Hence IEEE 802.15.4 spends more time for idle listening. In theory, no matter what topology, if BOPL is less than the active period, ADCF always consumes less energy than the IEEE 802.15.4 standard.

As shown in **Figure 3**, there are 7 sources with 1-hop traffic or 3 sources with multi-hop traffic. When Packet Interarrival Time decreases from 1.0 s to 0.1 s, the traffic load will increase. Therefore, End-to-end delay becomes larger. When Packet Interarrival Time is about 0.4 s, there are radical changes caused by buffer overflow.

With the current superframe (2 s) and active period configurations, the end-to-end delay for 1-hop traffic is about 1 s on average, including the time from packet generation to the scheduled data slot. ADCF has 0.06 s advantage than IEEE 802.15.4. If a packet arrives just before its scheduled data slot, the node can send this packet immediately. Otherwise, the node may send the packet in the next superframe. It also can be seen that ADCF saves about 25% end-to-end delay for multi-hop traffic. This is because some multi-hop traffic may be transmitted in one superframe as the same active period for all the mesh nodes. IEEE 802.15.4 works in a cluster-tree topology which may take several superframes from the source to the final destination. The average hop count is 3, so this end-to-end delay is about 2.7 s for ADCF and 3.5 s for IEEE 802.15.4. In some other topology cases, ADCF can be even more efficient since the 802.15.4 tree can be shortened thanks to the mesh links.

In the same simulation, when the CFDS buffers are available, both ADCF and IEEE 802.15.4 keep 100%

Table 1. Basic simulation parameters.

Parameter	Value
Scene area	$100 \times 100\ m^{-2}$
Transmission range	15 m
Number of nodes	14
Superframe period	2 s
Active period	0.245 s
Application payload	100 bits
CSMA buffer	0.5 k octets
CFDS buffer	1.5 k octets
Simulation duration	30 min
Simulation times	20

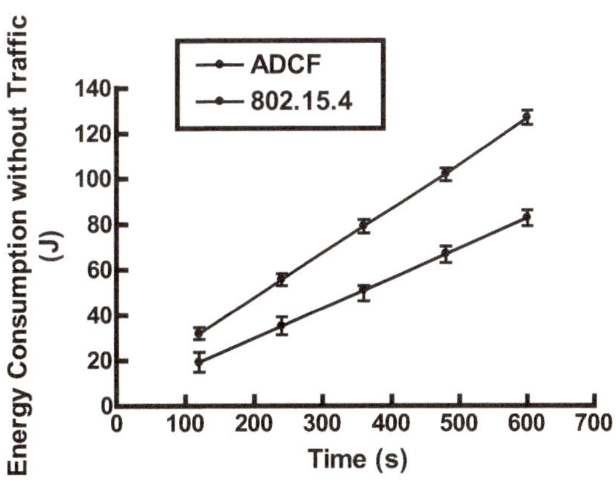

Figure 2. Energy consumption comparison.

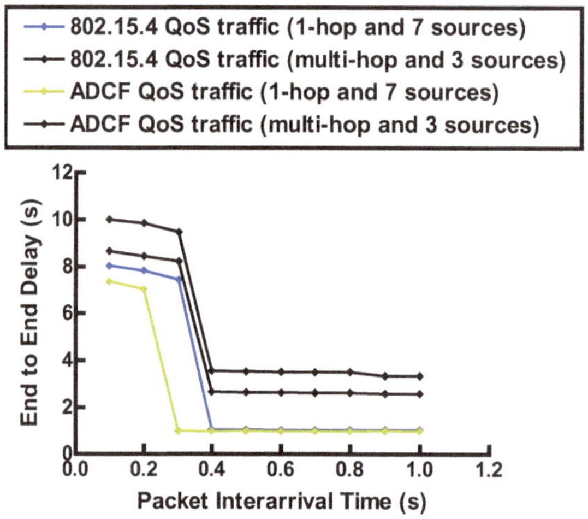

Figure 3. End-to-end delay comparison of traffic sent by CFDS.

packet success ratio, which indicates the percentage of successfully delivered packets.

By many other simulation works and results, we can conclude that ADCF satisfies our application request of delivering QoS messages with low energy consumption. Therefore, for the monitored people, environmental alerts, and remote assistant messages could be quickly sent to the destination without loss. The end-to-end delay is optimized, especially for the multi-hop alters traffic.

5. Prototype Implementation

Prototype implementation is a fundamental approach to verify a protocol and its performances. A platform WiNo [28] of emulation and rapid prototyping for WSN is utilized. WiNo is an open platform ready to accommodate protocols at MAC layer or network layer for sensor networks. It provides an open environment adapted to research projects, including the management of physical layer and the necessary tools to develop a full but very compact protocol stack using ANSI C-language. More accurately, a developer with WiNo can master not only the time access to the medium and the sleep-wake cycle, but also the CPU time and memory resource generally restricted by hardware in a WSN.

Currently, two types of nodes from Freescale, 13192-SARD and 1321x-SRB, are implemented with ADCF MAC. Both of them have the common features: 1) IEEE 802.15.4 2.4 GHz transceiver. 2) 8-bit MCU. 3) There are 4KB of RAM and 60KB of on-chip Flash. Two tools are mainly used for debugging, measuring and evaluating the protocol performance in real conditions: the date-logger console of nodes and SNA (Sensor Network Analyzer) [29]. Daintree's SNA is known as an expert tool providing comprehensive solution for IEEE 802.15.4 and ZigBee. It includes a protocol analyzer software fitted to sensor networks and a sensor network adapter hardware. The sensor network adapter hardware is a powerful node, which can capture all the packets of a network but do not participate in this network. By SNA software that includes a protocol decoder allowing user to drill down to packet, field and byte level, we can view all network nodes and interactions simultaneously, without disturbing the original network.

In part 5.1, two results obtained through practical measurements in our laboratory will be given. In part 5.2, the deployment of ADCF in a real application context using our smart home of Blagnac [30] will be presented.

5.1. Experimental Scenarios and Results

To build the multi-hop network, 8 nodes are gradually deployed in our laboratory. In each scenario we start each node at different time with a random MAC address. In fact, each multi-hop network is unique and has its own particularities. We do 20 times manipulations separately for the following two networks examples, as shown in **Figure 4** and **Figure 5**. In the first deployment, 8 nodes are disposed throughout the first floor of our laboratory. And in the second deployment, 4 nodes are disposed in the first floor, 3 nodes are disposed in the ground floor, 1 node is disposed on stairway. Even though the fixed positions, network topology may change due to the wireless

Figure 4. Network example 1.

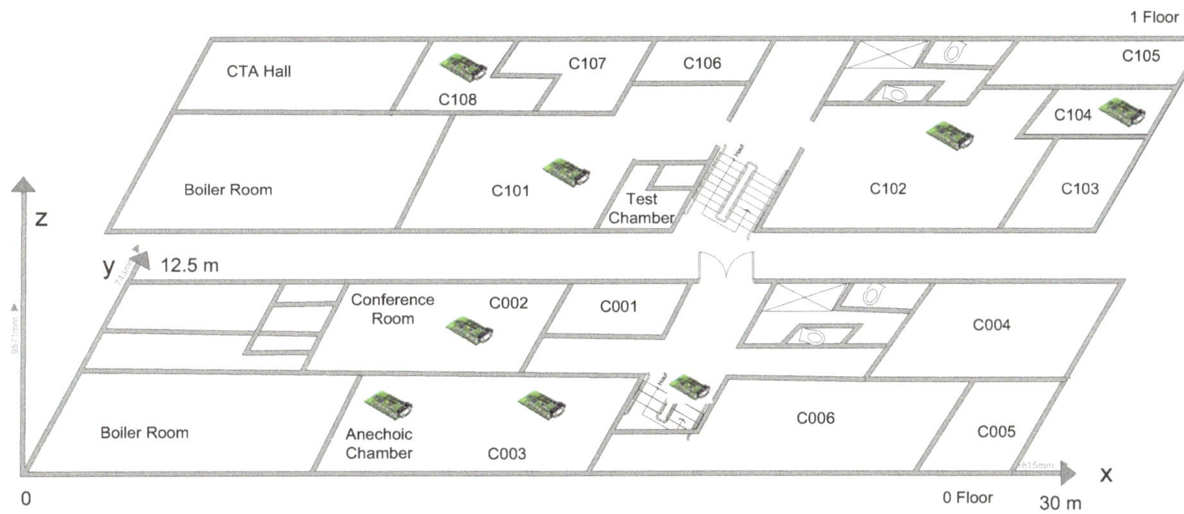

Figure 5. Network example 2.

links. Based on the SNA results, the maximum hop count is 3 in the two network examples. Some interesting results are shown in **Table 2**.

Here convergence time of the last node means the time duration from the start of the last node to a BOP well organized network in which each node has a collision-free beacon slot. In reality for the deployment of multi-hop topology, power-up of each node in the network is a process. At the same time, as in the mesh network each node can directly communicate with its neighbours, therefore network rebuilding may be triggered in this deployment process. We can see that this convergence time takes about 1 to 2 minutes on average. This value is acceptable considering the total network lifetime which is several days or weeks. In a normal situation where the nodes are fixed, the network topology may not be reorganized frequently. The reuse of CFBS confirms the successful achievement of the organized superframe and the synchronized multi-hop networks. In addition, this test highlights that asymmetric link occurs frequently in the real environment, but ADCF can work well without disturbing of asymmetric link problem.

5.2. Deployment of ADCF in a Smart Home

The smart home of Blagnac in France targets the elderly and the disabled living alone and provides them health

Table 2. Results for multi-hop network.

		Network example 1	Network example 2
	Average	123.99 s	97.74 s
Convergence time of the last node	Minimum	55.04 s	42.63 s
	Maximum	252.30 s	173.25 s
Reuse 1 CFBS		19/20	11/20
Reuse 2 CFBS		1/20	9/20
Asymmetric links		3/20	18/20

and medico-social assistance at home [31]. Many types of equipment, including wired products, mainly KNX-based, have been installed and are operational in this smart home. ADCF network fills some application gaps and provides an alternative wireless solution. Our testbed is also described in a video [32].

Six ADCF nodes were deployed in the smart home. From MAC point of view, these 6 nodes form a full mesh network. The nodes are connected with various sensors corresponding to the target of our application:

- Node @106 connected with a magnetic sensor can monitor the open/closure of the refrigerator,
- Node @107 connected with an infrared sensor can detect the motion of the person under his coverage area,
- Node @108 connected with an emergency button is worn by the user and may alert in case of fall, faintness, etc.
- Node @109 connected with a sensor carpet placed near the bed can detect the getting up of the user,
- Node @10A connected with a light can be switch on/off by an order of another ADCF node,
- Finally node @105 connected with screen is a sink for collecting and displaying all the network information. On the screen, we can see the superframe structure updating each second and log files from a web page [33].

From application point of view, the data exchanges are as following (**Figure 6**):

- Node @106 sends a message to node @105 when the fridge is opened or closed,
- Node @107 sends a message to node @105 when the person moves under the infrared sensor,
- Node @108 sends a message to node @105 when the person presses the emergency button,
- Node @109 sends a message to node @10A when the user arrives or leaves the carpet,
- Node @10A receives messages from node @109 and switches on the lighting when the user is on the carpet. It switches off the lighting when the user leaves the carpet. When the lighting is switched on/off, node @10A sends a message to node @105 containing a return state of lighting.
- Node @105 receives messages from node @106, @107, @108, @10A and logs actions in a journal. A voice synthesis software eSpeak [34] is also available and announces the reception of information from sensors (motion, carpet, etc.). Several scenarios using the sensor information have been implemented; for example, when the sink detects the opening of the fridge for more than 10 seconds, the user can hear a voice alert.

In the initialization stage or rebuilding stage, frames are transmitted immediately without any medium access control precaution. In working stage, frames are transmitted using CFDS.

Figure 7 shows a result of the above network deployment. From the ADCF journal and application journal displayed on the screen, we can see that each node can join or leave the network freely and the rest of the network works properly. Generally, the network rebuilding time is less than 10 s. At this moment, 6 nodes occupy 6 CFBS as they are all 1-hop neighbours. 5 CFDS are negotiated to transmit the corresponding application data. Thanks to a buzzer available on the nodes, which is activated for each data frame reception, we can verify that the bounded time is verified as expected. For example, end-to-end delay is always less than 2 s when the superframe is fixed as 1.5 s. It satisfies our application requirements. In addition, the superframe with its slot allocation is totally consistent with what we proposed in Section 3.

6. Conclusions and Perspectives

This paper presented an original MAC protocol ADCF based on the IEEE 802.15.4 standard. This protocol was designed to build and to maintain a wireless mesh sensor network providing QoS-guaranteed medium access and energy saving solution for home monitoring application. The work was organized in three phases: protocol

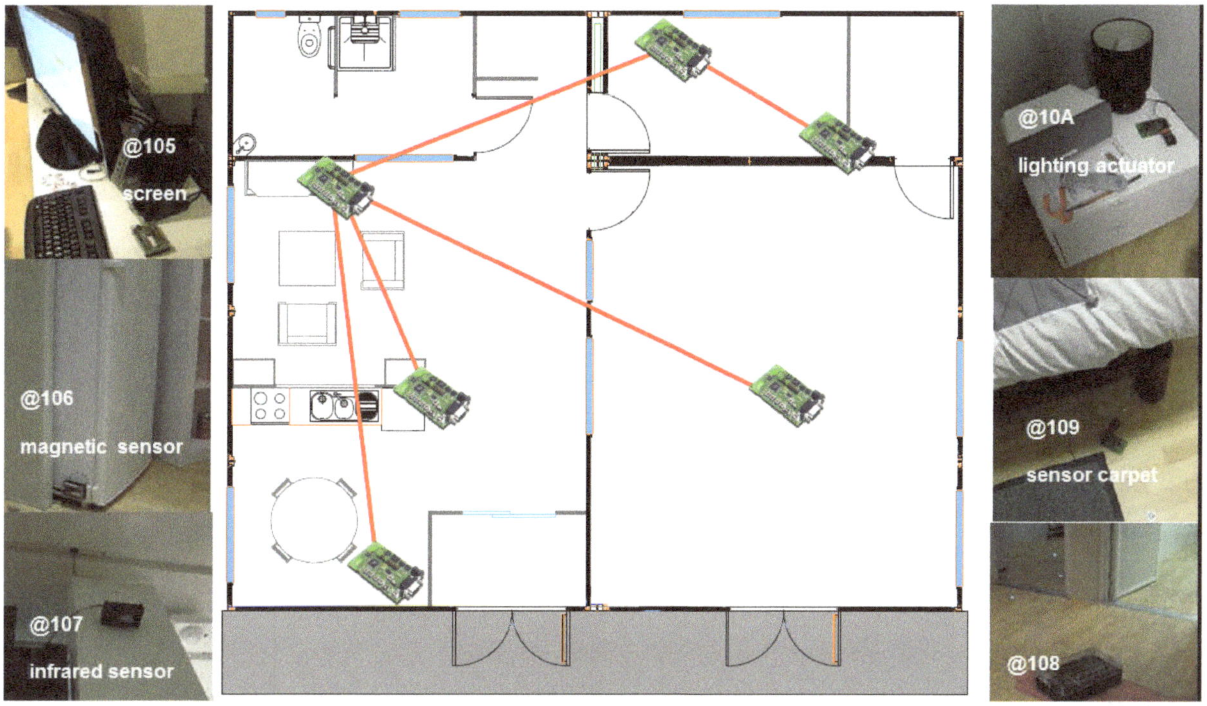

Figure 6. Six ADCF sensor nodes and the data transmission topology at application layer.

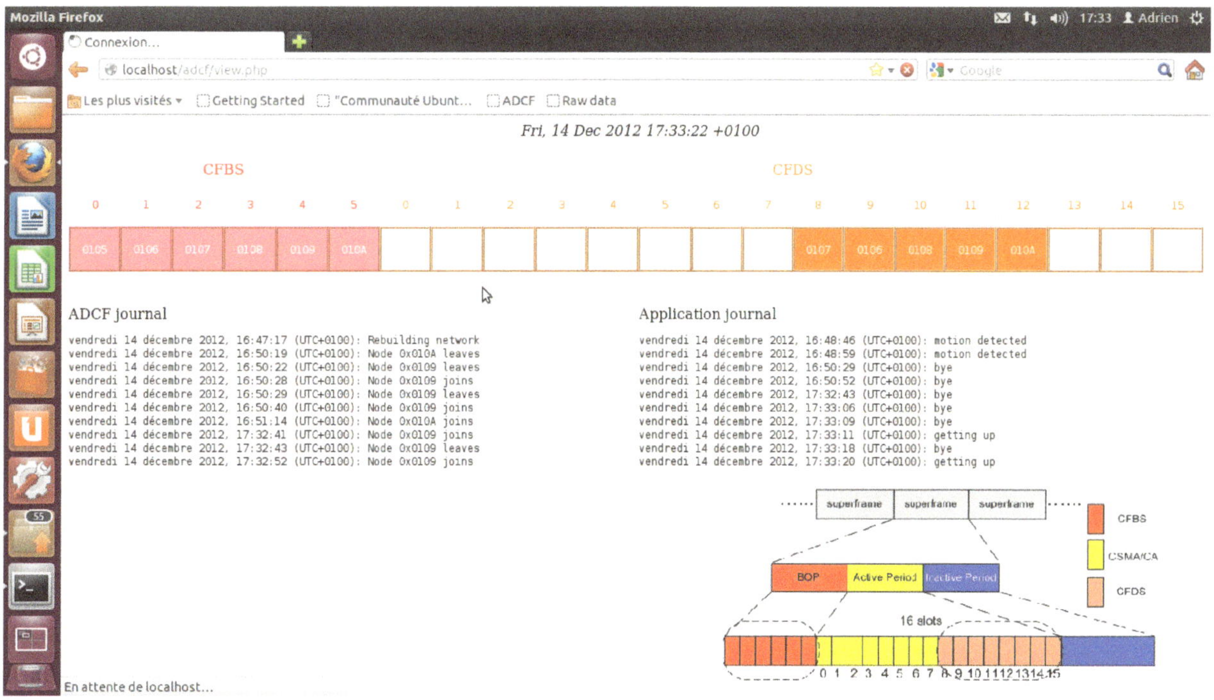

Figure 7. Example of a visual result.

presentation, implementation and results from simulation and from prototype. In conclusion, the contributions of this paper are:

1) The 2-hop CFBS and CFDS mechanisms were proposed and described. CFBS enables beacon collision to be avoided and therefore efficient multi-hop mesh network to be formed. Thanks to CFBS, CFDS enables the application data to be transmitted in a bounded time without loss. A smart repair protocol was also proposed to

improve the flexibility and robustness of a mesh network. It's efficient to manage BOPL according to the wireless topology changes.

2) We established a simulation model by OPNET. The simulation results show that ADCF satisfies our application requirements. For example of the specific topology, ADCF consumes less than 37% energy than IEEE 802.15.4. At the same time, end-to-end delay and packet success ratio show similar performance compared to the standard. Many other scenarios were simulated and the results may be good reference to prototype.

3) We developed some prototypes which were not considered in other works. By the results obtained through practical measurements, we can see: a mesh network of 8 nodes could successfully be built with an organized superframe where each node can choose a CFBS by ADCF. A full mesh network of 6 nodes was deployed in the smart home. This network can always converge in about 10 s with the suitable parameters configuration and, thanks to the buzzers we can hear the reception of an application data on the sink in 2 s at most, which confirms that ADCF is verified in a real environment as expected.

In the future, we will optimize the prototype such as enabling the evaluation of energy consumption of hardware and compare the results with the simulation. Another perspective is in the framework of cross-layer design, such as reuse the 2-hop NT, to reduce routing overhead and further optimize network performances. Finally, we will test various application scenarios and invite the elderly and the disabled to use the system in the smart home.

References

[1] Fnet, T. (2006) Counting on an Aging Population. *Computing in Science & Engineering*, **8**, 88-96. http://dx.doi.org/10.1109/MCSE.2006.107

[2] Boulanger, J.R. and Deroussent, C. (2008) Preliminary Based Service Evaluation for Elderly People and Healthcare Professional in Residential Home Care Units. *Second International Conference on the Digital Society* (*ICDS* 08), 93-101. http://dx.doi.org/10.1109/ICDS.2008.36

[3] Chen, S.-L., Lee, H.-Y., Chen, C.-A., *et al*. (2009) Wireless Body Sensor Network with Adaptive Low-Power Design for Biometrics and Healthcare Application. *IEEE Systems Journal*, **3**, 398-409. http://dx.doi.org/10.1109/JSYST.2009.2032440

[4] Otal, B., Alonso, L. and Verikoukis, C. (2009) Highly Reliable Energy-Saving Mac for Wireless Body Sensor Networks in Healthcare Systems. *IEEE Journal on Selected Areas in Communications*, **27**, 553-565. http://dx.doi.org/10.1109/JSAC.2009.090516

[5] Mamaghanian, H., Khaled, N., Atienza, D. and Vandergheynst, P. (2011) Compressed Sensing for Real-Time Energy-Efficient ECG Compression on Wireless Body Sensor Nodes. *IEEE Transactions on Biomedical Engineering*, **58**, 2456-2466. http://dx.doi.org/10.1109/TBME.2011.2156795

[6] Noury, N., Herve, T., Rialle, V., *et al*. (2000) Monitoring Behavior in Home Using a Smart Fall Sensor and Position Sensors. 1*st Annual International Conference on Microtechnologies in Medicine and Biology* (*ICMMB* 00), 607-610. http://dx.doi.org/10.1109/MMB.2000.893857

[7] Zatout, Y. and Campo, E. (2012) Using Wireless Technologies for Healthcare Monitoring at Home: A Survey. *IEEE* 14*th International Conference on e-Health Networking, Applications and Services* (*HealthCom* 12), 383-386. http://dx.doi.org/10.1109/HealthCom.2012.6379443

[8] Pomponio, L., Le Goc, M., Pascual, E. and Anfosso, A. (2011) Resident's Activity at Different Abstraction Levels: Proposition of a General Theoretical Framework. *IEEE 6th International Conference on Intelligent Data Acquisition and Advanced Computing Systems* (*IDAACS* 11), 540-545. http://dx.doi.org/10.1109/IDAACS.2011.6072825

[9] van den Bossche, A., Val, T. and Campo, E. (2007) Prototyping and Performance Analysis of a QoS MAC Layer for Industrial Wireless Network. 7*th International Conference on Fieldbuses and Networks in Industrial and Embedded Systems* (*IFAC* 07), **7**. http://dx.doi.org/10.3182/20071107-3-FR-3907.00028

[10] Mahfoudh, S. and Minet, P. (2009) Maximization of Energy Efficiency in Wireless Ad Hoc and Sensor Networks with SERENA. *Mobile Information Systems*, *Advances in Wireless Networks*, **5**, 33-52.

[11] IEEE 802.15.4 Standard (2006) Part 15.4: Wireless Medium Access Control (MAC) and Physical Layer (PHY) Specifications for Low-Rate Wireless Personal Area Networks (LR-WPANs). IEEE Standard for Information Technology, IEEE-SA Standards Board.

[12] Yoo, D.S., Park, S.S., Choi, S.S. and Park, S.H. (2008) Dynamic S-MAC Protocol for Wireless Sensor Networks Based on Network Traffic States. 14*th Asia-Pacific Conference on Communications* (*APCC* 2008), Tokyo, 14-16 October 2008, 1-5.

[13] Jeong Gil, K., Tsiftes, N., Dunkels, A. and Terzis, A. (2012) Pragmatic Low-Power Interoperability: ContikiMAC vs

TinyOS LPL. *9th Annual IEEE Communications Society Conference on Sensor, Mesh and Ad Hoc Communications and Networks* (*SECON* 2012), Seoul, 18-21 June 2012, 94-96. http://dx.doi.org/10.1109/SECON.2012.6276358

[14] Mihai, G., Alina, D.A. and Ion, B. (2010) Performance Analysis on T-MAC Protocol over a Body Area Network. *3rd International Symposium on Electrical and Electronics Engineering* (*ISEEE* 2010), Galati, 16-18 September 2010, 224-227. http://dx.doi.org/10.1109/ISEEE.2010.5628510

[15] Yang, O. and Heinzelman, W.B. (2012) Modeling and Performance Analysis for Duty-Cycled MAC Protocols with Applications to S-MAC and X-MAC. *IEEE Transactions on Mobile Computing*, **11**, 905-921. http://dx.doi.org/10.1109/TMC.2011.121

[16] ZigBee-Alliance. http://www.zigbee.org

[17] Koubaa, A., Cunha, A. and Alves, M. (2007) A Time Division Beacon Scheduling Mechanism for IEEE 802.15.4/ Zigbee Cluster-Tree Wireless Sensor Networks. *19th Euromicro Conference on Real-Time Systems* (*ECRTS* 2007), Pisa, 4-6 July 2007, 125-135. http://dx.doi.org/10.1109/ECRTS.2007.82

[18] Koubâa, A., Alves, M., Attia, M. and Van Nieuwenhuyse, A. (2007) Collision-Free Beacon Scheduling Mechanisms for IEEE 802.15.4/Zigbee Cluster-Tree Wireless Sensor Networks. *7th International Workshop on Applications and Services in Wireless Networks* (*ASWN* 2007), Santander, 24-25 May 2007.

[19] Dang, T., Devic, C., Livolant, E., Van Den Bossche, A. and Val, T. (2008) OCARI: Optimization of Communication for Ad Hoc Reliable Industrial Networks. *6th IEEE International Conference on Industrial Informatics* (*INDIN* 2008), Daejeon, 13-16 July 2008, 688-693. http://dx.doi.org/10.1109/INDIN.2008.4618189

[20] Alagha, K., Chalhoub, G., Guitton, A., Livolant, E., Mahfoudh, S., Minet, P., Misson, M., Rahme, J., Val, T. and van den Bossche, A. (2009) Cross-Layering in an Industrial Wireless Sensor Network: Case Study of OCARI. *Journal of Networks*, **4**, 411-420. http://dx.doi.org/10.4304/jnw.4.6.411-420

[21] Muthukumaran, P.S., de Paz, R., Špinar, R. and Pesch, D. (2010) MeshMAC: Enabling Mesh Networking over IEEE 802.15.4 through Distributed Beacon Scheduling. *AD HOC Networks*, **28**, 561-575. http://dx.doi.org/10.1007/978-3-642-11723-7_38

[22] Carballido Villaverde, B., De Paz Alberola, R., Rea, S. and Pesch, D. (2010) Experimental Evaluation of Beacon Scheduling Mechanisms for Multihop IEEE 802.15.4 Wireless Sensor Networks. *4th Conference on Sensor Technologies and Applications* (*SENSORCOMM* 2010), Venice, 18-25 July 2010, 226-231. http://dx.doi.org/10.1109/SENSORCOMM.2010.42

[23] de Paz Alberola, R., Villaverde, B.C. and Pesch, D. (2011) Distributed Duty Cycle Management for IEEE 802.15.4 Beacon-Enabled Wireless Mesh Sensor Networks. *IEEE 8th International Conference on Mobile Adhoc and Sensor Systems* (*MASS* 2011), Valencia, 17-22 October 2011, 721-726. http://dx.doi.org/10.1109/MASS.2011.78

[24] Lu, J., van den Bossche, A. and Campo, E. (2011) An Adaptive and Distributed Collision-Free MAC Protocol for Wireless Personal Area Networks. *6th International Symposium on Intelligent Systems Techniques for Ad hoc and Wireless Sensor Networks* (*IST-AWSN* 11), Niagara Falls, 19-21 September 2011, 798-803. http://dx.doi.org/10.1016/j.procs.2011.07.109

[25] Lu, J., van den Bossche, A. and Campo, E. (2012) Improving Robustness and Flexibility of MAC Layer for Guaranteed QoS Indoor Monitoring in Wireless Mesh Sensor Networks. *3rd International Conference on Wireless Communications in Unusual and Confined Areas* (*ICWCUCA* 12), Clermont Ferrand, 28-30 August 2012, 1-6. http://dx.doi.org/10.1109/ICWCUCA.2012.6402491

[26] www.open-zb.net/wpan_simulator.php

[27] Fourty, N., van den Bossche, A. and Val, T. (2012) An Advanced Study of Energy Consumption in an IEEE 802.15.4 Based Network: Everything but the Truth on 802.15.4 Node Lifetime. Elsevier, *Computer Communications*, **35**, 1759-1767.

[28] Van den Bossche, A. and Val, T. (2013) WiNo: Une plateforme d'émulation et de prototypage rapide pour l'ingénierie des protocoles en réseaux de capteurs sans fil. UbiMob, Nancy.

[29] http://www.daintree.net/sna/sna.php

[30] http://mi.iut-blagnac.fr

[31] Campo, E., Daran, X. and Redon, L. (2011) A Smart Home between Techniques Sciences and Human Sciences. *Une maison intelligente au carrefour des sciences technologiques et des sciences humaines*. 2nd Conference on Accessibility and Assistive Systems for Persons in Disabled Situation (*ASSISTH* 11), 33-42.

[32] http://irit.fr/~Adrien.Van-Den-Bossche/adcf_application_demonstration_video_v2.flv

[33] http://lab.iut-blagnac.fr/adcf/

[34] http://espeak.sourceforge.net/

5

A Packet-Interleaving Scheme Using RS Code for Burst Errors in Wireless Sensor Networks

Tsang-Ling Sheu, Yen-Hsi Kuo, Zi-Tsan Chou*

Department of Electrical Engineering, National Sun Yat-Sen University, Kaohsiung Taiwan
Email: sheu@ee.nsysu.edu.tw, ysk@atm.ee.nsysu.edu.tw, *ztchou@mail.ee.nsysu.edu.tw

Abstract

In this paper, we propose a packet-interleaving scheme (PIS) for increasing packet reliability under burst errors in wireless sensor networks (WSN). In a WSN, packet errors could occur due to weak signal strength or interference. These erroneous packets have to be retransmitted, which will increase network load substantially. The proposed PIS, encoding data using Reed-Solomon (RS) codes, can classify data into two different types: high-reliability-required (HRR) data and non-HRR data. An HRR packet is encoded with a short RS symbol, while a non-HRR packet with a long RS symbol. When an HRR and a non-HRR packet arrive at a sensor, they are interleaved on a symbol-by-symbol basis. Thus, the effect of burst errors (BE) is dispersed and consequently the uncorrectable HRR packets can be reduced. For the purpose of evaluation, two models, the uniform bit-error model (UBEM) and the on-off bit-error model (OBEM), are built to analyze the packet uncorrectable probability. In the evaluation, we first change the lengths of BE, then we vary the shift positions in a BE period, and finally we increase the number of correctable symbols to observe the superiority of the proposed PIS in reducing packet uncorrectable probability.

Keywords

WSN, RS Code, Burst Errors, Interleaving, Packet Uncorrectable Probability

1. Introduction

Along with the increasing requirements for living quality and home security, sensors have been widely deployed inside or outside a building to collect environmental information, such as temperature, humidity, image, motion

*Corresponding author.

picture, etc. To effectively deliver the collected data back to a control center for further analysis, a wireless sensor network (WSN) [1]-[3] is usually built. However, packet transmission over a WSN may encounter intermittent errors due to weak signals or interferences. The erroneous packets, if comprising of text or numbers, such as temperature or humidity, would require packet re-transmission, which surprisingly increases network load. Thus, the motivation of this paper is to increase transmission reliability over a WSN, which has recently attracted many researchers' attention.

Basically, previous researches on transmission reliability over a WSN can be divided into three major categories: reliable routing, information coding, and network coding. In the first category, to increase the transmission reliability after data are collected by a sensor node, relay nodes (RNs) are employed. For examples, H. Chebbo, *et al.* [4] modified IEEE 802.15.4 MAC frames. The authors added one bit in the frame control field, with which whether it is necessary to build a tree by RNs or just build a simple star, can be determined. Moreover, R. Sampangi, *et al.* [5] utilized RNs to divide sensors into several cluster networks. Since the distance from a sensor to its cluster head is reduced, the quality of data transmission is greatly improved. To protect the routing path, S. Kim, *et al.* [6] utilized both coding and retransmission schemes once the established path fails. However, in these schemes, it is inevitable that end-to-end packet delay will increase accordingly due to multiple-hop forwarding.

Thus, in the second category, instead of developing reliable routing, the authors switch their interests to information coding. For examples, E. Byrne, *et al.* [7] designed a coding scheme which can increase the probability of successful decoding based on graph theory. Y. Hamada, *et al.* [8] proposed a scheme to reduce packet error rate by using Luby Transform (LT) codes [9]. Their proposed schemes have achieved small complexity of $O(n)$; yet too many packets are required for retransmission when bit error rate is high. Thus, K. Ishibashi, *et al.* [10] proposed embedded forward error control (FEC) technique which utilizes RS (Reed Solomon) code to reduce packet error rate. Similarly, M. Busse, *et al.* [11] can recover lost chunk by using Fountain code and Raptor code. To increase data reliability and processing speed, K. Yu, *et al.* [12] designed new FEC which protects header and payload, respectively. Similarly, M. Srouji, *et al.* [13] proposed a reliable data transfer scheme which can adjust the lengths of redundancy code based on the successful receiving rate at the downstream node.

In the third category, to increase transmission reliability, researchers focus on network coding. For examples, to increase the probability of successfully recovering the transmitted data, G. Arrobo, *et al.* [14] proposed cooperative network coding. Similarly, A. Taparugssanagorn, *et al.* [15] proposed unequally error protection (UEP) scheme. In UEP, data are classified into two types: medical and non-medical. Only medical data are EX-ORed, while non-medical data are not. However, in UEP, it is not guaranteed that a downstream sensor node can always decode data successfully. Thus, I. Salhi, *et al.* [16] proposed another reliable coding scheme where a relay node can produce much better packet combination through the exchange of neighboring table and topology interference information. Finally, Z. Kiss, *et al.* [17] tried to combine two coding techniques, RS coding and network coding. Once a relay node receives data packets transmitted from all the downstream sensor nodes, it begins to chop every data packet into smaller segments. Each segment is encoded into two symbols using RS codes. The two symbols are then converted to two checking blocks before data transmission. A receiving node can therefore increase the probability of correct data by decoding these two checking blocks.

Unlike the previous research work, where the authors focused on reliable routing, information coding, and network coding, in this paper, we propose a packet interleaving scheme (PIS) to reduce the impact of burst errors (BE) on high-reliability-required (HRR) data in WSNs. BE is defined as a sequence of transmitted bits which may encounter errors in continuous bit pattern. Although Reed-Solomon (RS) codes [18], which encode data in a block-by-block fashion, may correct bit errors in a certain constraints, it may not be economically worthy in dealing with burst errors when the number of consecutive bit errors exceeds a threshold. In other words, it makes the overhead of redundancy codes become unacceptable. Hence, to increase packet correctable probability in a WSN, the proposed PIS first classifies the collected data into two different types: HRR data and non-HRR data. An HRR packet is encoded with a short RS symbol, while a non-HRR packet with a long RS symbol. When an HRR and a non-HRR packet arrive at a sensor, they are interleaved on a symbol-by-symbol basis. The noticeable benefit from the packet interleaving is that the burst errors are dispersed and the uncorrectable probability of HRR packets is significantly reduced.

The remainder of this paper is organized as follows. In Section 2, we introduce RS codes employed in a WSN. In Section 3, the proposed PIS and its operations are described. In Section 4, an analytical model is built using two bit-error models, the uniform bit-error model (UBEM) and the on-off bit-error model (OBEM). Finally, conclusions are drawn in Section 5.

2. Reed-Solomon Codes

2.1. Forward Error Correction

Forward Error Correction (FEC) codes have been widely used in today's unreliable communication system, such as wireless networks, to reduce bit error rate (BER). The basic principle behind FEC is to generate redundancy code by encoding the user data; the former is then appended to the latter before transmission. At the receiver, user data together with the redundancy code are examined so that a certain pattern of bit errors can be corrected. FEC is composed of two categories, convolution code (CC) and block code (BC) [3]. The former has been applied to the transmission for continuous data, such as video streaming. Encoding process in CC requires relatively larger memory to store the previous, the current, and the upcoming information. As shown in **Figure 1(a)**, each symbol in the k symbols, $(m_0, m_1, \cdots, m_{k-1})$ is Exclusive-ORed with the previous or the upcoming symbols. As a result, n symbols $(c_0, c_1, \cdots, c_{n-1})$ are produced from k symbols. BC, on the other hand, has been used in encoding discrete data on the basis of block by block. Thus, BC is more applicable to a sensor network, where data, such as temperature, moisture, etc., are collected discretely in a block size. The basic idea behind BC is to encode k-symbol information with a division operation. Redundancy codes with n-k symbols, $(r_0, r_1, \cdots, r_{n-k-1})$, are generated after the division operation. Before the transmission to an upper stream node, the n-k symbols are attached to the original k symbols, as shown in **Figure 1(b)**.

2.2. RS Encoding

1) It is well known that Reed-Solomon (RS) code is one of the most popular block codes. RS Code is usually represented by $GF(2^m)$, where GF is the Galois Field [19] and m denotes the number of bits in a symbol. Alternatively, RS code can be represented by (n, k), where $n = 2^m - 1$ denotes the total number of symbols encoded, k is number of symbols for user data, and (n-k) denotes the number of redundancy symbols. Hence, $t = \frac{n-k}{2}$ denotes the number of erroneous symbols that can be corrected.

2) Let us briefly review the operations of RS codes. Before a message is transmitted using RS encoding, the message is segmented into a number of m-bit symbols. The k m-bit symbols can be represented by a polynomial: $M(X) = \sum_{i=0}^{k-1} m_i X^i = m_{k-1}X^{k-1} + m_{k-2}X^{k-2} + \cdots + m_1 X^1 + m_0$. A generator polynomial, $G(X)$, is defined as $G(X) = \prod_{i=0}^{2t-1}(X - \alpha_i) = (X - \alpha_0)(X - \alpha_1)\cdots(X - \alpha_{2t-1})$, which is used to calculate a parity polynomial, $P(X)$; *i.e.*, $P(X) = M(X)X^{2t} \bmod G(X)$. Finally, codeword, $C(X)$, is calculated by $C(X) = M(X)X^{2t} + P(X)$. The coefficients of polynomial $C(X)$ are the final encoded data to be transmitted.

3) For example, a RS code has $m = 3$ and $k = 3$ or (n, k) = (7, 3); *i.e.*, a message is segmented into a num-

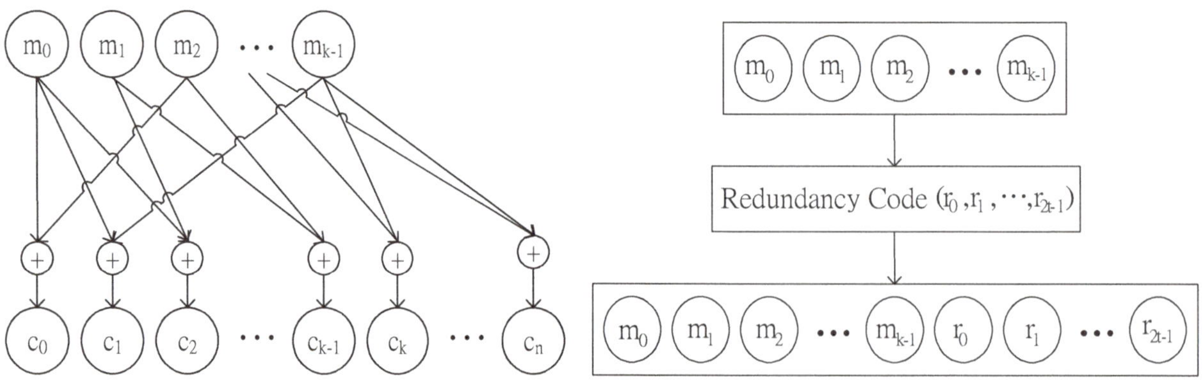

Figure 1. Convolution code versus block code. (a) Convolution Code; (b) Block Code.

ber of 3-bit symbols. The 8 different symbols and their binary representations are shown in **Table 1**.

We assume the original user data is 9 bits: 001 010 011. Since a symbol consists of 3 bits and $t=(n-k)/2=(7-3)/2=2$. From **Table 1**, the three symbols are represented by $\alpha_1\alpha_2\alpha_3$. Thus, we have $M(X)=\sum_{i=0}^{2} m_i X^i = \alpha_1 X^2 + \alpha_2 X^1 + \alpha_3$ and $G(X)=\prod_{i=0}^{3}(X-\alpha_0)(X-\alpha_1)(X-\alpha_2)(X-\alpha_3)=X^4+\alpha_3X^3+\alpha_0X^2+\alpha_1X+\alpha_3$.

Parity polynomial is computed as $P(X)=M(X)X^4 \bmod G(X)=0X^3+0X^2+\alpha_0X+\alpha_3$. Thus, $C(X)=M(X)X^4+P(X)=\alpha_1X^6+\alpha_2X^5+\alpha_4X^4+\alpha_0X^3+\alpha_0X^2+\alpha_1X+\alpha_4$. From **Table 1**, we know the encoded data to be transmitted are 001 010 100 000 000 001 100.

2.3. RS Decoding

Due to noise or interference, data may encounter bit errors over wireless link. Let *R*(*X*) be the data received, *C*(*X*) be the data transmitted, and *E*(*X*) be the data in error. Thus, we have, $R(X)=C(X)+(X)$. To determine whether or not *R*(*X*) requires any corrections, we have to compute *E*(*X*) by their syndromes: $(S_1,S_2,\cdots,S_{2t})$. From Equation (1), if $S_i=0$, then $E(\alpha_i)=0$, which implies $R(\alpha_i)=C(\alpha_i)$; *i.e.*, no bit errors occur. Otherwise, bit errors occur and *E*(*X*) is computed.

$$S_i=R(\alpha_i)=C(\alpha_i)+E(\alpha_i),\quad i=1,2,\cdots,2t \tag{1}$$

If bit errors do occur, error locations and error values are required computation. Assume there are v errors appear at the locations, $X^{l1},X^{l2},\cdots,X^{lv}$. That is, $E(X)=X^{l1}+X^{l2}+\cdots+X^{lv}$ and their syndromes can be expressed as in Equation (2).

$$\begin{cases} S_1=E(\alpha_1)=\alpha_1^{l1}+\alpha_1^{l2}+\cdots+\alpha_1^{lv} \\ S_2=E(\alpha_2)=\alpha_2^{l1}+\alpha_2^{l2}+\cdots+\alpha_2^{lv} \\ \vdots \\ S_{2t}=E(\alpha_{2t})=\alpha_{2t}^{l1}+\alpha_{2t}^{l2}+\cdots+\alpha_{2t}^{lv} \end{cases} \tag{2}$$

Next, if we can find $(\alpha^{l1},\alpha^{l2},\cdots,\alpha^{lv})$ such that all syndromes are equal to zero, then the errors can be located from the exponents of α. Let $\beta^i=\alpha^{li},\ \forall i=1,2,\cdots,v$. The syndromes can be expressed as in Equation (3).

$$\begin{cases} S_1=\beta_1^1+\beta_1^2+\cdots+\beta_1^v=\sum_{i=1}^{v}\beta_1^i \\ S_2=\beta_2^1+\beta_2^2+\cdots+\beta_2^v=\sum_{i=1}^{v}\beta_2^i \\ \vdots \\ S_{2t}=\beta_{2t}^1+\beta_{2t}^2+\cdots+\beta_{2t}^v=\sum_{i=1}^{v}\beta_{2t}^i \end{cases} \tag{3}$$

Table 1. An example of RS code with 8 symbols.

Symbols	$b_2b_1b_0$
α_0	000
α_1	001
α_2	010
α_3	011
α_4	100
α_5	101
α_6	110
α_7	111

Thus, the polynomial of error locations can be expressed as shown in Equation (4).

$$\sigma(X) = \prod_{i=1}^{v}\left(1+\beta^{i}\right) = \sigma_v X^v + \sigma_{v-1}X^{v-1} + \cdots + \sigma_1 X + 1 \tag{4}$$

To compute σ_i, we let $X = \left(\beta^i\right)^{-1}$ and substitute it into Equation (4). Next, we let $\sigma(X) = 0$, which results to Equation (5), where $i = 1, 2, \cdots, v$ and $l = 1, 2, \cdots, 2t - v$. From Equation (5), we can derive Equation (6).

$$\sigma_v \left(\beta^i\right)^l + \sigma_{v-1}\left(\beta^i\right)^{l+1} + \cdots + \left(\beta^i\right)^{l+v} = 0 \tag{5}$$

$$\sum_{i=1}^{v}\left(\sigma_v \left(\beta^i\right)^l + \sigma_{v-1}\left(\beta^i\right)^{l+1} + \cdots + \left(\beta^i\right)^{l+v}\right) = 0 \tag{6}$$

Expanding Equation (6), the result is shown in Equation (7).

$$\sum_{i=1}^{v}\sigma_v \left(\beta^i\right)^l + \sum_{i=1}^{v}\sigma_{v-1}\left(\beta^i\right)^{l+1} + \cdots + \sum_{i=1}^{v}\left(\beta^i\right)^{l+v} = 0 \tag{7}$$

From Equation (3) and Equation (7), we can derive Equation (8).

$$\sigma_v S_l + \sigma_{v-1} S_{l+1} + \cdots + S_{l+v} = 0 \tag{8}$$

Substituting $l = 1, 2, \cdots, 2t - v$ to Equation (8), we have derive Equation (9).

$$\begin{cases} S_1\sigma_v + S_2\sigma_{v-1} + \cdots + S_v\sigma_1 = -S_{v+1} \\ S_2\sigma_v + S_2\sigma_{v-1} + \cdots + S_{v+1}\sigma_1 = -S_{v+2} \\ \vdots \\ S_{2t-v}\sigma_v + S_{2t-v+1}\sigma_{v-1} + \cdots + S_{2t-1}\sigma_1 = -S_{2t} \end{cases} \tag{9}$$

Equation (9) can be expressed by a matrix equation as shown in Equation (10). By applying the operation of inverse matrix, we can compute σ_i.

$$\begin{bmatrix} S_1 & S_2 & \cdots & S_v \\ S_2 & S_3 & \cdots & S_{v+1} \\ \vdots & \vdots & \ddots & \vdots \\ S_v & S_{v+1} & \cdots & S_{2t-1} \end{bmatrix} \begin{bmatrix} \sigma_v \\ \sigma_{v-1} \\ \vdots \\ \sigma_1 \end{bmatrix} = \begin{bmatrix} -S_{v+1} \\ -S_{v+2} \\ \vdots \\ -S_{2v} \end{bmatrix} \tag{10}$$

Substituting σ_i to Equation (4), we can find β^i; *i.e.*, the error locations. To compute error values, we define them as $e'_v, e'_{v-1}, \cdots, e'_1$ as shown in Equation (11).

$$R(X) = C(X) + E'(X) = C(X) + e'_v X^v + e'_{v-1}X^{v-1} + \cdots + e'_1 X \tag{11}$$

Let $X = \alpha_1, \alpha_2, \cdots, \alpha_v$. By substituting X to Equation (11), we have derive Equation (12).

$$\begin{cases} e'_v\alpha_1^v + e'_{v-1}\alpha_1^{v-1} + \cdots + e'_1\alpha_1 = S_1 \\ e'_v\alpha_2^v + e'_{v-1}\alpha_2^{v-1} + \cdots + e'_1\alpha_2 = S_2 \\ \vdots \\ e'_v\alpha_v^v + e'_{v-1}\alpha_v^{v-1} + \cdots + e'_1\alpha_v = S_v \end{cases} \tag{12}$$

Since Equation (12) has v equations and v unknown, it is easy to compute v. Finally, errors can be corrected by adding error values v to their corresponding positions. At last, let us give an example. We assume the received data = 001 010 000 011 000 001 011, which can be represented by a polynomial $R(X) = \alpha_1 X^6 + \alpha_2 X^5 + \alpha_0 X^4 + \alpha_3 X^3 + \alpha_0 X^2 + \alpha_1 X + \alpha_3$. We then calculate the 4 syndromes, $S_1 = R(\alpha_1) = \alpha_2$, $S_2 = R(\alpha_2) = \alpha_1$, $S_3 = R(\alpha_3) = \alpha_6$, and $S_4 = R(\alpha_4) = \alpha_6$. By substituting them into Equation (10), we obtain the error locations as shown in Equation (13).

$$\begin{bmatrix} \alpha_3 & \alpha_5 \\ \alpha_5 & \alpha_6 \end{bmatrix} \begin{bmatrix} \sigma_2 \\ \sigma_1 \end{bmatrix} = \begin{bmatrix} \alpha_6 \\ 0 \end{bmatrix} \tag{13}$$

$$\begin{bmatrix}\sigma_2\\ \sigma_1\end{bmatrix}=\begin{bmatrix}\alpha_1 & \alpha_0\\ \alpha_0 & \alpha_5\end{bmatrix}\begin{bmatrix}\alpha_6\\ 0\end{bmatrix}=\begin{bmatrix}\alpha_0\\ \alpha_6\end{bmatrix} \tag{14}$$

By multiplying an inverse matrix on both sides in Equation (13), we have derive Equation (14). Next, if we substitute σ_2 and σ_1 to an error-location polynomial, we have $\sigma(X)=\alpha_0X^2+\alpha_6X+\alpha_0$. Now, we can derive $\sigma(\alpha_0)=\alpha_6$, $\sigma(\alpha_1)=\alpha_2$, $\sigma(\alpha_2)=\alpha_6$, $\sigma(\alpha_3)=0$, $\sigma(\alpha_4)=0$, $\sigma(\alpha_5)=\alpha_2$, and $\sigma(\alpha_6)=\alpha_0$. Hence, the two error locations are found, which show us the third and the fourth symbols are in error. To compute these two error values, we need to compute S_1 and S_2 first, as shown in Equation (15). Thus, the two error values can be computed from Equation (16).

$$\begin{bmatrix}\alpha_1^3 & \alpha_1^4\\ \alpha_2^3 & \alpha_2^4\end{bmatrix}\begin{bmatrix}e_1'\\ e_2'\end{bmatrix}=\begin{bmatrix}S_1\\ S_2\end{bmatrix} \tag{15}$$

$$\begin{bmatrix}e_1'\\ e_2'\end{bmatrix}=\begin{bmatrix}\alpha_2 & \alpha_5\\ \alpha_0 & \alpha_4\end{bmatrix}\begin{bmatrix}\alpha_2\\ \alpha_1\end{bmatrix}=\begin{bmatrix}\alpha_3\\ \alpha_3\end{bmatrix} \tag{16}$$

Since $E(X)=\alpha_3X^4+\alpha_3X^3$, the received data are corrected as, $C'(X)=R(X)+E(X)=\alpha_1X^6+\alpha_2X^5+\alpha_3X^4+\alpha_0X^3+\alpha_0X^2+\alpha_1X+\alpha_3$, which in binary representation is 001 010 011 000 000 001 011.

3. Packet Interleaving Scheme

3.1. WSN with Multi-Hop Tree Structure

In a wireless sensor network (WSN), a coordinator is the sink which gathers all the data collected from other distant sensor nodes. To facilitate data gathering from all the sensor nodes, it is very constructive that the coordinator and the sensor nodes will collaborate to build a multi-hop tree structure (MTS), as shown in **Figure 2**. In an MTS-based WSN, data collected by a sensor node will be forwarded hop-by-hop to the coordinator. Thus, by fully utilizing a branch node of the MTS, in this paper, we propose a packet interleaving scheme (PIS) based on RS codes to reduce the impact of burst errors on packet uncorrectable probability.

3.2. Packet Interleaving

In the proposed PIS, packets collected by a sensor node are classified into two different types: high-reliability required (HRR) packet and non-HRR packet. An HRR packet is defined as a packet which requires for retransmission, if uncorrectable burst errors exist. Payload in an HRR packet consists of numerical data, such as temperature, moisture, luminance, etc. This type of packet has relatively shorter data length (usually, a couple of bytes) and each uncorrectable HRR packet requires for retransmission. Hence, it is better to employ a shorter-length symbol (in this paper, we use $m = 4$) to encode an HRR packet with shorter data length. On the other

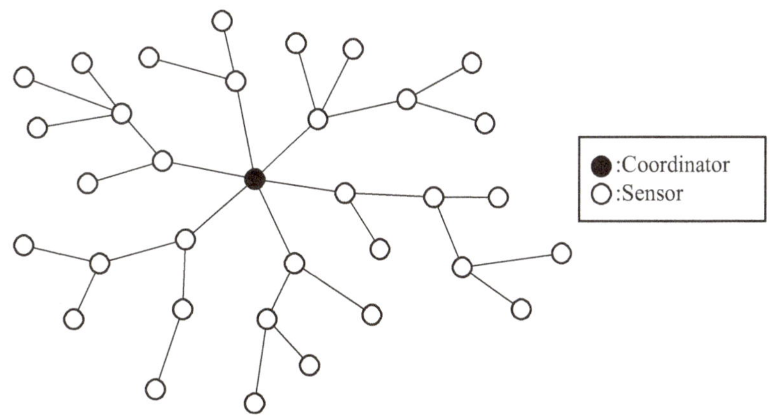

Figure 2. WSN with multi-hop tree structure.

hand, payload in a non-HRR packet consists of non-numerical data, such as video, audio, etc. This type of packet has relatively longer data length (usually, in the order of kilo bytes) and each uncorrectable non-HRR packet may not require for retransmission. Thus, it is better to employ longer-length symbol (in this paper, we use m = 8) to encode a non-HRR packet with longer data length.

As it is illustrated in **Figure 3**, an HRR packet is encoded with a shorter-length RS symbol (i.e., m = 4). First, an M-byte MAC-layer header and payload is converted to $\frac{(M\times 8)}{4}=2M$ symbols on the basis of 4 bits per symbol. Thus, we have the length of a codeword is n, where $n=2^m-1=2^4-1=15$, the number of symbols for user data in a codeword is k, where $k=n-2\times t=15-2\times t$, and the number of symbols for redundancy code in a codeword is $2\times t$. Let $N=\left\lceil \frac{2M}{k} \right\rceil$, which denotes the number of codeword required for encoding the M-byte packet (header plus payload). The total redundancy code (or FCS) is therefore equal to $N\times 2\times t$ symbols, or $N\times 2\times t\times 4\ (=N\times t\times 8)$ bits.

Similarly, as it is illustrated in **Figure 4**, a non-HRR packet is encoded with a longer-length symbol (m = 8). An M-byte MAC-layer header and payload is converted to $\frac{(M\times 8)}{8}=M$ symbols on the basis of 8 bits per symbol. Since the length of a codeword, $n=2^m-1=2^8-1=255$ bytes, which is much greater than the maximum length of a packet, which is 127 bytes in a WSN, a codeword is sufficiently enough to encode a MAC-layer packet in a WSN. Thus, the length of redundancy code (or FCS) is equal to $2\times t$ bytes.

When both an HRR and a non-HRR packet are received by a branch node in a tree-structured WSN, these two packets are interleaved on a symbol-by-symbol basis, as shown in **Figure 5**. The interleaved packet is then forwarded to an upper stream node, which performs decoding and correction process. However, as shown in **Figure 6**, an interleaved packet may not be correctable, if it encounters burst errors where the number of total errors is greater than t. In the proposed PIS, by separating the interleaved single packet back to their original two packets, each individual packet may become correctable. This is because the number of errors in each separated packet is highly possible to be smaller than t.

4. Mathematical Model and Analysis

In RS codes, whether or not a coded packet is correctable is mainly determined by the number of erroneous symbols. In this section, to study the robustness and the error-correction capability of the proposed PIS, we build two mathematical models for comprehensive numerical simulations. The first one is referred to as uniform bit error model (UBEM), while the second one is referred to as on-off bit error model (OBEM). The first model assumes the errors occur evenly on the coded packets. The second model assumes the errors may occur continuously

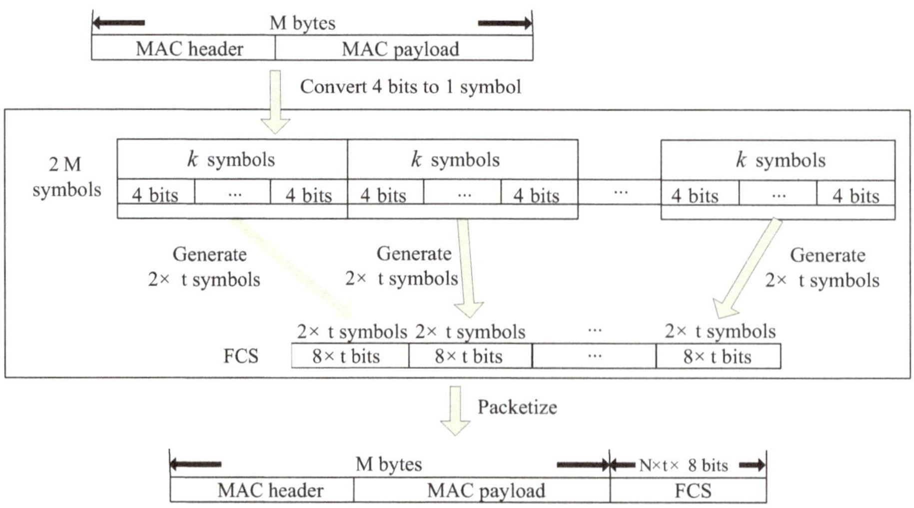

Figure 3. An HRR packet encoded with a shorter-length symbol (m = 4).

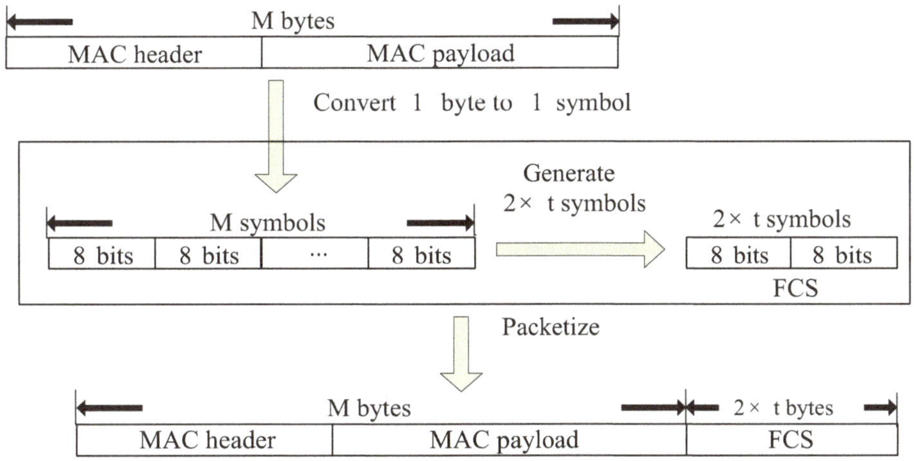

Figure 4. A non-HRR packet encoded with a longer-length symbol ($m = 8$).

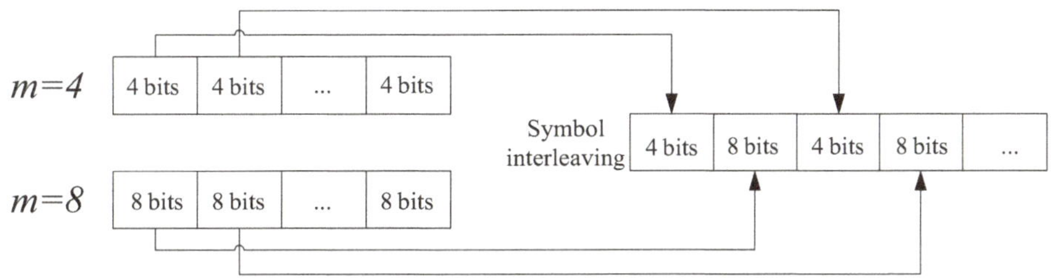

Figure 5. Packet interleaving on a symbol-by-symbol basis.

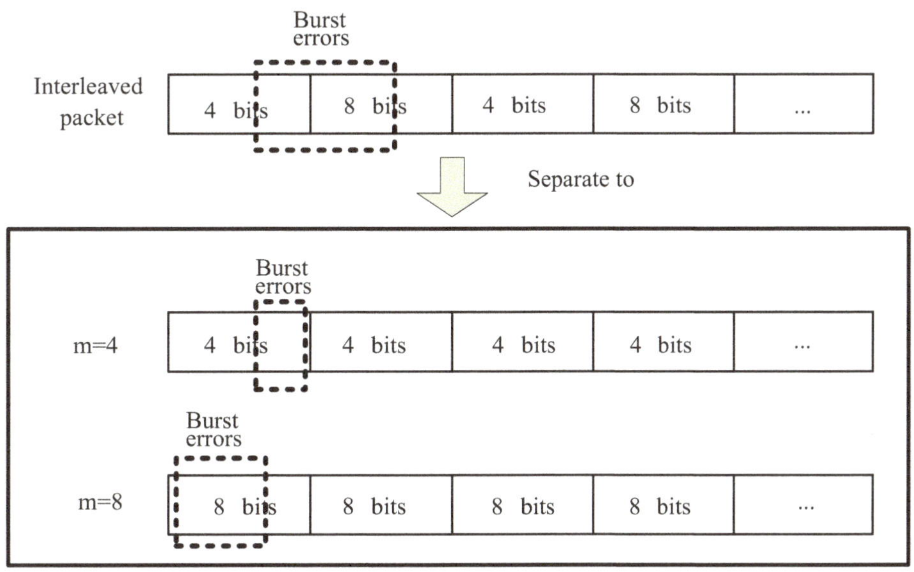

Figure 6. Burst errors dispersed on two symbols.

in a burst length.

4.1. UBEM

Let P_{UB_be} and P_{UB_se} denote the probability of bit errors and the probability of symbol errors, respectively. Since any bit errors occur in a symbol may result in a symbol error and each symbol has m bits, we can derive

P_{UB_se} directly from P_{UB_be}, as shown in Equation (17).

$$P_{UB\_se} = 1 - \left(1 - P_{UB\_be}\right)^{m} \quad (17)$$

Next, let us define two more parameters, N_S and N_C. The first parameter denotes the number of symbols in a codeword and the second parameter denotes the number of codeword in a packet. Hence, the uncorrectable probability of a codeword $\left(P_{UB\_cuc}\right)$, can be derived as shown in Equation (18).

$$P_{UB\_cuc} = 1 - \sum_{i=0}^{t}\left(\binom{N_s}{i} \times \left(P_{UB\_se}\right)^{i} \times \left(1 - P_{UB\_se}\right)^{N_S - i}\right) \quad (18)$$

In Equation (18), we know in a codeword if the number of symbol errors is smaller than t, then the codeword is correctable. Thus, the uncorrectable probability of a codeword can be summed up from $i = 0$ to t, since there are $\binom{N_s}{i}$ different types of errors. Next, let us define P_{UB_puc} as the packet uncorrectable probability. Since there are N_C codeword in a packet, we can derive P_{UB_puc} as shown in Equation (19).

$$P_{UB\_puc} = 1 - \left(1 - P_{UB\_cuc}\right)^{N_C} \quad (19)$$

4.2. OBEM

The on-off bit error model (OBEM) is illustrated in **Figure 7**. All the parameters used in the analysis are defined in **Table 2**. A burst error (BE) period is defined as two consecutive bit error intervals where high bit errors ap-

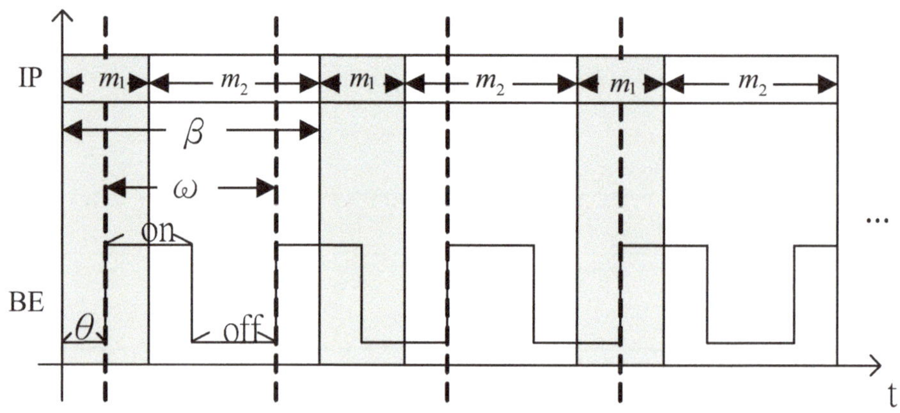

Figure 7. On-Off bit error model.

Table 2. Parameters used in OBEM.

m_1	Number of bits in a symbol of HRR packet
m_2	Number of bits in a symbol of non-HRR packet
β	Packet interleaved length in bits $(m_1 + m_2)$
on	Length of high bit errors (in bits)
σ	Error probability of high bit errors
off	Length of low bit errors (in bits)
ρ	Error probability of low bit errors
ω	Burst length (*on* + *off*) (in bits)
θ_0	Length of initial right-shift position (in bits)

pear first and then followed by low bit errors. Notice that θ_0 is defined as the length of right-shift position for an initial BE period; $\theta_0 = 0$ implies that no gap exists between the beginning of an interleaved packet and the beginning of the first BE period.

First, we define P_{OB_se} as the symbol-error probability in OBEM. An interleaved packet (IP) and a burst error (BE) may have different lengths; here we assume the former has a length of β bits $(\beta = m_1 + m_2)$ and the later has a length of ω bits $(\omega = on + off)$. Since every symbol in an IP may encounter different positions of bit errors, we have to analyze the bit error positions of a symbol before we can compute the symbol-error probability. To compute the error probability of the αth symbol, we define (i) m_{1s} = the distance between the first bit of m_1 and the first bit of an IP, and (ii) m_{1e} = the distance between the last bit of m_1 and the first bit of an IP. Similarly, we define m_{2s} and m_{2e} for m_2. Thus, m_{1s}, m_{1e}, m_{2s}, and m_{2e} can be computed as shown in Equation (20), (21), (22), and (23), respectively.

$$m_{1s} = \alpha \times \beta \tag{20}$$

$$m_{1e} = \alpha \times \beta + m_1 - 1 \tag{21}$$

$$m_{2s} = \alpha \times \beta + m_1 \tag{22}$$

$$m_{2e} = (\alpha + 1) \times \beta - 1 \tag{23}$$

For simplicity, the four parameters, m_{1s}, m_{1e}, m_{2s}, and m_{2e} are generalized to m_{ij}, where $i = 1,2$ and $j = s,e$. Let θ denote the length of right-shift position between an IP and a BE at the αth symbol. Let $\Omega_{m_{ij}}$ denote the length of right-shift position for m_{1s}, m_{1e}, m_{2s}, and m_{2e} at the αth symbol. Thus, we can compute θ and $\Omega_{m_{ij}}$ as shown in Equation (24) and (25), respectively.

$$\theta = \left(\theta_0 + \left(\left\lceil \frac{\text{header length}}{\omega} \right\rceil \times \omega - \text{header length} \right) \right) \ mod\ \omega \tag{24}$$

$$\Omega_{m_{ij}} = \left(\left\lfloor \frac{m_{ij} + 1}{\omega} \right\rfloor - 1 \right) \times \omega + \theta \tag{25}$$

After we found the right-shift position between an IP and a BE, we can categorize the symbol errors into four cases, as illustrated from **Figure 8(a)** to **Figure 8(d)**. Case 1 in **Figure 8(a)** shows whether or not a symbol may occupy one BE or two BE periods, their start bit and stop bit of a symbol all appear at the high-bit-error interval. Case 2 in **Figure 8(b)** shows only the start bit of a symbol appears at the high-bit-error interval. Case 3 in **Figure 8(c)** shows whether or not a symbol may occupy one BE or two BE periods, neither the start bit nor the stop bit appear at the high-bit-error interval. Finally, Case 4 in **Figure 8(d)** shows only the stop bit of a symbol appears at the high-bit-error interval. Actually, the four different cases of symbol errors can be constrained by eight inequalities with four parameters, m_{ij}, $\Omega_{m_{ij}}$, *on*, and ω. These eight inequalities (two inequalities for each case) are shown in **Table 3**. Once we identify the four cases of symbol errors, we can compute the symbol-error probability. Since one codeword consists of N_s symbols, we define $P_{OB\_se}(\alpha), \alpha = 0,1,\cdots,N_s - 1$, as the symbol-error probability of the αth symbol. Let σ denote the error probability of high bit errors and ρ denote the error probability of low bit errors. Let N_σ represent the number of bits with errors in a symbol and N_ρ represent the number of bits without errors in a symbol. We can compute $P_{OB\_se}(\alpha)$ as shown in Equation (26).

$$P_{OB\_se}(\alpha) = 1 - (1 - \sigma)^{N_\sigma} \times (1 - \rho)^{N_\rho} \tag{26}$$

Now, we can compute N_σ and N_ρ for case 1 as shown in Equation (27) and (28), case 2 as shown in Equation (29) and (30), case 3 as shown in Equation (31) and (32), and case 4 as shown in Equation (33) and (34), respectively.

Case 1:

$$N_\rho = \left(\left\lfloor \frac{m_{ie} + 1}{\omega} \right\rfloor - 1 \right) - \left(\left\lfloor \frac{m_{is} + 1}{\omega} \right\rfloor - 1 \right) \times Off = \left(\left\lfloor \frac{m_{ie} + 1}{\omega} \right\rfloor - \left\lfloor \frac{m_{is} + 1}{\omega} \right\rfloor \right) \times Off \tag{27}$$

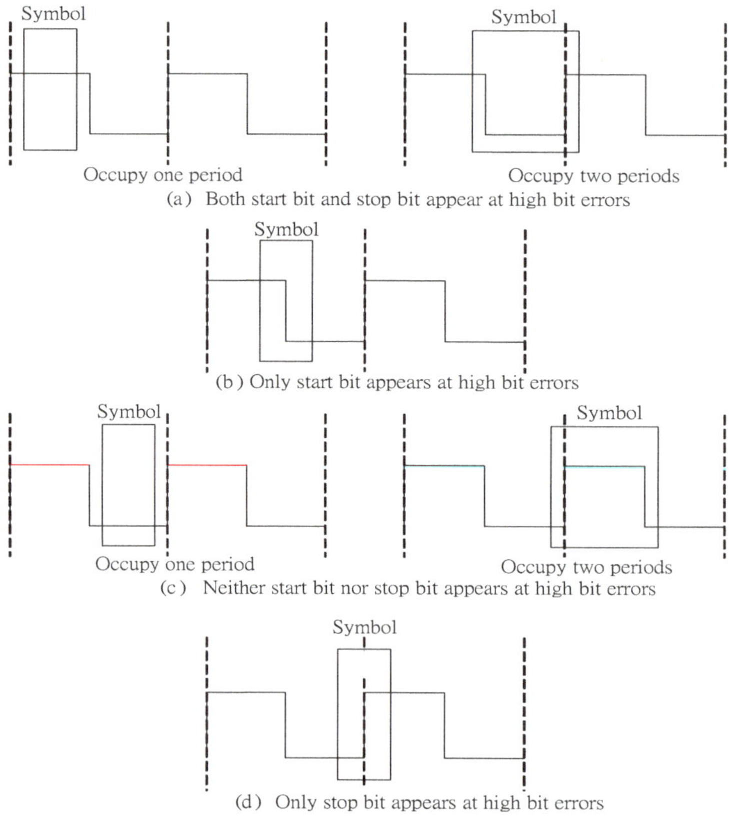

Figure 8. Four cases of symbol errors in terms of the start and stop bits.

Table 3. Four cases of symbol errors with 8 constraints.

Case	Conditions
1	$\Omega_{m_{is}} \le m_{is} < \Omega_{m_{is}} + On$ and $\Omega_{m_{ie}} \le m_{ie} < \Omega_{m_{ie}} + On$
2	$\Omega_{m_{is}} \le m_{is} < \Omega_{m_{is}} + On$ and $\Omega_{m_{ie}} + On \le m_{ie} < \Omega_{m_{ie}} + \omega$
3	$\Omega_{m_{is}} + On \le m_{is} < \Omega_{m_{is}} + \omega$ and $\Omega_{m_{ie}} + On \le m_{ie} < \Omega_{m_{ie}} + \omega$
4	$\Omega_{m_{is}} + On \le m_{is} < \Omega_{m_{is}} + \omega$ and $\Omega_{m_{ie}} \le m_{ie} < \Omega_{m_{ie}} + On$

$$N_\sigma = m_i - N_\rho = m_i - \left(\left\lfloor \frac{m_{ie}+1}{\omega} \right\rfloor - \left\lfloor \frac{m_{is}+1}{\omega} \right\rfloor \right) \times Off \tag{28}$$

Case 2:

$$\begin{aligned} N_\sigma &= \left(\left\lfloor \frac{m_{ie}+1}{\omega} \right\rfloor - 1 \right) - \left(\left\lfloor \frac{m_{is}+1}{\omega} \right\rfloor - 1 \right) \times On + \Omega_{m_{is}} + On - m_{is} \\ &= \left(\left\lfloor \frac{m_{ie}+1}{\omega} \right\rfloor - \left\lfloor \frac{m_{is}+1}{\omega} \right\rfloor \right) \times On + \left(\left\lfloor \frac{m_{is}+1}{\omega} \right\rfloor - 1 \right) \times \omega + \theta + On - m_{is} \\ &= \left\lfloor \frac{m_{ie}+1}{\omega} \right\rfloor \times On + \left(\left\lfloor \frac{m_{is}+1}{\omega} \right\rfloor - 1 \right) \times Off + \theta - m_{is} \end{aligned} \tag{29}$$

$$N_\rho = m_i - N_\sigma = m_i - \left\lfloor \frac{m_{ie}+1}{\omega} \right\rfloor \times On - \left(\left\lfloor \frac{m_{is}+1}{\omega} \right\rfloor - 1 \right) \times Off - \theta + m_{is} \tag{30}$$

Case 3:

$$N_\sigma = \left(\left\lfloor \frac{m_{ie}+1}{\omega} \right\rfloor - 1\right) - \left(\left\lfloor \frac{m_{is}+1}{\omega} \right\rfloor - 1\right) \times On = \left(\left\lfloor \frac{m_{ie}+1}{\omega} \right\rfloor - \left\lfloor \frac{m_{is}+1}{\omega} \right\rfloor\right) \times On \tag{31}$$

$$N_\rho = m_i - N_\sigma = m_i - \left(\left\lfloor \frac{m_{ie}+1}{\omega} \right\rfloor - \left\lfloor \frac{m_{is}+1}{\omega} \right\rfloor\right) \times On \tag{32}$$

Case 4:

$$\begin{aligned} N_\rho &= \left(\left(\left\lfloor \frac{m_{ie}+1}{\omega} \right\rfloor - 1\right) - \left(\left\lfloor \frac{m_{is}+1}{\omega} \right\rfloor - 1\right) - 1\right) \times Off + \Omega_{m_{is}} + \omega - m_{is} \\ &= \left(\left\lfloor \frac{m_{ie}+1}{\omega} \right\rfloor - \left\lfloor \frac{m_{is}+1}{\omega} \right\rfloor - 1\right) \times Off + \left(\left\lfloor \frac{m_{is}+1}{\omega} \right\rfloor - 1\right) \times \omega + \theta + \omega - m_{is} \\ &= \left\lfloor \frac{m_{is}+1}{\omega} \right\rfloor \times On + \left(\left\lfloor \frac{m_{ie}+1}{\omega} \right\rfloor - 1\right) \times Off + \theta - m_{is} \end{aligned} \tag{33}$$

$$N_\sigma = m_i - N_\rho = m_i - \left\lfloor \frac{m_{is}+1}{\omega} \right\rfloor \times On - \left(\left\lfloor \frac{m_{ie}+1}{\omega} \right\rfloor - 1\right) \times Off - \theta + m_{is} \tag{34}$$

By substituting N_σ and N_ρ back to Equation (26), we can now summarize $P_{OB\_se}(\alpha)$ for the four different cases of symbol errors as shown in **Table 4**.

After we compute the symbol-error probability for the four different cases, our next step is to derive the uncorrectable probability of a codeword and the uncorrectable probability of a packet. We know there are N_C codeword in a packet and the uncorrectable probabilities of the N_C codeword are all different. Let $P_{OB\_cuc}(l), l = 0, 1, \cdots, N_c - 1$, denote the uncorrectable probability of the lth codeword and let P_{OB_puc} denote the uncorrectable probability of a packet. By using the combination theory of probabilities, we can derive $P_{OB\_cuc}(l)$ and P_{OB_puc} as shown in Equation (35) and (36), respectively.

$$P_{OB\_cuc}(l) = 1 - \sum_{i=0}^{t} \left(\sum_{j=0}^{\tau_i - 1} \left(\prod_{k=0}^{\lambda-1} \left(\binom{N_k}{r_k^{ij}} \times \left(P_{OB\_se}^k\right)^{r_k^{ij}} \times \left(1 - P_{OB\_se}^k\right)^{N_k - r_k^{ij}} \right) \right) \right) \tag{35}$$

$$P_{OB\_puc} = 1 - \prod_{l=0}^{N_C - 1} \left(1 - P_{OB\_cuc}(l)\right) \tag{36}$$

5. Numerical Simulations

In OBEM, to study the influences of the four parameters, (i) the length of high bit errors (*i.e.*, the *On* period), (ii) the length of low bit errors (*i.e.*, the *Off* period), (iii) the right-shift position (*i.e.*, θ), and (iv) the number of correctable symbols (*i.e.*, *t*), on the packet uncorrectable probability (*i.e.*, P_{OB_puc}), we perform comprehensive numerical simulation. Table V shows the parameters and their setting used in the simulation. From **Table 5**, we can observe that an *On-Off* period is fixed to 12 bits. Within the 12 bits, the *On* period is varied among three different lengths; shorter length (4 bits), equal length (6 bits), and longer length (8 bits). The length of *Off* period is varied from the other directions (*i.e.*, from longer to shorter length).

Table 4. Four cases of symbol-error probabilities.

Case	$P_{OB\_se}(\alpha)$
1	$1-(1-\sigma)^{m_i - \left(\left\lfloor \frac{m_{ie}+1}{\omega} \right\rfloor - \left\lfloor \frac{m_{is}+1}{\omega} \right\rfloor\right) \times Off} \times (1-\rho)^{\left(\left\lfloor \frac{m_{ie}+1}{\omega} \right\rfloor - \left\lfloor \frac{m_{is}+1}{\omega} \right\rfloor\right) \times Off}$
2	$1-(1-\sigma)^{\left\lfloor \frac{m_{ie}+1}{\omega} \right\rfloor \times On + \left(\left\lfloor \frac{m_{is}+1}{\omega} \right\rfloor - 1\right) \times Off + \theta - m_{is}} \times (1-\rho)^{m_i - \left\lfloor \frac{m_{ie}+1}{\omega} \right\rfloor \times On - \left(\left\lfloor \frac{m_{is}+1}{\omega} \right\rfloor - 1\right) \times Off - \theta + m_{is}}$
3	$1-(1-\sigma)^{\left(\left\lfloor \frac{m_{ie}+1}{\omega} \right\rfloor - \left\lfloor \frac{m_{is}+1}{\omega} \right\rfloor\right) \times On} \times (1-\rho)^{m_i - \left(\left\lfloor \frac{m_{ie}+1}{\omega} \right\rfloor - \left\lfloor \frac{m_{is}+1}{\omega} \right\rfloor\right) \times On}$
4	$1-(1-\sigma)^{m_i - \left\lfloor \frac{m_{is}+1}{\omega} \right\rfloor \times On - \left(\left\lfloor \frac{m_{ie}+1}{\omega} \right\rfloor - 1\right) \times Off - \theta + m_{is}} \times (1-\rho)^{\left\lfloor \frac{m_{is}+1}{\omega} \right\rfloor \times On - \left(\left\lfloor \frac{m_{ie}+1}{\omega} \right\rfloor - 1\right) \times Off + \theta - m_{is}}$

Table 5. Parameter settings.

Parameter	Setting
m_1	4 bits
m_2	8 bits
on	4/6/8 bits
off	8/6/4 bits
σ	0.1
ρ	0.0001

Prior to studying the influences of the three different *On-Off* lengths on packet uncorrectable probability, we illustrate three scenarios in **Figure 9**. When the length of *On* period is shorter, as shown in **Figure 9(a)**, an HRR packet encoded with a shorter length of symbol (m = 4, dark-shaded portion) completely falls in the *Off* period, while a non-HRR packet encoded with a longer length of symbol (m = 8, light-shaded portion) just has a few bits fall in the *Off* period. When the length of the *On* period is increased to 6 bits, which is equal to the length of the *Off* period, as shown in **Figure 9(b)**, half of the HRR packet falls in the *On* period. The error-bit coverage for an HRR packet becomes even worse when the *On* period is increased to 8 bits, as shown in **Figure 9(c)**. As it can be seen, a whole symbol of an HRR packet is now covered by the *On* period, while only half of a symbol of a non-HRR packet is covered by the *On* period.

Impact of Correctable Symbols

First, we are interested in studying the impact of increasing the number of correctable symbols when the length of high bit errors is shorter than the length of low bit errors; *i.e.*, *On* = 4 bits and *Off* = 8 bits. As shown in Figures 10(a) to 10(e), we observe the variations of packet uncorrectable probability $\left(P_{OB\_puc}\right)$ when the right-shift position (θ) is increased from 0 to 11 bits and the number of correctable symbols (t) is increased from 1 to 5.

From **Figure 10**, we can observe that both the packet uncorrectable probabilities $\left(P_{OB\_puc}\right)$ of HRR and non-HRR curves drop off very quickly, when the number of correctable symbols (t) is increased from 1 to 5. Additionally, we can observe that he curves of P_{OB_puc} with PIS (the two dashed lines) are much lower than the curves of P_{OB_puc} without using PIS (the two solid lines). The improvement of P_{OB_puc} by using PIS is more significant when t is small, which is quite beneficial for reducing packet overhead, since the number of redundancy bits in RS codes can be shorter. Another noticeable phenomenon is that although when θ is smaller than 7, the curves of HRR with PIS are completely inverted to the curves of non-HRR with PIS, the curves of HRR with PIS do drop to zero when θ is larger than 7. In other words, on average we have come out a result; *i.e.*, the packet uncorrectable probability of HRR with PIS is significantly lower than that of non-HRR with PIS.

From **Figure 11(a)** to **Figure 11(e)**, we show the packet uncorrectable probabilities when the *on* period (6 bits) is equal to the *Off* period (6 bits). It is not surprising to notice that, for the corresponding t values (t = 1 to 5), the four curves in **Figure 11** are much higher than those four curves in **Figure 10**. The reason is quite straightforward, since the period of high bit errors (*i.e.*, the *On* period) is increased from 4 bits to 6 bits. However, on average the proposed PIS can effectively reduce packet uncorrectable probability, as it is compared to the curves without using PIS. In particular, we can observe that, no matter what the value t is, the curves of HRR with PIS can be reduced to near zero when θ is larger than 7. Additionally, we observe that a side effect could occur. That is, the curves of non-HRR with PIS cannot be reduced very significantly, as they are compared to those curves in **Figure 10**. This is because no matter how we adjust the right-shift position (θ), a non-HRR packet will be affected by the *on* period since the latter has been increased from 4 bits to 6 bits.

From **Figure 12(a)** to **Figure 11(e)**, we show the packet uncorrectable probabilities when the *on* period (8 bits) is longer than the *Off* period (4 bits). It is interesting to notice that when t is larger than 3 and θ is smaller than 6, the packet uncorrectable probabilities of HRR with PIS are higher than those curves without PIS. Of course, when t is smaller than 3 and θ is larger than 6, the situations are completely inverted. Hence, from **Figure 12**, we have discovered that when the period of high bit errors exceeds the length of an HRR symbol (m = 4) and approaches to a non-HRR symbol (m = 8), it is better to encode packets with a small value of t; otherwise, there

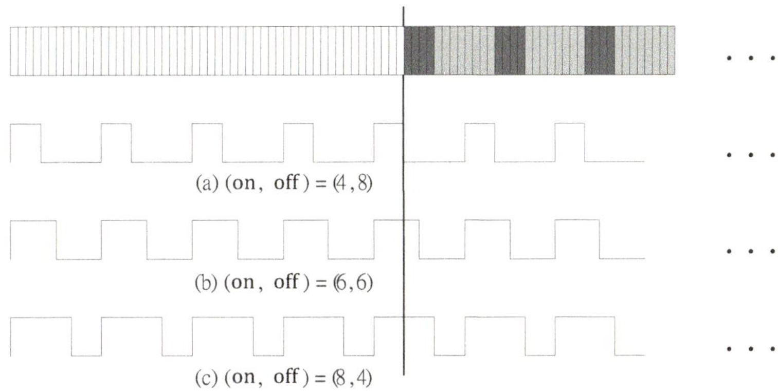

Figure 9. Influences of three *On-Off* lengths on the packet errors.

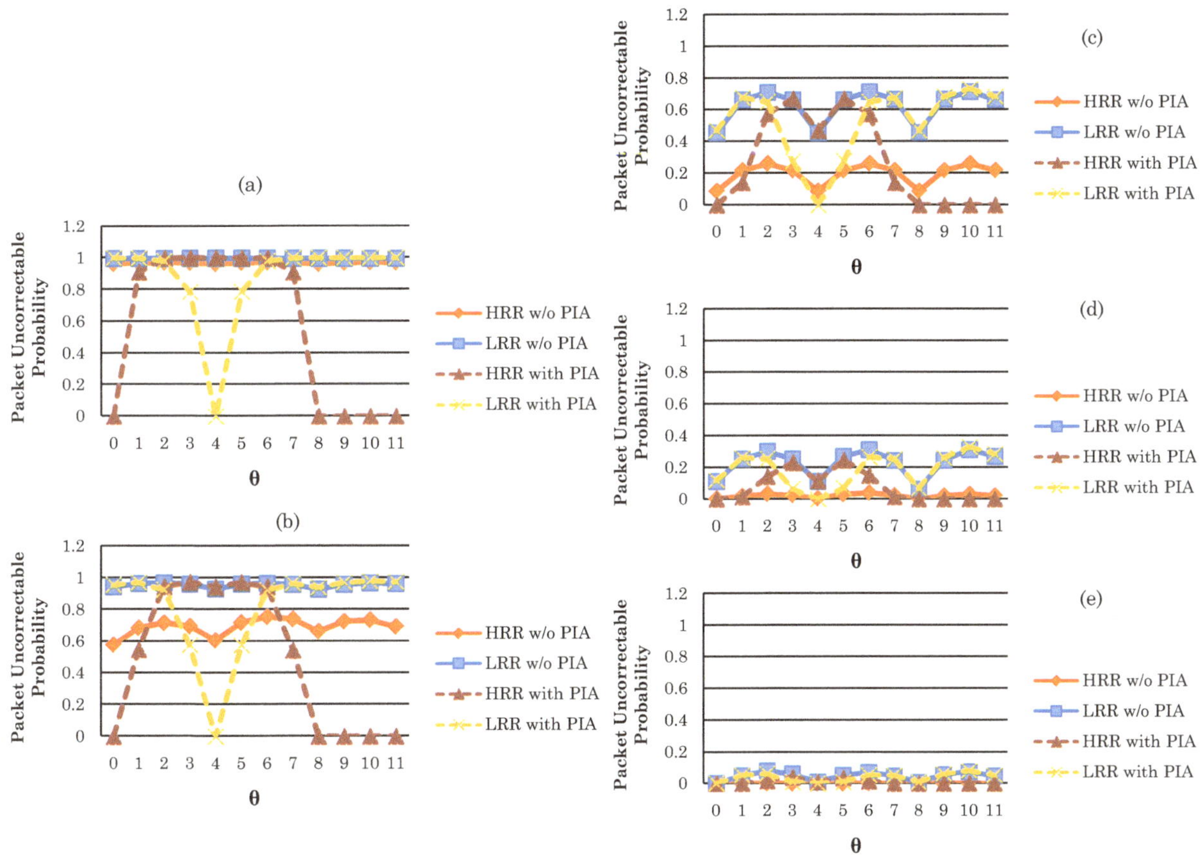

Figure 10. Variations of packet uncorrectable probability (*on* is smaller than *off*).

is no advantage achieved by using the proposed PIS.

From **Figure 13(a)** to **Figure 13(c)**, we show the comparisons in packet uncorrectable probabilities between UBEM and OBEM. First, we observe that in UBEM it does not make any big difference whether or not using the proposed PIS, since in UBEM every bit receives the same error probability. Second, we observe that packet uncorrectable probability in UBEM is relatively higher than that in OBEM. Notice that the curves of UBEM and the curves of OBEM vary along with the following two parameters: (i) when the number of correctable symbols (t) increases from 1 to 5, the curves of UBEM dissever very quickly from those of OBEM; and (ii) when the right-shift position (θ) increases from zero to 8 bits, the gap between these two models becomes smaller. Since OBEM is more reactive to a real word than UBEM, it is rewarding to know that the proposed PIS can reduce packet uncorrectable probability in OBEM more significantly than that in UBEM. Another noticeable re-

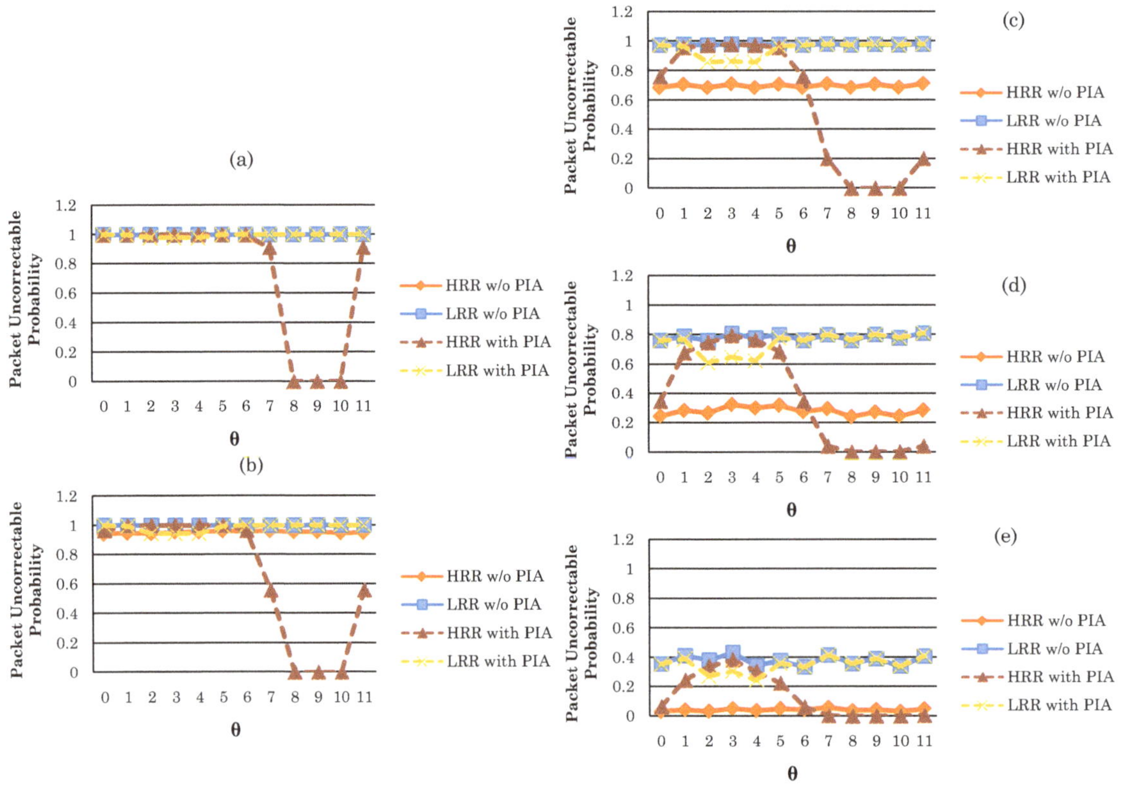

Figure 11. Variations of packet uncorrectable probability (*on* is equal to *off*).

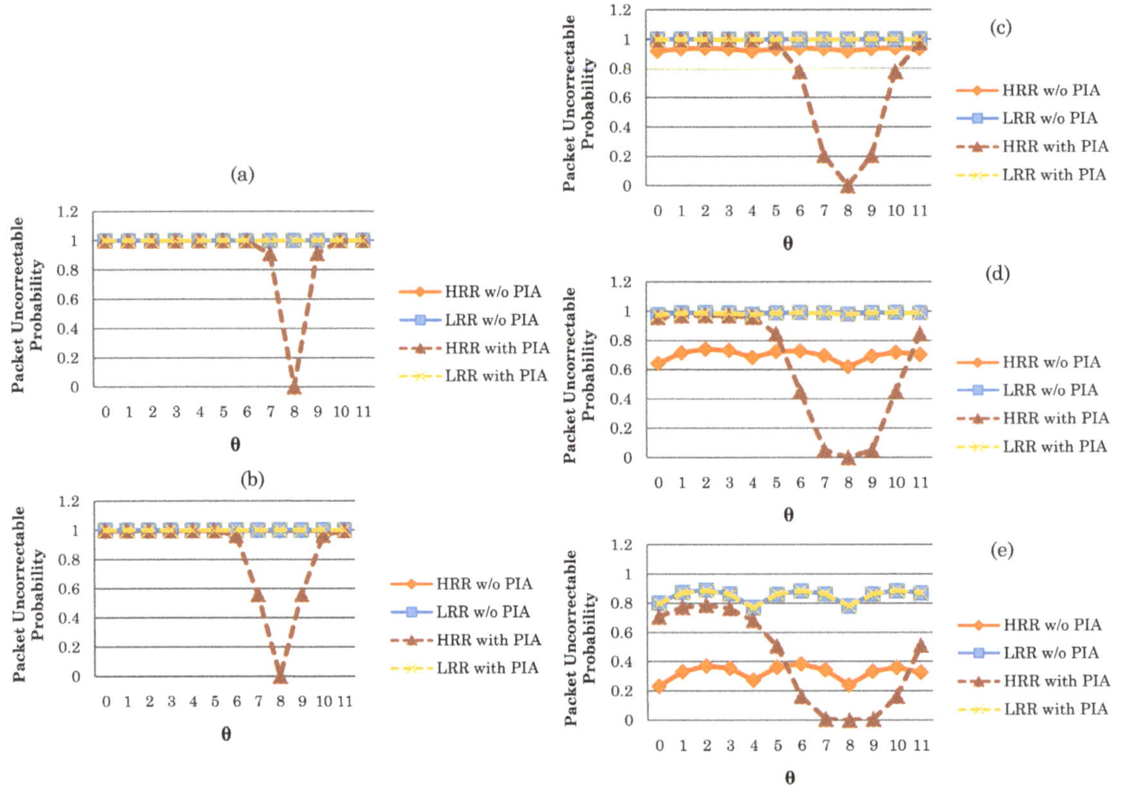

Figure 12. Variations of packet uncorrectable probability (*on* is larger than *off*).

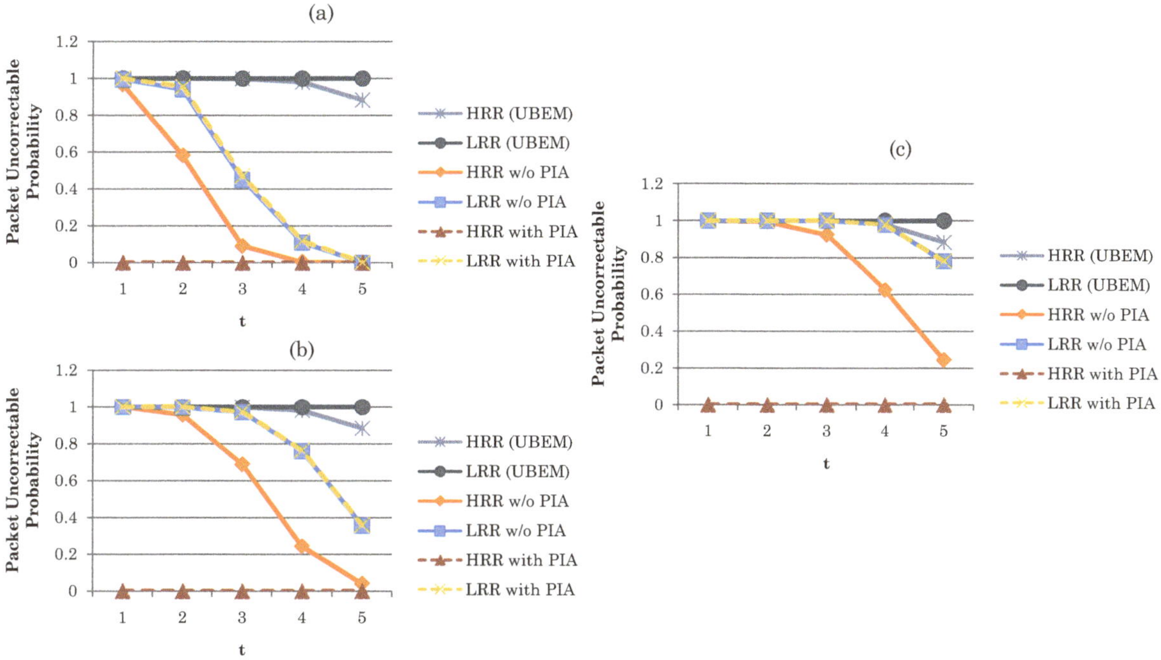

Figure 13. Packet uncorrectable probability in UBEM vs in OBEM. (a) (On, Off, θ) = (4, 8, 0); (b) (On, Off, θ) = (6, 6, 8); (c) (On, Off, θ) = (4, 8, 8)

sult is that no matter how we increase the period of high bit errors (*i.e.*, we increase the *on* period from 4 bits in 13(a) to 6 bits in 13(b), and finally to 8 bits in 13(c)), HRR with PIS in OBEM always exhibits the lowest packet uncorrectable probability (near zero, in some cases). The relatively lower packet uncorrectable probability for HRR packets has demonstrated that the proposed PIS can successfully protect HRR packets from burst errors, while at the same time it does not sacrifice non-HRR packets from large uncorrectable bit errors.

6. Conclusion

In this paper, we have presented a packet interleaving scheme (PIS) using RS code to reduce packet uncorrectable probability under burst errors in wireless sensor networks. In PIS, the collected data are classified into HRR and non-HRR. An HRR packet is encoded with a shorter RS symbol, while a non-HRR packet with a longer RS symbol. One of the contributions of this paper is right in that an HRR and a non-HRR packet are interleaved on a symbol-by-symbol basis, which effectively disperses the burst errors among an interleaved packet. Consequently, the packet uncorrectable probability of an HRR packet can be significantly reduced. In the performance evaluation, we built two mathematical models, UBEM and OBEM. From the thorough numerical simulations, we have demonstrated that, no matter how we adjust the period of high bit errors, the proposed PIS behaves more resilient to burst errors in OBEM than in UBEM. Finally, we have revealed that, by carefully adjusting the period of high bit errors and the right-shift positions, the proposed PIS can reduce the uncorrectable probability of HRR packets to near zero.

Acknowledgements

The authors would like to thank the reviewers, whose valuable comments helped improve the presentation of this paper. Additionally, the authors would like to thank Prof. Tsang-Ling Sheu's assistant, Miss Yi-Ying Ke for her cautious redrawing of the figures which helped improve the quality of the paper. This study is supported under the three grant numbers: (1) NSC 102-2221-E-110-036-MY2 of MoST, Taiwan, (2) 103-EC-17-A-03-S1-214, of MOEA, Taiwan, and (3) NHRI-EX103-10142EI, Taiwan.

References

[1] Barac, F., Yu, K., Gidlund, M., Akerberg, J. and Bjrkman, M. (2012) Towards Reliable and Lightweight Communica-

tion in Industrial Wireless Sensor Networks. *Proceedings of the INDIN* 2012: *IEEE* 10*th International Conference on Industrial Informatics*, Beijing, 25-27 July 2012, 1218-1224. http://dx.doi.org/10.1109/indin.2012.6300846

[2] Marinkovic, S. and Popovici, E. (2009) Network Coding for Efficient Error Recovery in Wireless Sensor Networks for Medical Applications. *Proceedings of the* 1*st International Emerging Network Intelligence*, Sliema, 11-16 October 2009, 15-20. http://dx.doi.org/10.1109/emerging.2009.22

[3] Sklar, B. (2001) Digital Communications: Fundamentals and Applications. 2nd Edition, Prentice Hall, Upper Saddle River.

[4] Chebbo, H., Abedi, S., Lamahewa, T.A., Smith, D.B., Miniutti, D. and Hanlen, L. (2010) Reliable Body Area Networks Using Relays: Restricted Tree Topology. *Proceedings of the* 2012 *International Conference on Computing, Networking and Communications* (*ICNC*), Maui, 30 January-2 February 2010, 82-88. http://dx.doi.org/10.1109/iccnc.2012.6167540

[5] Sampangi, R.V., Urs, S.R. and Sampalli, S. (2011) A Novel Reliability Scheme Employing Multiple Sink Nodes for Wireless Body Area Networks. *Proceedings of the* 2011 *IEEE Symposium on Wireless Technology and Applications* (*ISWTA*), Langkawi, 25-28 September 2011, 162-167. http://dx.doi.org/10.1109/iswta.2011.6089401

[6] Kim, S., Fonseca, R. and Culler, D. (2004) Reliable Transfer on Wireless Sensor Network. *Proceedings of the* 1*st Annual IEEE Communications Society Conference on Sensor and Ad Hoc Communications and Networks*, Santa Clara, 4-7 October 2004, 449-459.

[7] Byrne, E., Manada, A., Marinkovic, S. and Popovici, E. (2011) A Graph Theoretical Approach for Network Coding in Wireless Body Area Networks. *Proceedings of the* 2011 *IEEE International Symposium on Information Theory Proceedings* (*ISIT*), Saint Petersburg, 31 July-5 August 2011. http://dx.doi.org/10.1109/isit.2011.6034156

[8] Hamada, Y., Takizawa, K. and Ikegami, T. (2012) Highly Reliable Wireless Body Area Network Using Error Correcting Codes. *Proceedings of the* 2012 *IEEE Radio and Wireless Symposium*, Santa Clara, 15-18 January 2012, 231-234. http://dx.doi.org/10.1109/rws.2012.6175351

[9] Luby, M. (2002) LT Codes. *Proceedings of the* 43*rd Annual IEEE Symposium on Foundations of Computer Science*, Vancouver, 16-19 Novenber 2002, 271-280. http://dx.doi.org/10.1109/sfcs.2002.1181950

[10] Ishibashi, K., Ochiai, H. and Kohno, R. (2008) Embedded Forward Error Control Technique (EFECT) for Low-Rate but Low Latency Communications. *IEEE Transactions on Wireless Communications*, **7**, 1456-1460. http://dx.doi.org/10.1109/TWC.2008.060573

[11] Busse, M., Haenselmann, T. and Effelsberg, W. (2007) Energy-Efficient Data Dissemination for Wireless Sensor Networks. *Proceedings of the* 5*th Annual IEEE International Conference on Pervasive Computing and Communications Workshops*, White Plains, 19-23 March 2007. http://dx.doi.org/10.1109/percomw.2007.44

[12] Yu, K., Barac, F., Gidlund, M., Akerberg, J. and Bjorkman, M. (2012) A Flexible Error Correction Scheme for IEEE 802.15.4-Based Industrial Wireless Sensor Networks. *Proceedings of the* 2012 *IEEE International Symposium on Industrial Electronics* (*ISIE*), Hangzhou, 28-31 May 2012, 1172-1177. http://dx.doi.org/10.1109/isie.2012.6237255

[13] Srouji, M.S., Wang, Z. and Henkel, J. (2011) RDTS: A Reliable Erasure-Coding Based Data Transfer Scheme for Wireless Sensor Networks. *Proceedings of the* 2011 *IEEE* 17*th International Conference on Parallel and Distributed Systems*, Tainan, 7-9 December 2011, 481-488. http://dx.doi.org/10.1109/icpads.2011.104

[14] Arrobo, G.E. and Gitlin, R.D. (2011) Improving the Reliability of Wireless Body Area Networks. *Proceedings of the* 33*rd Annual International Conference of the IEEE Engineering in Medicine and Biology Society* (*EMBC*), Boston, 30 August-3 September 2011, 2192-2195. http://dx.doi.org/10.1109/iembs.2011.6090413

[15] Taparugssanagorn, A., Ono, F. and Kohno, R. (2010) Network Coding for Non-Invasive Wireless Body Area Networks. *Proceedings of the* 2010 *IEEE* 21*st International Symposium on Personal*, *Indoor and Mobile Radio Communications Workshops*, Istanbul, 26-30 September 2010, 134-138. http://dx.doi.org/10.1109/pimrcw.2010.5670413

[16] Salhi, I., Ghamri-Doudane, Y., Lohier, S. and Roussel, G. (2011) Reliable Network Coding for ZigBee Wireless Sensor Networks. *Proceedings of the* 8*th IEEE International Conference on Mobile Ad-Hoc and Sensor Systems*, Valencia, 17-22 October 2011, 135-137.

[17] Kiss, Z.I., Polgar, Z.A., Stef, M.P. and Bota, V. (2012) Network Coding Solution for Improving Transmission Reliability in Wireless Sensor Networks Employed in Industrial Monitoring. *Proceedings of the* 35*th International Conference on Telecommunications and Signal Processing* (*TSP*), Prague, 3-4 July 2012, 190-195. http://dx.doi.org/10.1109/tsp.2012.6256280

[18] Reed, I.S. and Solomon, G. (1960) Polynomial Codes over Certain Finite Fields. *Journal of the Society of Industrial and Applied Mathematics*, **8**, 300-304. http://dx.doi.org/10.1137/0108018

[19] Lidl, R. and Niederreiter, H. (1997) Finite Fields. Cambridge University Press.

6

A Novel Low-Complexity Low-Latency Power Efficient Collision Detection Algorithm for Wireless Sensor Networks

Fawaz Alassery, Walid K. M. Ahmed, Mohsen Sarraf, Victor Lawrence

Electrical and Computer Engineering Department, Stevens Institute of Technology, Hoboken, NJ, USA
Email: falasser@stevens.edu

Abstract

Collision detection mechanisms in Wireless Sensor Networks (WSNs) have largely been revolving around direct demodulation and decoding of received packets and deciding on a collision based on some form of a frame error detection mechanism, such as a CRC check. The obvious drawback of full detection of a received packet is the need to expend a significant amount of energy and processing complexity in order to fully decode a packet, only to discover the packet is illegible due to a collision. In this paper, we propose a suite of novel, yet simple and power-efficient algorithms to detect a collision without the need for full-decoding of the received packet. Our novel algorithms aim at detecting collision through fast examination of the signal statistics of a short snippet of the received packet via a relatively small number of computations over a small number of received IQ samples. Hence, the proposed algorithms operate directly at the output of the receiver's analog-to-digital converter and eliminate the need to pass the signal through the entire. In addition, we present a complexity and power-saving comparison between our novel algorithms and conventional full-decoding (for select coding schemes) to demonstrate the significant power and complexity saving advantage of our algorithms.

Keywords

Wireless Sensor Networks, Wireless Sensor Protocols, Collision Detection Algorithms, Power-Efficient Techniques, Low Complexity Algorithms

1. Introduction

Most research activities in WSNs focus on maximizing the network lifetime and minimizing the power consumption since they are powered by using finite energy sources (e.g., batteries). In this regard, some efforts deal with the routing schemes in order to rout a packet through most energy efficient links from a source node to a destination ([1]-[3]), while other researches extensively explore MAC schemes ([4]-[6]) which efficiently reduce packets collisions. However, MAC layer schemes intrinsically cannot eliminate all kind of collisions, because of hidden nodes problems, as well as collisions when multiple nodes sense the medium free at the same time followed by transmitting their packets. Hence the collision may occur at the receiver where it is difficult to distinguish between desirable and interferers signals.

Authors in [7], investigate the effect of interference signals on decoding power. They suggest adapting the decoder power based on the communication range. That is the decoder power needed to be increased, while transmitter power needs to be decreased for short rang communication systems. Authors in [8], design LDPC decoder architecture for low power WSNs. They suggest different LDPC codes and analyze the energy saving for encoded communication system. Their analysis shows how the decoder power levels affect the Bit Error Rate (BER). Author in [9], investigate the trade-off between the transmission power and decoding power in WSNs by employing convolutional codes with a specific ECC complexity in order to extend the network lifetime. In [10], the authors studied the relationship between the number of received bits and the decoder power consumption using LDPC codes in WSNs. Their analysis shows a large improvement in the network lifetime up to four times with LDPC codes which is more efficient than the convolutional and block codes. A power management technique at the receiver side in WSNs has been presented in [11]. Authors used rateless codes to minimize the power consumption, and their analytical results showed up to 80% of energy that is saved in comparison with IEEE 802.15.4 physical layer standard. Some efforts (e.g. [12]) focused on the actual design of LDPC decoder where early stopping methods are proposed in order to reduce the number of unnecessary iterations when decoding received packets. Such method is efficient in low SNR but it has a limitation when SNR is high.

Error correction schemes in wireless communication systems increase the reliability between a transmitter and receiver by reducing the probability of error. Reducing the probability of error can be achieved by increasing the transmit power or using a complex decoder that consume too much power to decode every received packet correctly. However, in limited power recourses networks such as WSNs, such increase in the transmit power as well as decoding power are not efficient since it contradicts with the design objective of WSNs which aims at energy-efficient solutions. Hence, in WSNs a fundamental trade-off exists between the transmitter and receiver power that should be considered to enhance the network lifetime.

One of the main sources of overhead power consumption in WSNs is collision detection. When multiple sensors transmit at the same time, their transmitted packets collide at the central node (the receiver) [13]. Authors in [14] use out of band control channel to indicate the transmission status (*i.e.* active state) for sensors which have packets ready to be transmitted. Sensors sense the control channel to detect collision. However, such technique is not accurate to detect collisions that may occur at the receiver. In addition, current collision detection mechanisms have largely been revolving around direct demodulation and decoding of received packets and deciding on a collision based on some form of a frame error detection mechanism, such as a CRC check [15]. The obvious drawback of full decoding of a received packet is the need to expend a significant amount of energy and processing complexity in order to fully-decode a packet, only to discover the packet is invalid and corrupted due to collision. So, decoding of corrupted packets becomes useless and provides the main cause of unnecessary power consumption.

In this paper we pose the following questions: *Can we propose a power-efficient technique to detect packets' collisions at the receiver side of WSNs without the need for full-decoding of received packets*? *Further*, *can we eliminate the need to pass corrupted packets through the entire demodulator/decoder*? From the perspective of achieving an efficient collision detection scheme at the receiver-side of WSNs we propose a suite of novel, yet simple and power-efficient algorithms to detect a collision without the need for full-decoding of the received packet. Our novel approach aims at detecting collision through fast examination of the signal statistics of a short snippet of the received packet via a relatively small number of computations over a small number of received IQ samples. Hence, operating directly at the output of the receiver's analog-to-digital-converter (ADC) and eliminating the need to pass the signal through the entire demodulator/decoder line-up. **Figure 1** illustrates where we apply our proposed scheme. In addition, we present a complexity and power-saving comparison between our novel Statistical Discrimination (*SD*) algorithms and conventional Full-Decoding (*FD*) algorithm (*i.e.* Soft Output Viterbi Algorithm) to demonstrate the significant power and complexity saving advantage of our scheme. Accordingly, our novel *SD* scheme has the following advantages:

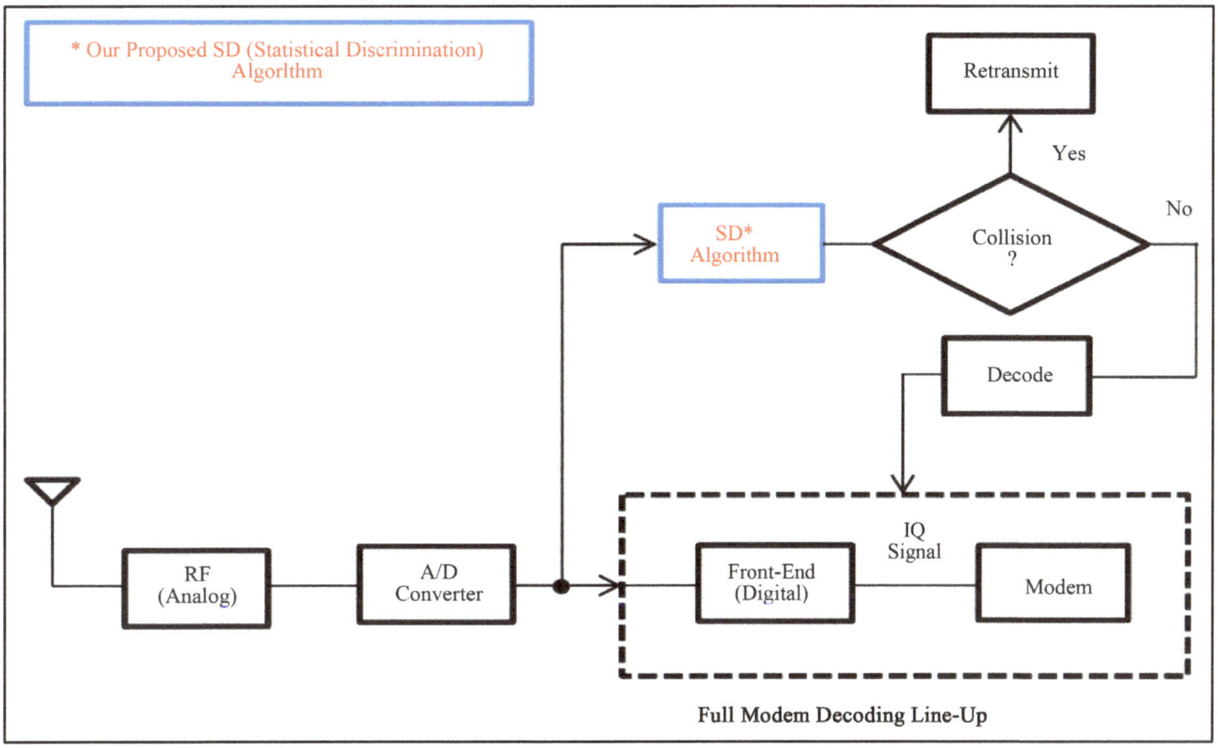

Figure 1. Block diagram for a receiver's line-up.

- The *SD* scheme not only reduces processing complexity and hence power consumption, but it also reduces the latency incurred to detect a collision since it operates on only a small number of samples-that may be chosen to be in the beginning of a received packet-instead of having to buffer and process the entire packet as is the case with a Full-Decoding (*FD*) algorithms.
- The *SD* scheme does not require any special pilot or training patterns. It operates directly on the (random) data, *i.e.*, the received packet as is.
- With a relatively short measurement period, the *SD* scheme can achieve low False-Alarm and Miss probabilities. It achieves a reliable collision-detection mechanism at the receiver-side of WSNs in order to minimize the receiption power consumption.
- The *SD* scheme can be turned over various design parameters in order to allow a system designer multiple degrees of freedom for design trade-offs and optimization.

The remainder of this paper is organized as follows. Section 2 describes our system. In section 3, we explain the proposed algorithms and show how to select a system threshold level. In section 4, we evaluate the power saving based on our proposed algorithms. In addition, we compare the computational complexity of our algorithms against commonly used decoding techniques (e.g., Soft Output Viterbi Algorithm, or SOVA). In section 5, we provide analysis and numerical empirical characterization to provide some quantitative theoretical framework and shed some light on the behavior of the various system factors and parameters involved in our proposed algorithms. In section 6, we present performance results, and finally in section 7 we provide the conclusion for this paper.

2. System Description

Figure 2 depicts an example of a WSN where a number of intermediate sensors are deployed arbitrarily to perform certain functionalities including sensing and/or collecting data and then communicating such information to a central sensor node (a receiver). The central node may process and relay the aggregate information to a backbone network.

As seen in **Figure 2**, there are *N* wireless sensors that communicate to the central node, where at any point in time, multiple packets may accidentally arrive simultaneously and cause a collision. Without loss of generality, we shall assume for the sake of argument that one sensor is denoted a "desirable" sensor, while the rest of the

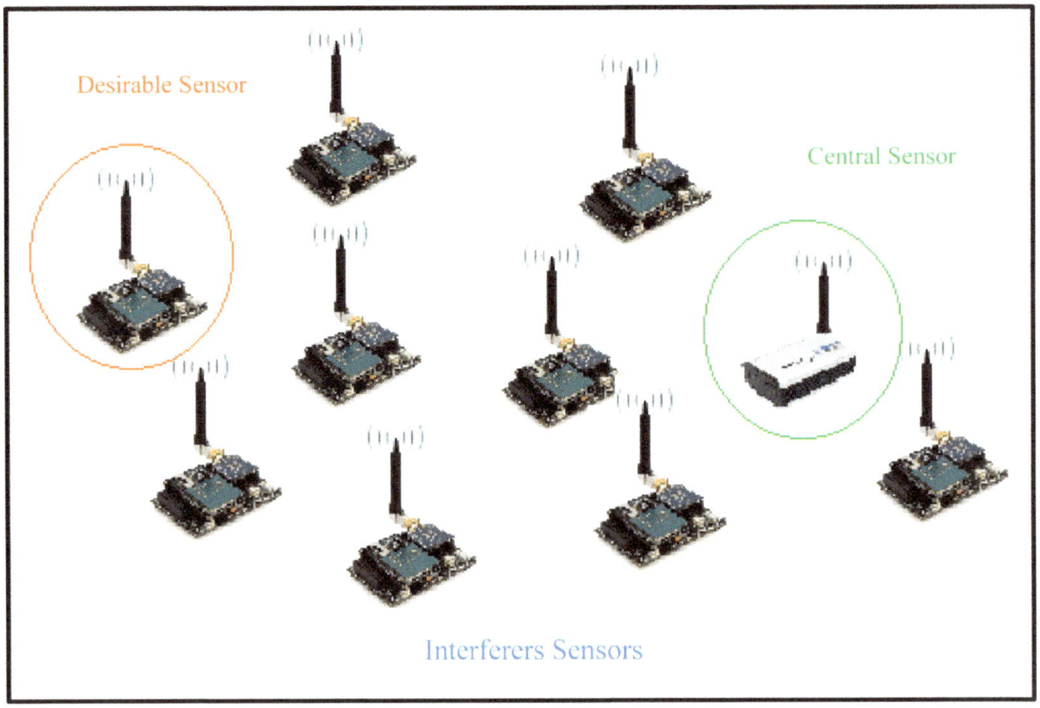

Figure 2. Wireless Sensor Network (WSN) with one desirable sensor, multiple interferer sensors and a central sensor (a receiver).

colliding sensors become "interferers". We assume the maximum number of sensors *i.e.*, $N = 30$. This number can be tuned as required is order to meet designers' requirements.

A commonly accepted model for packet arrivals, *i.e.*, a packet is available at a sensor and ready to be transmitted, is the well-known Bernoulli-trial-based arrival model, where at any point in time, the probability that a sensor has a packet ready to transmit is α.

Upon the receipt of a packet, the central node processes and evaluates the received packet and makes a decision on whether the packet is a collision-free (good) or has suffered a collision (bad). In this paper, we propose a suite of fast collision detection algorithms where the central node evaluates the statistics of the received signal's IQ samples at the output of the receiver's ADC directly using a simple *SD* scheme (as will be explained in more detail in the following sections), saving the need to expend power and time on the complex modem line-up processing (e.g., demodulation and decoding). If the packet passes the *SD* test, it is deemed collision-free and undergoes all the necessary modem processing to demodulate and decode the data. Otherwise, the packet is deemed to have suffered a collision, which in turn triggers the central node to issue a NACK message per the mechanism and rules mandated by the specific multiple-access scheme employed in the network.

It is noted that the actual design details and choice of the multiple access mechanism, e.g., slotted or un-slotted Aloha, are beyond the scope of this paper and irrelevant to the specifics of the technique proposed herein.

3. Algorithm Description

As mentioned earlier, our proposed algorithms are based upon evaluating the statistics of the received signal at the receiver ADC output via the use of a simple statistical discrimination metrics calculation that are performed on a relatively small portion of the received IQ packet samples. The resulting metrics values are then compared with a pre-specified threshold level to determine if the statistics of the samples of received signals reflect an acceptable signal-to-interference-plus-noise ratio (*SINR*) from the decoding mechanism perspective. If so, the packed is deemed collision-free and qualifies for further decoding. Otherwise, the packet is deemed to have suffered a collision with other interferer(s) and is rejected without expending any further processing/decoding energy. A repeat request may then be issued so the transmitting sensors to re-try depending on the MAC scheme. In other words, the idea is to use a fast and simple calculation to determine if the received signal strength (RSS) is indeed due to a single transmitting sensor that is strong enough to achieve an acceptable *SINR* at the central node's receiver, or the RSS is rather due to the superposition of the powers of multiple colliding packets, hence the associated *SINR* is less than acceptable to the decoding mechanism.

Let's define the kth received signal (complex-valued) IQ sample at the access node as:

$$y_k = x_{0,k} + \sum_{m=1}^{N-1} x_{m,k} + n_k$$

where $y_k = y_{k,I} + jy_{k,Q}$, $j = \sqrt{-1}$,

$$x_{0,k} = x_{0,k,I} + jx_{0,k,Q}$$

Is a complex-valued quantity that represents the kth IQ sample component contributed by the desired sensor, and

$$x_{m,k} = x_{m,k,I} + jx_{m,k,Q};\ m = 1,\cdots,N-1$$

Is the kth IQ sample component contributed by the mth interfering (colliding) sensor? Finally, $n_k = n_{k,I} + jn_{k,Q}$ Is a complex-valued Additive-White-Gaussian Noise (AWGN) quantity (e.g., thermal noise).

We propose three time-averaging statistical discrimination (*SD*) metrics that are applied to the envelope value, $|y_k| = \sqrt{y_{k,I}^2 + y_{k,Q}^2}$, of the received IQ samples at the central node as follows:

1) Entropy (Logarithmic) based metric:

$$\text{Log}_{\text{metric}} = \left\langle \log_{10} |y_k| \right\rangle = \frac{1}{K}\sum_{k=0}^{K-1} \log_{10} |y_k| \tag{1}$$

2) Moment based metric[1]:

$$\text{Moment}_{\text{metric}} = \left\langle |y_k^t| \right\rangle = \frac{1}{K}\sum_{k=0}^{K-1} |y_k^t|\ ;\ t = 3,4,5,\cdots \tag{2}$$

3) Signal Dynamic-Range Maximum-Minimum based metric:

$$\text{MaxMin}_{\text{metric}} = \frac{\max\limits_{k=0,\cdots,K} \{|y_k|\}}{\min\limits_{k=0,\cdots,K} \{|y_k|\}} \tag{3}$$

The computed statistical discrimination metric is then compared with a pre-specified threshold value that is set based on a desired signal-to-interference-plus-noise ratio (*SINR*) cut-off assumption[2], $SINR_{cut\_off}$. That is (and as will be described in more detail later in this paper), a system designer pre-evaluates the appropriate threshold value that corresponds to the desired $SINR_{cut\_off}$. If the *SD* metric value is higher than the threshold value, then the *SD* metric value reflects a *SINR* that is less than $SINR_{cut\_off}$ and the packet is deemed not usable, and vice-versa. Accordingly, a "False-Alarm" event occurs if the received *SINR* is higher than $SINR_{cut\_off}$ but the *SD* algorithm erroneously deems the received *SINR* to be less than $SINR_{cut\_off}$. On the other hand, if the *SD* algorithm deems the *SINR* to be higher than $SINR_{cut\_off}$ while it is actually less than $SINR_{cut\_off}$, a "Miss" event is encountered. Miss and False-Alarm probabilities directly impact the overall system performance as will be discussed in the following sections. Therefore, it is desired to minimize such probabilities as much as possible.

Threshold Selection

The decision threshold is chosen based on evaluating the False-Alarm and Miss probabilities and choosing the threshold values that satisfy the designer's requirements of such quantities. For example, we generate, say, a 100,000 Monte-Carlo simulated snapshots of interfering sensors (e.g., 1 - 30 sensors with random received powers to simulate various path loss amounts) where for each snapshot we compute the statistical discrimination

[1]We have found that odd-valued moment ranks (t) give better discrimination. Clearly, the second moment cannot be used as it represents the received signal power. Hence, it does not really bear any statistical discrimination information.

[2]In order to have a threshold setting that is independent of the absolute level of the received signal power (hence independent of path loss, receiver gain …etc.) the collected IQ samples of the measurement period may first be normalized to unity power.

value for the received *SINR* and compare it with various threshold levels, determine if there is a corresponding False-Alarm or Miss event and record the counts of such events. At the end of the simulations the False-Alarm and Miss probabilities are computed and plotted versus the range of evaluated threshold values, which in-turn, enables the designer to determine a satisfactory set point for the threshold.

4. Power Saving Analysis

To analyze the power saving of our proposed *SD* system we introduce the following computational complexity metrics:

$$F_B = S + P_{miss} F \tag{4}$$

$$F_G = S + (1 - P_{FA}) F \tag{5}$$

In above formulas, S is the number of computational operations incurred in our proposed scheme, while F is the number of computational operations incurred in a full-decoding scheme, P_{miss} and P_{FA} are the probabilities of Miss and False-Alarm events respectively. Hence, F_B represents the computational complexity for the case where the central node (the receiver) makes a wrong decision to fully-decode the received packet (*i.e.*, declared as a collision-free packets) while the packet should has been rejected (*i.e.*, due to collision). On the other hand, F_G is the computational complexity for the case where the central node makes a correct decision to fully decode received packet.

In addition, and for the comparison purposes, we introduce the following formulae in order to compare the computational complexity saving achieved by our proposed *SD* approach (*i.e.*, T_{SD}) over the *FD* approach (*i.e.*, T_{FD}):

$$T_{SD} = F_B P_{collision} + F_G P_{no\_collosion} \tag{6}$$

$$T_{FD} = F \tag{7}$$

In above formulae, $P_{collision}$ and $P_{no\_collosion}$ are the probabilities of collision and no-collision events respectively. $P_{collision}$ and $P_{no\_collosion}$ have been obtained via Monte-Carlo simulation as follows: A random number of interfering sensors (maximum of 30 sensors) is generated per a simulation snapshot, where each sensor is assumed to have a randomly received power level at the access node (to reflect a random path loss/location effect). The generation of the interfering sensors is based on a Bernoulli trial model where it is assume that the probability of a packet available for transmission at a sensor (hence the existence/generation of the sensor for the snapshot at hand) is equal to α. If the total *SINR* is found to be worse than the cut-off limit, a collision is assumed and vice-versa. For our numerical example in this section we used $\alpha = 0.3$ and $SINR_{cut\_off} = 5$ dB. Also, we typically generate more than 100,000 snapshots in order to achieve a reliable estimate of the collision probabilities. For the aforementioned choices of α and $SINR_{cut\_off}$, we found the collision probabilities to be $P_{collision} = 0.3649$ and $P_{no\_collosion} = 0.6351$.

Comparing with Full-Decoding Algorithms

In order to assess the computational complexity of our *SD* scheme, we first quantize our metrics calculation in order to define fixed-point and bit-manipulation requirement of such calculations. We also assume a look-up table (LUT) approach for the logarithm calculation. Note that the number of times the algorithm needs to access the LUT equals the number of IQ samples involved in the metric calculation. Thus, our algorithm only needs to perform addition operations as many times as the number of samples. Hence, if the number of bits per LUT word/entry is equal to M at the output of the LUT, our algorithm needs as many M-bit addition operations as the number of IQ samples involved in the metric calculation.

As a case-study, we compare the complexity of our *SD* scheme with the complexity of a *FD* algorithm assuming a Soft Output Viterbi Algorithm (SOVA). SOVA has been an attractive choice for WSNs [16]. Authors in [17] measure the computational complexity of SOVA (per information bit of the decoded codeword) based on the size of the encoder memory. It has been shown in [17] that for a memory length of λ, the total computational complexity per information bit can be estimated as:

$$F_{\mathrm{SOVA}} = 3 \times 2^{\lambda} + 9(\lambda + 1) + 16 \tag{8}$$

In contrast, our *SD* system does not incur such complexity related to the size of the encoder memory. In addition, our *SD* system avoids other complexities required by a full decoding such as time and frequency synchronization, Doppler shift correction, fading and channel estimation, etc., since our *SD* scheme operates directly at the IQ samples at the output of the ADC "as is". Finally, the *FD* approach requires buffering and processing of the entire packet/codeword while our *SD* scheme needs only to operate on a short portion of the received packet.

Now let's compute the computational complexity for our *SD* approach using the logarithmic (entropy) metric. Let's assume that the IQ ADCs each is *D* bits. Also, let's assume a $(\cdot)^2$ operation is done through a LUT approach to save multiplication operations. In addition, let's also assume that the square-root, $\sqrt{\cdot}$, is also done through a LUT approach. Hence, each of the I^2 and Q^2 operations consume of the order of *D* bit-comparison operations to address the $(\cdot)^2$ LUT. Then, if the output of the LUT is *G* bits, it follows that we need about *G* bit additions for an I^2+Q^2 operation. Let's assume that the $\sqrt{\cdot}$ LUT has *G* bits for input addressing and *K* output bits. Then, we need about *G*+1 bit-comparison operations to address the $\sqrt{\cdot}$ LUT. Let's assume a $\log(\cdot)$ is also done through a *K*-bit-input/*L*-bit-output LUT. Hence, a $\log\left(\sqrt{I^2+Q^2}\right)$ costs about *K* bit-comparison operations to address the $\log(\cdot)$ LUT. Finally, for simplicity, let's assume that a bit comparison operation costs as much as a bit addition operation[3]. Accordingly, the total number of operations needed to compute the $\log(\cdot)$ for one IQ sample is:

$$2D+G+(G+1)+K=2D+2G+1+K \tag{9}$$

If we assume the IQ over-sampling rate (OSR) to be *Z* (*i.e.*, we have *Z* samples per information symbol), then we need about $Z\times L$ bit additions to add the $Z\log(\cdot)$ values for every information symbol. Hence, for one information symbol, we need a total of:

$$(2D+2G+1+K)\times Z+Z\times L=(2D+2G+1+K+L)Z \tag{10}$$

Now if we assume an *M*-ary modulation (*i.e.*, $\log_2(M)$ information bits are mapped to one symbol), then the computational complexity per information bit can be computed as:

$$S\,/\,\text{Info Bit}=\frac{(2D+2G+1+K+L)Z}{\log_2(M)} \tag{11}$$

For example, in order to show the complexity saving of our *SD* algorithm, let's assume a QPSK modulation scheme (*M* = 4). Also, let's assume *Z* = 2 (2 samples per symbol), and *D* = *G* = *K* = *L* = 10 bits, which represents a good bit resolution. Also, let's assume a memory size of $\lambda=6$ for the SOVA decoder. Using the formulae (8), it follows the SOVA *FD* algorithm costs 271 operations per an information bit while our Entropy (Logarithmic) *SD* algorithm based on formula (11) costs only 61 operations per an information bit, which represents a 77% saving on the computational complexity.

In addition, in a no-collision event, the *SD* algorithm check would represent a processing overhead. Nonetheless, our *SD* scheme still provides a significant complexity saving over the *FD* scheme as demonstrated by the following example. **Table 1** in **Appendix A** shows the probability of Miss and False-Alarm to be 0.0762 and 0.0684, respectively for QPSK and a 50 bits measurement period[4]. Now, based on formulae (4) and (5), F_B and F_G (per information bit) for our *SD* algorithm will equal:

$$F_B=S+P_{miss}F=61+0.0762\times 271=82 \text{ Operations/Info Bit}$$

$$F_G=S+(1-P_{FA})F=61+(1-0.0684)\times 271=314 \text{ Operations/Info Bit}$$

For the comparison purposes between our *SD* algorithm and SOVA *FD* algorithm, formulae (6) and (7) are used to find the computational complexity when no-collision is detected:

$$\begin{aligned} T_{SD} &= F_B P_{collision} + F_G P_{no\_collision} \\ &= 82\times 36.49\% + 314\times 63.51\% \\ &= 230 \text{ Operations/Info Bit} \end{aligned}$$

[3]Similar assumptions were made in [17].

[4]The measurement period is 50 bits and the modulation scheme is QPSK, so the number of symbols is 25.

$$T_{FD} = F = 271 \text{ Operations/Info Bit}$$

Hence, the complexity savings (in number of operations per information bit) becomes:

$$\Delta_{SD}\% = (T_{FD} - T_{SD})/T_{FD} = (271 - 230)/271 = 15.12\%$$

Note that the above complexity saving calculation, in fact, represents a lower bound on the saving since the above calculation did not take into account the modem line-up operational complexity in order to demodulate and receive the bits in their final binary format properly (*i.e.*, synchronization, channels estimation, etc.).

The performance of our algorithms can be tuned as desired by a system designer. Appendix A provides performance comparisons for various examples where the system designer may choose to reduce the measurement period (e.g., to 25 or 50 bits) at the expense of increasing the Miss and False-Alarm probabilities, or may increase the throughput by using a longer estimation period in order to improve the accuracy of the statistical discriminator performance and reduce the Miss and False-Alarm probabilities (*i.e.* our system throughput (δ) is defined as $\delta = (1 - P_{FA})_{SD}$; Where P_{FA} denotes the False-Alarm probability).

5. Empirical Characterization

In this section, we attempt at empirically characterizing the statistics of various key quantities considered and encountered in this work, in an attempt to shed some light onto the behavior of such quantities and pave the way for some analytic mathematical tractability.

5.1. Statistics of the IQ Signal Envelope

In order to obtain reliable statistics, we have simulated different scenarios that reflect reasonably realistic assumptions[5]. For example, in our simulations, we assume that packets are generated at the various sensors using a Bernoulli trial model. That is, the probability of a packet available for transmission at a sensor is equal to α. We also generate random number of sensors per a network snapshot that are placed at random locations and distances from the central node in order to reflect various/random path loss situations[6]. The individual received sensor and noise components at the access node, as well as the total received signal (the superposition of the sensor received signals plus AWGN) are always normalized properly to reflect the correct *SINR* assumption.

In general, the parameters covered in this investigation include:

- Number of sensors[7].
- $SINR_{cut-off}$ level. For our simulations, we typically assumed $SINR_{cut-off} = 5$ dB[8].
- Sensitivity (tolerance) around the $SINR_{cut-off}$. That is, if the received *SINR* is within, for example, 1 dB, 1.5 dB, –2 dB, –10 dB or etc. around $SINR_{cut-off}$ (5 dB), we denote such *SINR* tolerance level as Δ_{SINR}.
- Probability of transmission per sensor.
- Modulation scheme.
- Measurement duration.
- *SD* metric choice.

As seen from the above simulation setting list, the simulations are always run assuming a fixed *SINR* value, in order to enforce a collision, or a no-collision event for the entire simulation session. Accordingly, the statistical analysis and characterization in this section are evaluated conditional on a collision or no-collision event in order to isolate the statistical characteristics of the metrics from the collision statistics, which can be dependent upon the MAC mechanism and other system parameters such as the specifics of the path loss distributions encountered by the sensors, which would affect the level of the received *SINR* …etc.

In general, we have found that the Normal (Gaussian) distribution has the closest fit to the actual (simulated) PDF of received signal envelope when $SINR \geq 0$ dB. For $SINR < 0$ dB, however, the Rayleigh distribution seems to be a better fit. We qualify the fitting accuracy of a distribution using the least-mean-square error (LMSE) criterion. Accordingly, the Normal and Rayleigh distributions have exhibited the minimum LMSE in comparison with other distributions as seen in **Figure 3** and **Figure 4** (such as 5th degree polynomial fit, the Weibull

[5]We simulate 100,000 - 1,000,000 snapshots per case.
[6]We simulate the random path-loss effect by simply generating random received power levels from the various sensors at the access node.
[7]In our analysis, we limit the maximum number of sensors in a simulation snapshot to 30 sensors.
[8]A 5 dB *SINR* seems a reasonable assumption based on typical coding requirements in wireless systems [18].

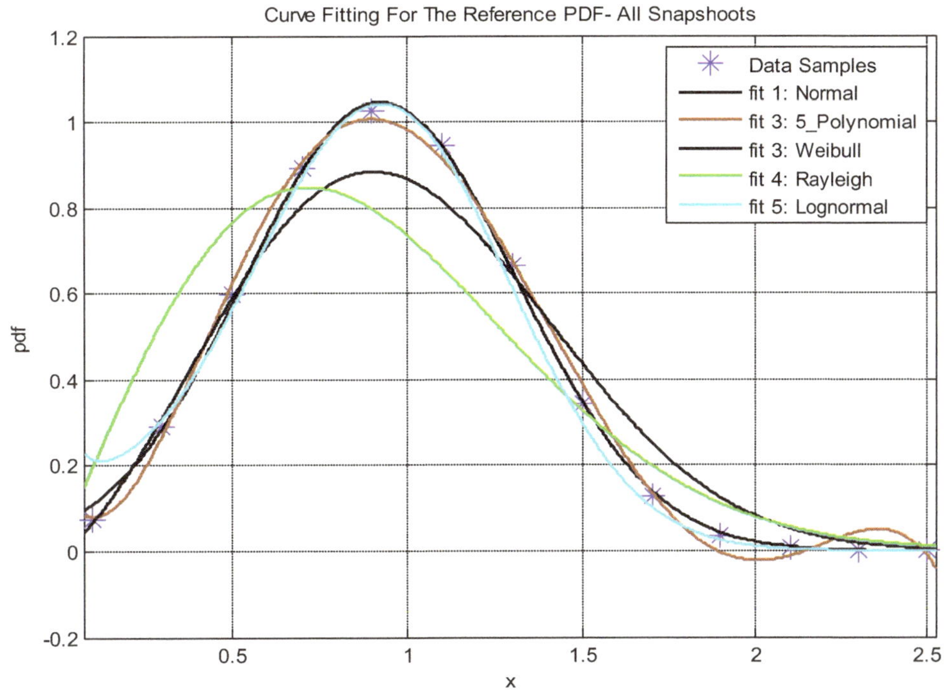

Figure 3. A curve-fitting comparison of various statistical distributions overlaid on the actual PDF for the IQ signal envelope as obtained from Monte-Carlo simulations: Logarithmic metric, $SINR = 4$ dB.

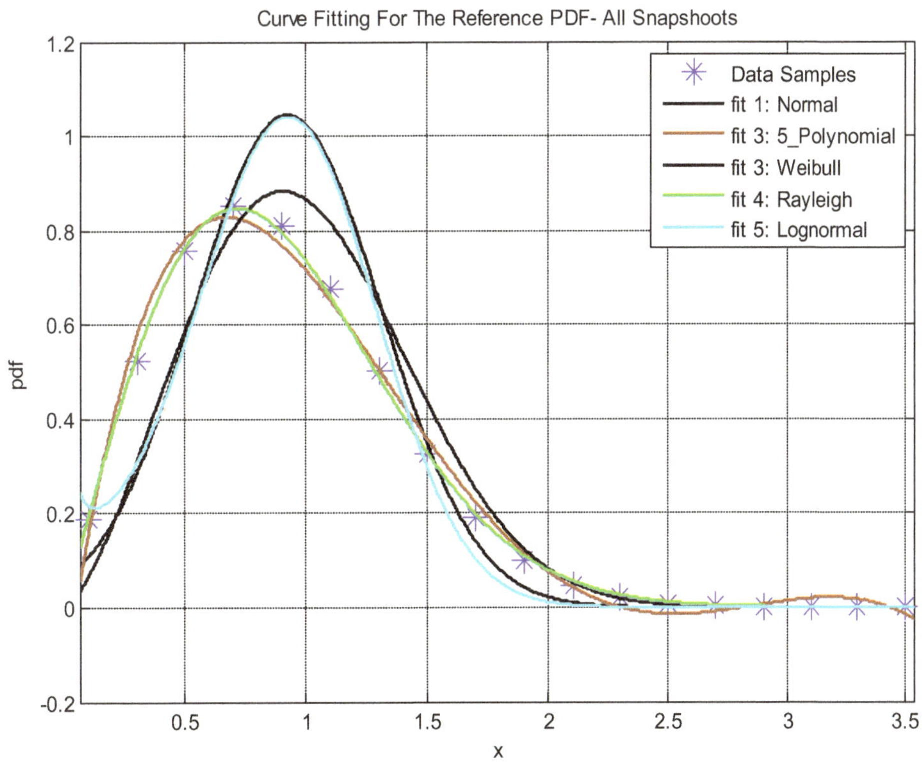

Figure 4. A curve-fitting comparison of various statistical distributions overlaid on the actual PDF for the IQ signal envelope as obtained from Monte-Carlo simulation: Logarithmic metric, $SINR = -6$ dB.

distribution and the Log-normal distribution).

For example, in **Figure 3**, the normal distribution with mean $(\mu = 0.9223)$, variance $(\sigma^2 = 1.047)$ resulted in a LMSE = 0.0024 and exhibited the closest fit to the actual (simulated) PDF of the received signal envelope. The choice of parameters for this example has been as follows:

- Maximum number of sensors is 30 (*i.e.*, the number of simultaneous sensors existing in the network per a simulation snapshot is between 2 and 30 sensors).
- $SINR_{cut-off} = 5\text{ dB}$, $SINR = 4\text{ dB}$ $(\Delta_{SINR} = -1\text{ dB})$.
- Probability of a packet available for transmission at a sensor is 0.3 (*i.e.*, theBernoullitrial model probability is $\alpha = 0.3$).
- Modulation scheme is 8PSK.
- Measurement period is equal to 50 information bits.
- Logarithmic (entropy) metric.

In **Figure 4**, the Rayleigh distribution achieved a LMSE = 0.0032 and exhibited the closest fit to the PDF of received signal envelop. Again, the choice of parameters in this figure is assumed as follows:

- Maximum number of sensors is 30.
- $SINR_{cut-off} = 5\text{ dB}$, $SINR = -6\text{ dB}$ $(\Delta_{SINR} = -11\text{ dB})$.
- $\alpha = 0.3$.
- Modulation scheme is 8 PSK.
- Measurement period is equal to 50 information bits.
- Logarithmic (entropy) metric.

Figure 5 and **Figure 6** show similar examples for a 3rd moment *SD* metric. **Figure 5** shows how the Normal distribution continues to have the closest fit and achieves a LMSE = 0.0036, while in **Figure 6** the Rayleigh distribution has the best fit with LMSE = 0.0041. The choice of parameters follows:

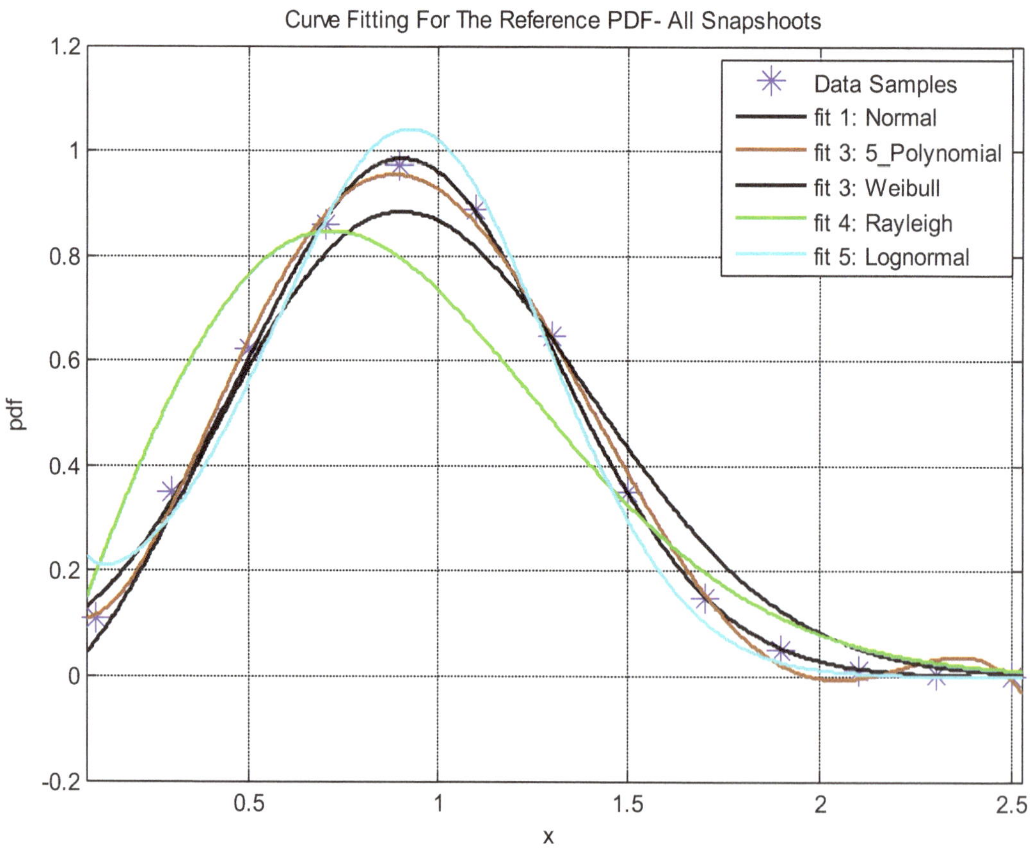

Figure 5. A curve-fitting comparison of various statistical distributions overlaid on the actual PDF for the IQ signal envelope as obtained from Monte-Carlo simulations: 3rd moment metric, *SINR* = 3 dB.

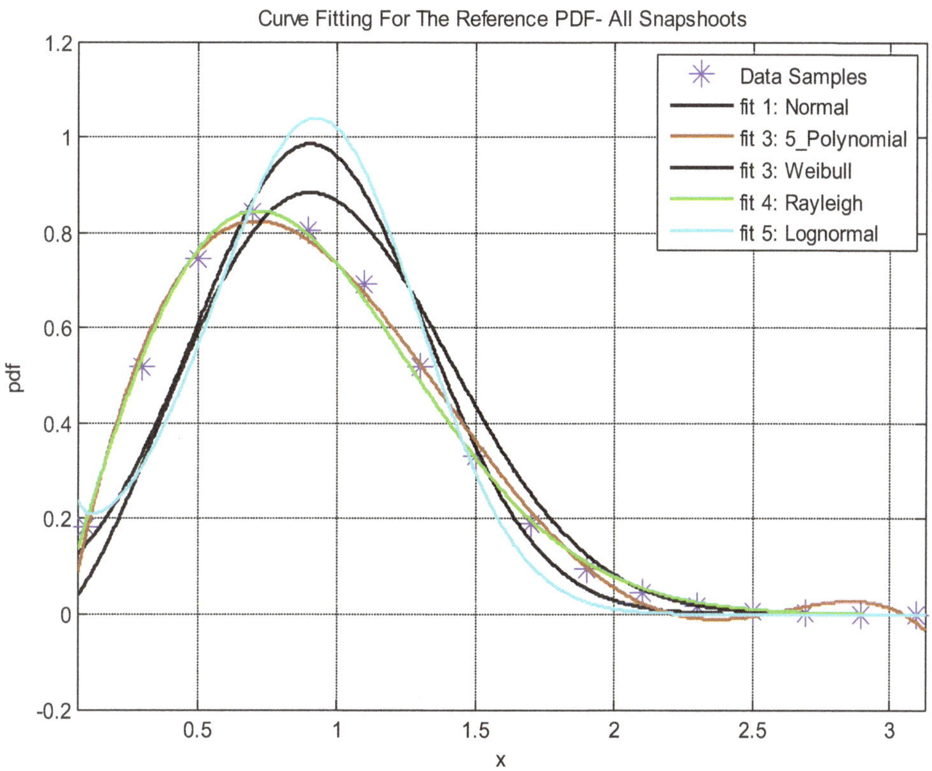

Figure 6. A curve-fitting comparison of various statistical distributions overlaid on the actual PDF for the IQ signal envelope as obtained from Monte-Carlo simulation: 3rd moment metric, $SINR = -4$ dB.

- Maximum number of sensors is 30.
- $SINR_{cut-off} = 5\text{ dB}$, $SINR = 3$ dB $(\Delta_{SINR} = -2\text{ dB})$ for **Figure 5**, while $SINR_{cut-off} = 5\text{ dB}$, $SINR = -4$ dB $(\Delta_{SINR} = -9\text{ dB})$ for **Figure 6**.
- $\alpha = 0.2$.
- Modulation scheme is QPSK.
- Measurement period is 200 information bits.
- 3rd moment metric.

Figure 7 and **Figure 8** show corresponding examples for the MAX-MIN based metric. Again, the Normal and Rayleigh distributions have best fits with LMSE = 0.0039 and LMSE = 0.0220, respectively. Our choice of parameters follows:

- Maximum number of sensors is 30.
- $SINR_{cut-off} = 5\text{ dB}$, $SINR$= 3 dB $(\Delta_{SINR} = -2\text{ dB})$ for **Figure 7**, while $SINR_{cut-off} = 5\text{ dB}$, $SINR = -5$ dB $(\Delta_{SINR} = -10\text{ dB})$ for **Figure 8**.
- $\alpha = 0.4$.
- Modulation scheme is 16 PSK.
- Measurement period is 1000 information bits.

MAX-MIN based metric.

5.2. Statistics of the *SD* Metrics

In general, the ensemble (overall) average (mean) of the first moment of the IQ envelope of the received signal, as well as the second moment (*i.e.*, the power of received signal) is a function of the received *SINR*. In the following, we plot the ensemble averages of the first and second moments (in **Figure 9** and **Figure 10** respectively) of the IQ envelope quantity viruses the corresponding first and second moment values that correspond to the best fitting distribution (*i.e.*, Normal and Rayleigh PDFs as pointed out above). The parameters in **Figure 9** and

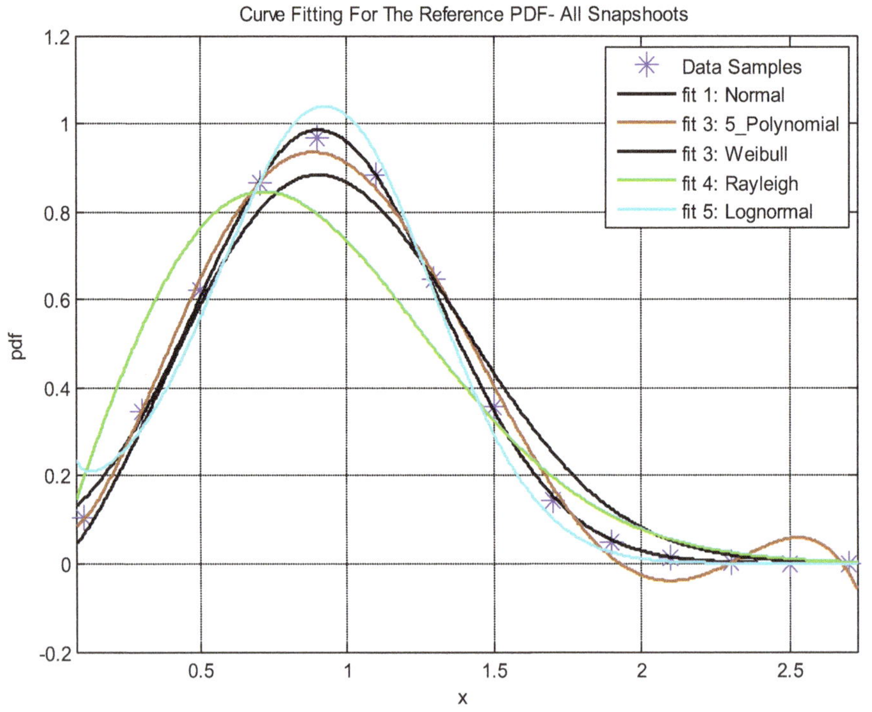

Figure 7. A curve-fitting comparison of various statistical distributions overlaid on the actual PDF for the IQ signal envelope as obtained from Monte-Carlo simulations: MAX-MIN based metric, $SINR = 3$ dB.

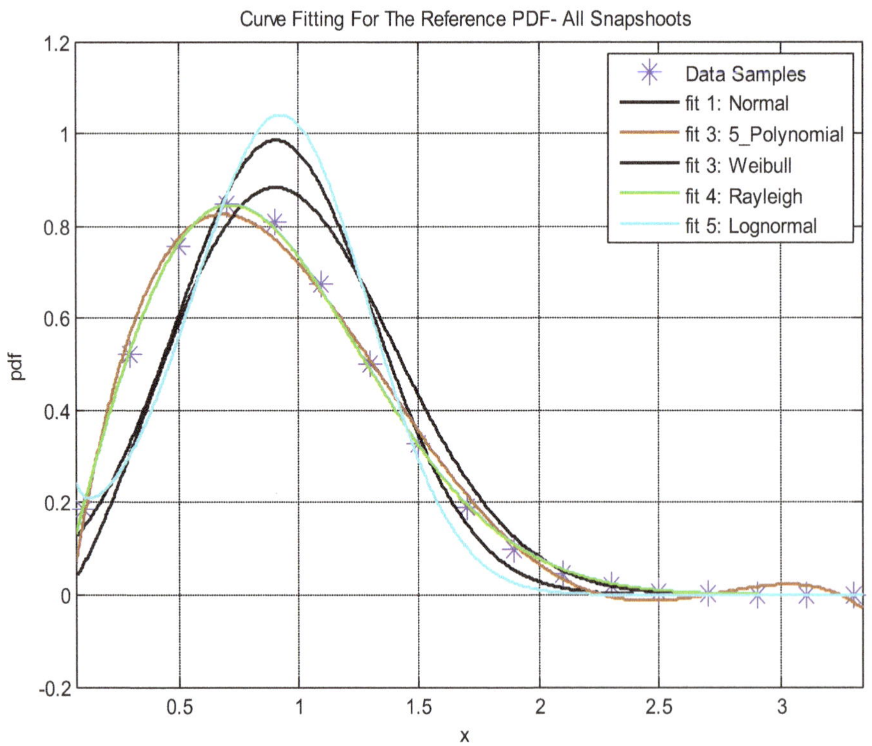

Figure 8. A curve-fitting comparison of various statistical distributions overlaid on the actual PDF for the IQ signal envelope as obtained from Monte-Carlo simulations: MAX-MIN based metric, $SINR = -5$ dB.

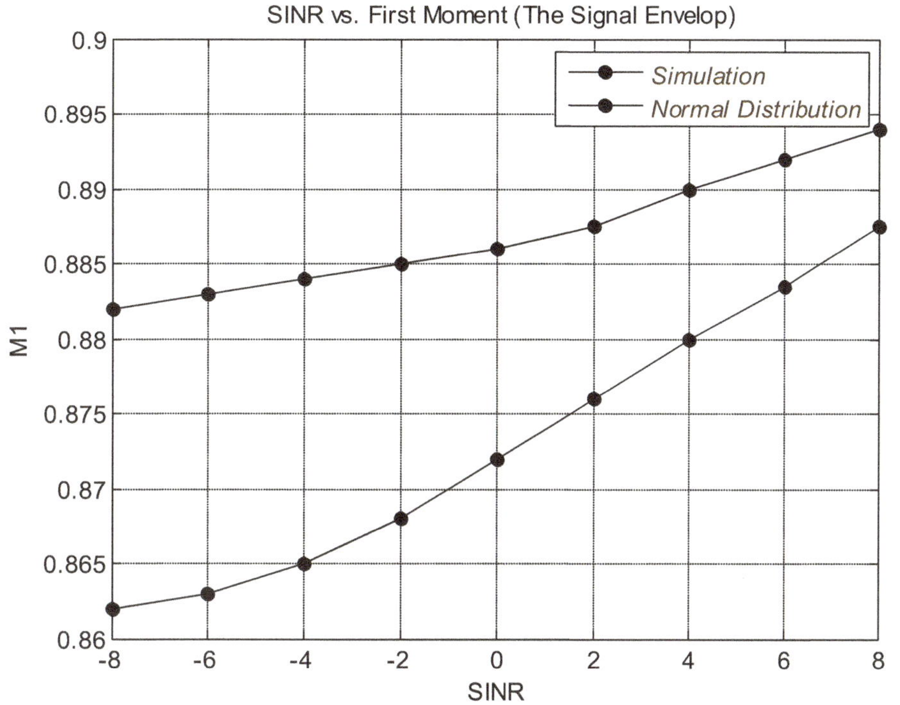

Figure 9. The mean μ for the received signal envelope for the simulation data samples & curve fitting distribution vs. *SINR*.

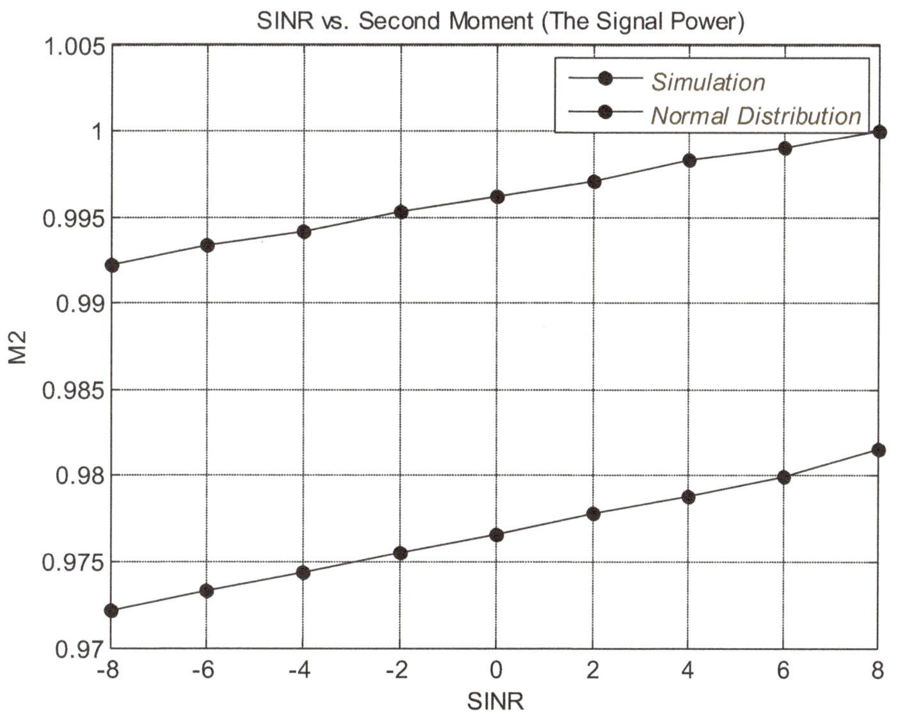

Figure 10. The second moment for the received signal power for the simulation data samples & curve fitting distribution vs. *SINR*.

Figure 10 are assumed as follows:

- Maximum number of sensors is 30.

- $SINR_{cut-off} = 5$ dB , $SINR = [-8$ dB, 8 dB], *i.e.*, the interval between −8 dB and 8 dB.
- $\alpha = 0.3$.
- Modulation scheme is 8 PSK.
- Measurement period is 200 information bits.

Logarithmic metric.

In addition, we have found that the normal distribution has the best fit to the simulated PDFs for the Logarithmic, Moment and MAX-MIN based metrics[9]. The corresponding normal curve fittings are shown in **Figures 11-13** for logarithmic, moment and MAX-MIN based metrics respectively.

Based on the normal PDF fit [19], one can calculate the False-Alarm and Miss probabilities as follows: If we assume a pre-defined threshold level (γ), then it can be shown that:

$$P_{FA} = \left(x > \gamma \middle| \left(SINR = SINR_{operating}\,; SINR_{operating} > SINR_{cut-off}\right)\right)$$
$$= \int_{\gamma}^{\infty} \frac{1}{\sigma(SINR)\sqrt{2\pi}} e^{\frac{-(x-\mu(SINR))^2}{2\sigma^2(SINR)}} dx \tag{12}$$

$$P_{Miss} = \left(x < \gamma \middle| SINR = SINR_{operating}\,; SINR_{operating} < SINR_{cut-off}\right)$$
$$= \int_{-\infty}^{\gamma} \frac{1}{\sigma(SINR)\sqrt{2\pi}} e^{\frac{-(x-\mu(SINR))^2}{2\sigma^2(SINR)}} dx \tag{13}$$

It should be noted that direction of the metric threshold-crossing versus *SINR*, *i.e.*, whether the metric being greater than or less than the threshold is an indicative of *SINR* being greater than or less than the cut-off *SINR* (*i.e.*, a collision or not event) is easily seen by inspecting the numerical behavior of the metric, which has been strictly consistent. Also, it should be noted that as indicated by Equations (12) and (13) above, the means (and variances) of the curve-fitting Gaussian PDFs used in approximating the False-Alarm probability versus the Miss probability are of generally different values that are functions of the operating *SINR*[10] since these PDFs are computed under disjoint conditions (*i.e.*, *SINR* greater than or less than the cut-off), as demonstrated, for example,

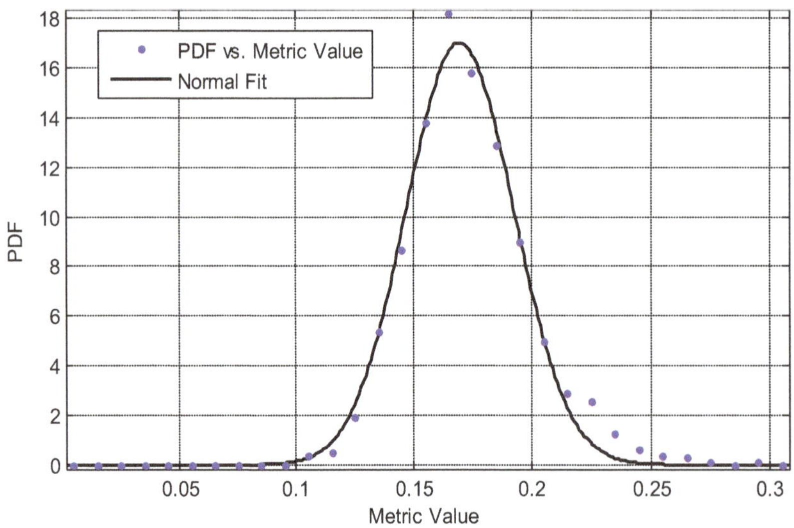

Figure 11. The PDF (simulation versus fitted) of the metric value, when treated as a random variable (over snapshots): Logarithmic metric.

[9]It is worth noting that the definitions of the logarithmic and the moment metrics (see Equation (1) and (2)) as sums of many random variables (*i.e.*, the IQ sample envelope log or moment values), which more or less have the same variance levels, indicates that by the central-limit theorem, one should expect the PDFs of these metrics to converge asymptotically to the Gaussian distribution, as the number of summed samples (terms) increases.

[10]As noted earlier in this section, the statistical analysis and characterization have been defined conditional on the collision or no-collision state.

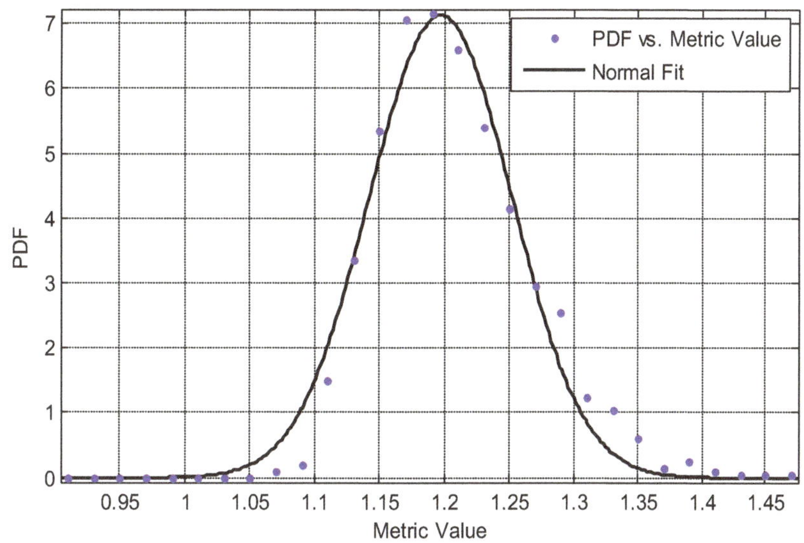

Figure 12. The PDF (simulation versus fitted) of the metric value, when treated as a random variable (over snapshots): 3rd moment metric.

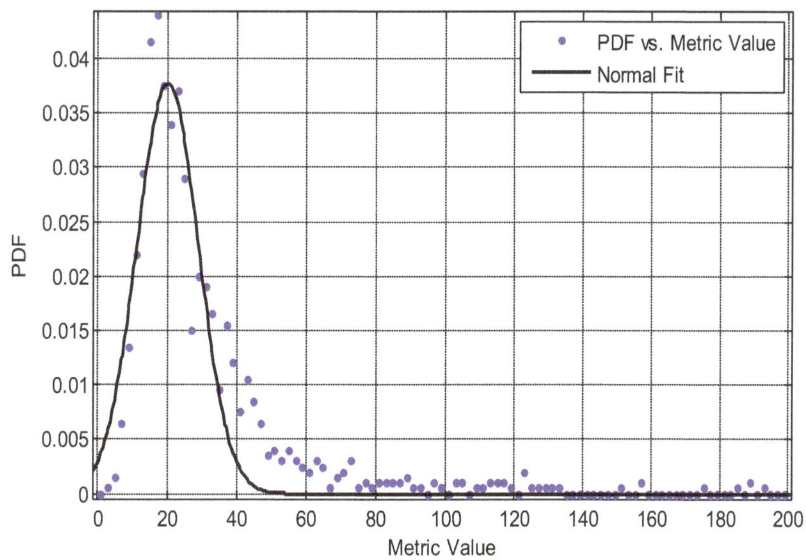

Figure 13. The PDF (simulation versus fitted) of the metric value, when treated as a random variable (over snapshots): MAX-MIN based metric.

in **Figure 9** and **Figure 10**. Clearly, P_{FA} and P_{Miss} are not complimentary (*i.e.*, do not necessarily add up to unity).

Figures 14-19 compare the simulated versus the empirically derived mathematical results for the False-Alarm and the Miss probabilities, for the Logarithmic, Moment and MAX-MIN based metrics. Our choice of parameters in these figures is as follows:

For **Figure 14**:

- Maximum number of sensors is 30.
- $SINR_{cut-off} = 5$ dB, $SINR = 6.5$ dB $\left(\Delta_{SINR} = 1.5 \text{ dB}\right)$.
- $\alpha = 0.3$.
- Modulation scheme is QPSK.
- Measurement period is 50 information bits.
- Logarithmic metric.

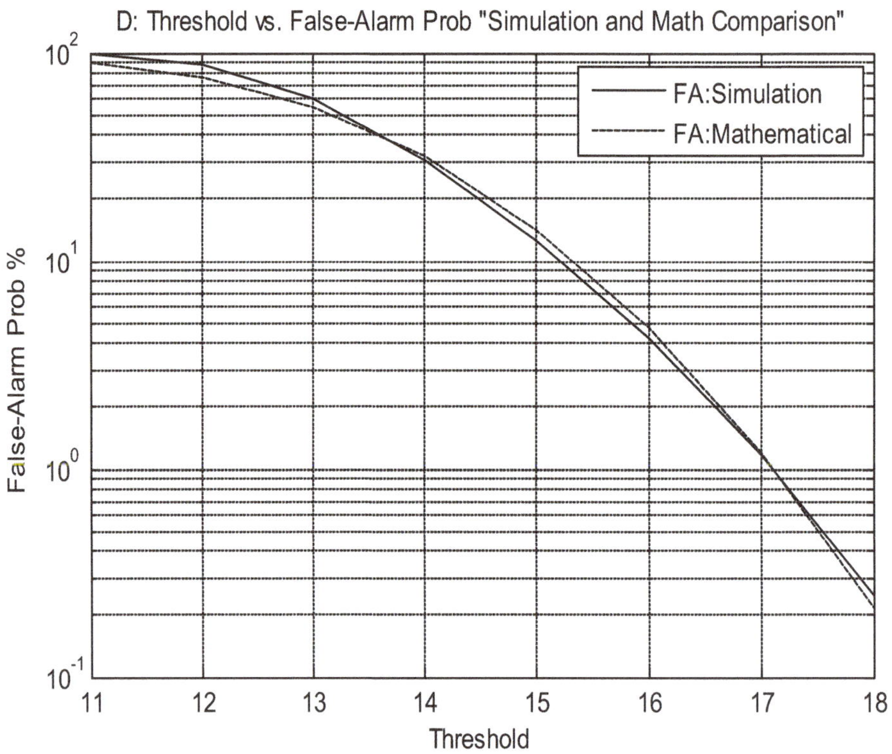

Figure 14. Comparison of False-Alarm probabilities for simulation and mathematical calculations: Logarithmic metric.

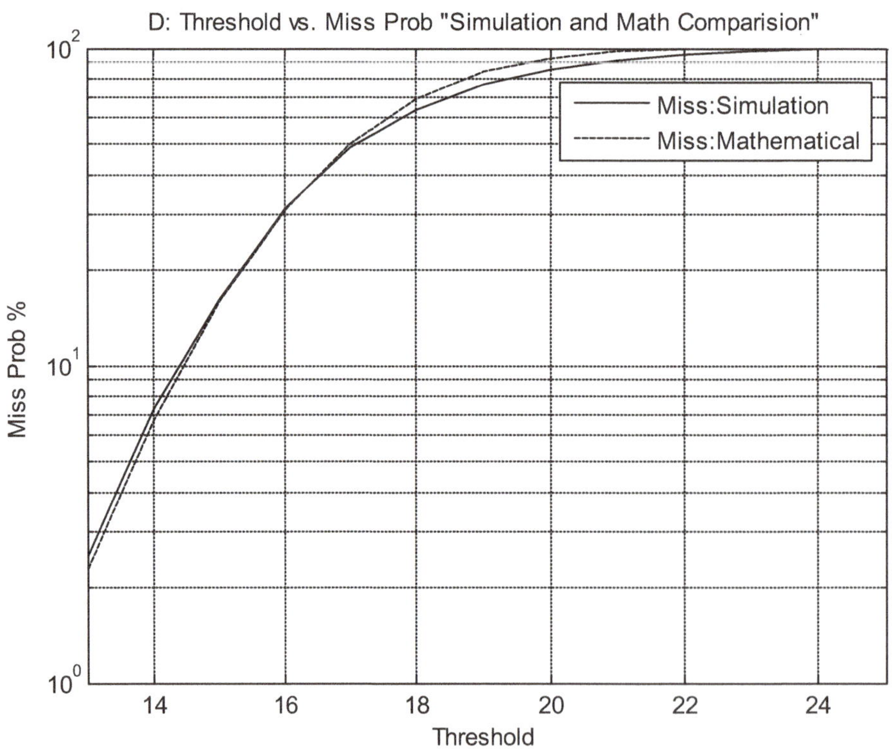

Figure 15. Comparison of Miss probabilities for simulation and mathematical calculations: Logarithmic metric.

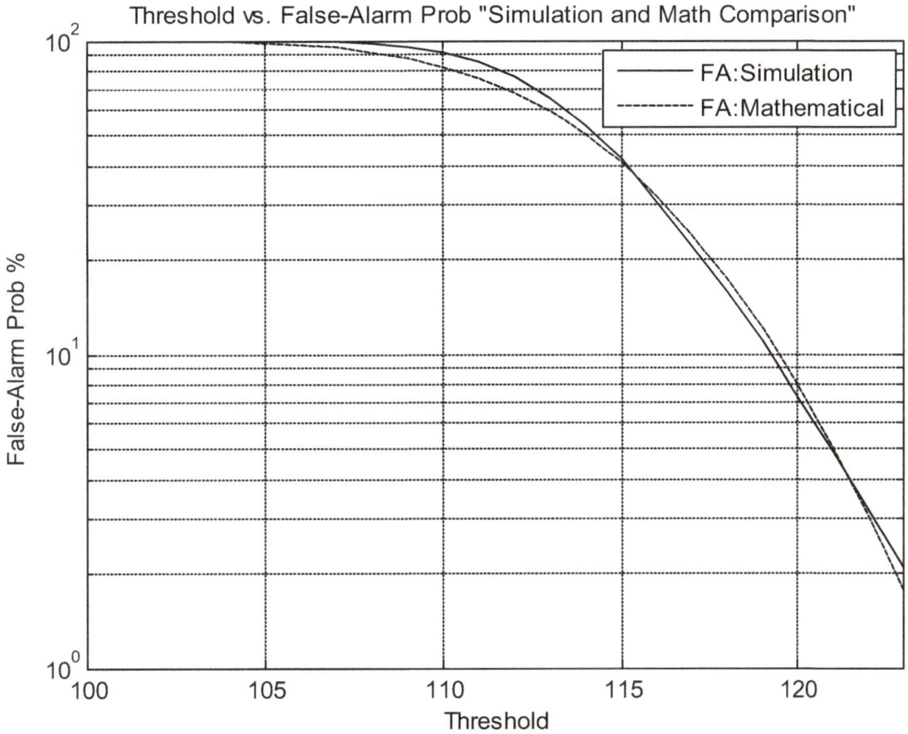

Figure 16. Comparison of False-Alarm probabilities for simulation and mathematical calculations: 3rd moment metric.

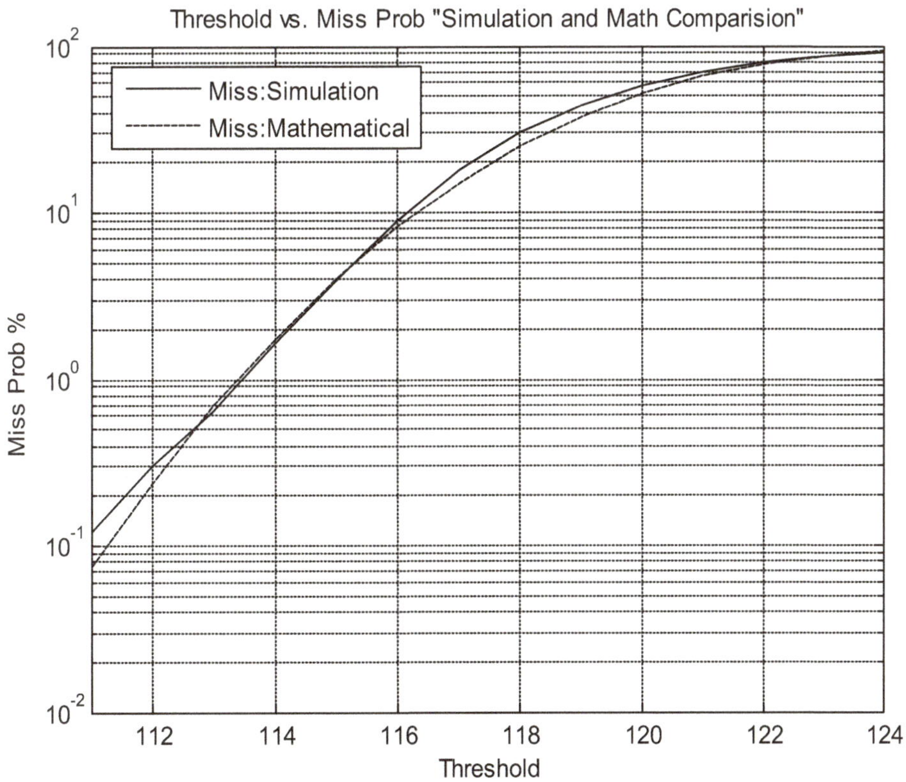

Figure 17. Comparison of Miss probabilities for simulation and mathematical calculations: 3rd moment metric.

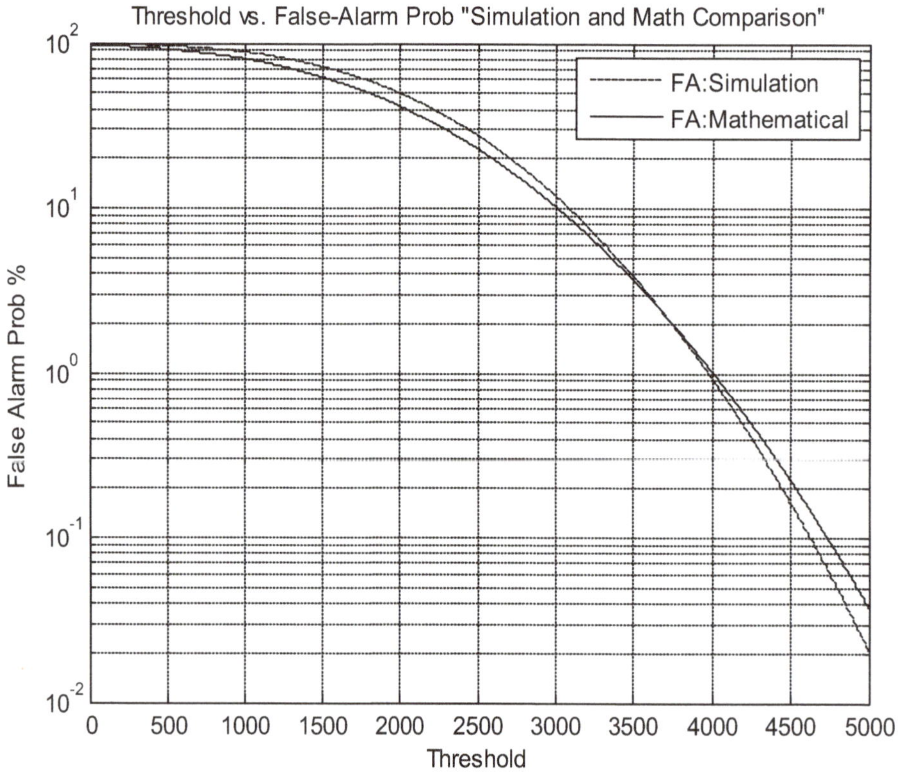

Figure 18. Comparison of False-Alarm probabilities for simulation and mathematical calculations: MAX-MIN based metric.

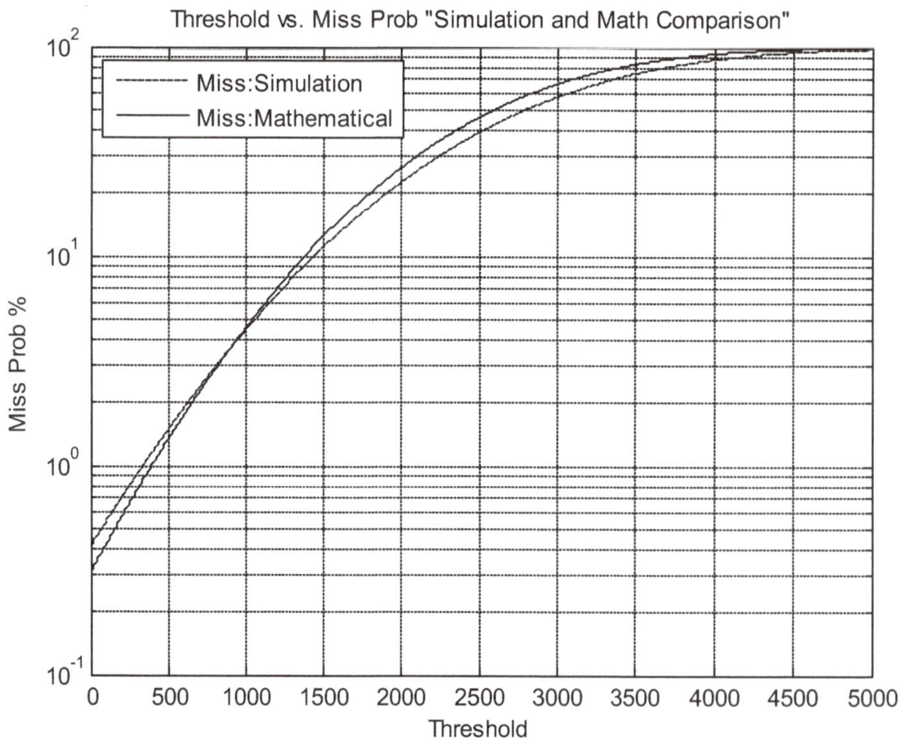

Figure 19. Comparison of Miss probabilities for simulation and mathematical calculations: MAX-MIN based metric.

For **Figure 15**:
- Maximum number of sensors is 30.
- $SINR_{cut-off} = 5$ dB, $SINR = 3.5$ dB $(\Delta_{SINR} = -1.5 \text{ dB})$.
- $\alpha = 0.25$.
- Modulation scheme is QPSK.
- Measurement period is 25 information bits
- Logarithmic metric.

For **Figure 16**:
- Maximum number of sensors is 30.
- $SINR_{cut-off} = 5$ dB, $SINR$= 6 dB $(\Delta_{SINR} = 1 \text{ dB})$.
- $\alpha = 0.2$.
- Modulation scheme is 8PSK.
- Measurement period is 25 information bits.
- 3rd moment metric.

For **Figure 17**:
- Maximum number of sensors is 30.
- $SINR_{cut-off} = 5$ dB, $SINR = 4$ dB $(\Delta_{SINR} = -1 \text{ dB})$.
- $\alpha = 0.25$.
- Modulation scheme is QPSK.
- Measurement period is 50 information bits.
- 3rd moment metric.

For **Figure 18**:
- Maximum number of sensors is 30.
- $SINR_{cut-off} = 5$ dB, $SINR$= 6.5 dB $(\Delta_{SINR} = 1.5 \text{ dB})$.
- $\alpha = 0.4$.
- Modulation scheme is 16 PSK.
- Measurement period is 500 information bits.
- MAX-MIN based metric.

For **Figure 19**:
- Maximum number of sensors is 30.
- $SINR_{cut-off} = 5$ dB, $SINR$= 3.5 dB $(\Delta_{SINR} = -1.5 \text{ dB})$.
- $\alpha = 0.4$.
- Modulation scheme is 16 PSK.
- Measurement period is 500 information bits.

MAX-MIN based metric.

6. Performance Evaluation

In this section we provide numerical performance evaluation of our proposed statistical discrimination algorithms for various system design scenarios and parameter choices. We also consider three modulation schemes, namely, QPSK, 8 PSK and 16 PSK. As pointed out in previous sections, without loss of generality and for the sake of a case study, we assume that a typical error correcting decoding scheme can successfully decode a packet with a satisfactory bit-error rate (BER) as long as the received signal-to-interference-plus-noise ratio (*SINR*) is higher than 5 dB (*i.e.*, $SINR_{cut-off} = 5$ dB), since a 5 dB *SINR* seems a reasonable assumption based on typical coding requirements in wireless systems [18]. Although the majority of the numerical results presented in this section are focused on the example of $SINR_{cut-off} = 5$ dB, we also show some example results for $SINR_{cut-off} = 7$ dB (**Figure 20**) and $SINR_{cut-off} = 10$ dB (**Figure 21**), to demonstrate the ability of our technique to work reliably with various *SINR* requirements.

We also evaluate the sensitivity of our proposed discriminators to the *SINR* deviation from the 5 dB cut-off point. That is, since the thresholds designed for the discriminators are pre-set based on studying (e.g., simulating) the statistics of the IQ signal envelope assuming "cut-off" *SINR* of 5 dB, it is important to investigate if the algorithm would still work reliably if the signal's *SINR* is offset by a $\pm\Delta$dB (e.g. $\Delta_{SINR} = \pm 1.5$ dB means the *SINR* = 6.5 dB for calculating False-Alarm probabilities, and the *SINR* = 3.5 dB for calculating Miss probabilities

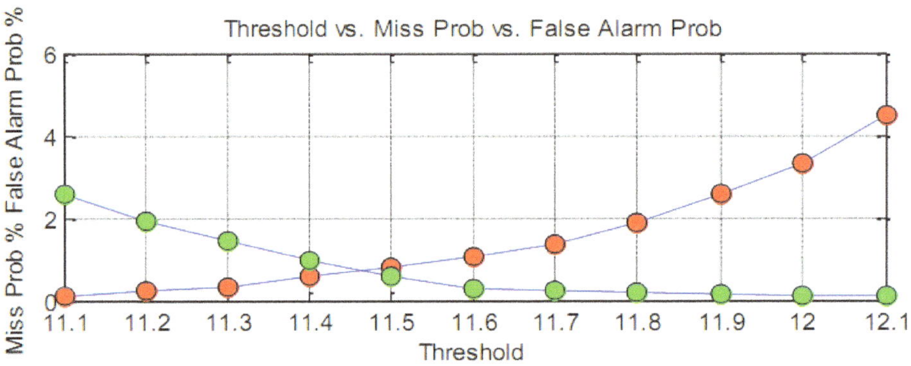

Figure 20. Miss probability = 1.01% (red) vs. False-Alarm probability = 0.93% (green) vs. threshold = 11.5, $\Delta_{SINR} = \pm 1.5$ dB, $SINR_{cut-off} = 7$ dB, logarithmic metric, 16PSK, measurement period = 250 bits.

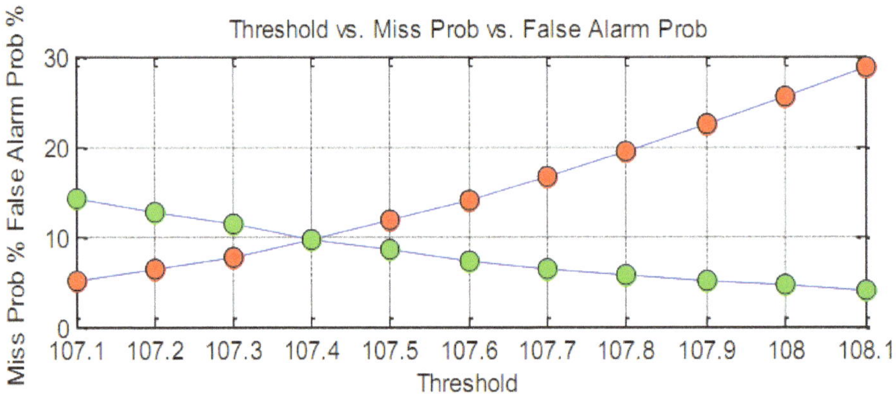

Figure 21. Miss probability = 10.00% (red) vs. False-Alarm probability = 10.00% (green) vs. threshold = 107.4, $\Delta_{SINR} = \pm 1$ dB, $SINR_{cut-off} = 10$ dB, 3rd moment metric, 8PSK, measurement period = 100 bits.

when $SINR_{cut-off}$ is 5 dB). In addition, we evaluate various measurement periods (number of information bits and number of samples per symbol, *i.e.*, over-sampling rate), as well as various levels of quantization of the *SD* metric computation to evaluate the performance of our algorithms in fixed-point implementation.

We typically generate 100,000 simulation snapshots where each snapshot generates a random number of interferers up to 30 sensors with random power assignments. **Figure 22** shows a flowchart for our simulation setup and procedure.

Figures 23-25 show the Miss (red points) and False-Alarm (green points) probabilities versus the choice of the metric comparison threshold level (*i.e.*, above which we decide the packet is valid (collision-free) and vice-versa) for the entropy (logarithmic) metric, the 3rd moment metric, the MAX-MIN based metric, and for QPSK, 8 PSK and 16 PSK modulation schemes, respectively (The choice of system parameters is defined in the caption of the corresponding figure). As shown in the figures, the intersection point of the red and green curves, can be a reasonable point to choose the threshold level in order to have a reasonable (or balanced) consideration of the Miss and False-Alarm probabilities, but certainly a designer can refer to **Appendix A** to choose an arbitrarily different point for a different criterion of choice. For example, **Figure 26** shows how the throughput of our proposed metrics may improve to 99.00% if a system designer sets the threshold at 15.2 or higher since this threshold results in a low False-Alarm probability of 0.01. Also, more results for logarithmic, moment, MAX-MIN based metrics are available in **Appendix A**.

7. Conclusion

In this paper we propose a novel simple power-efficient low-latency collision detection scheme for WSNs and

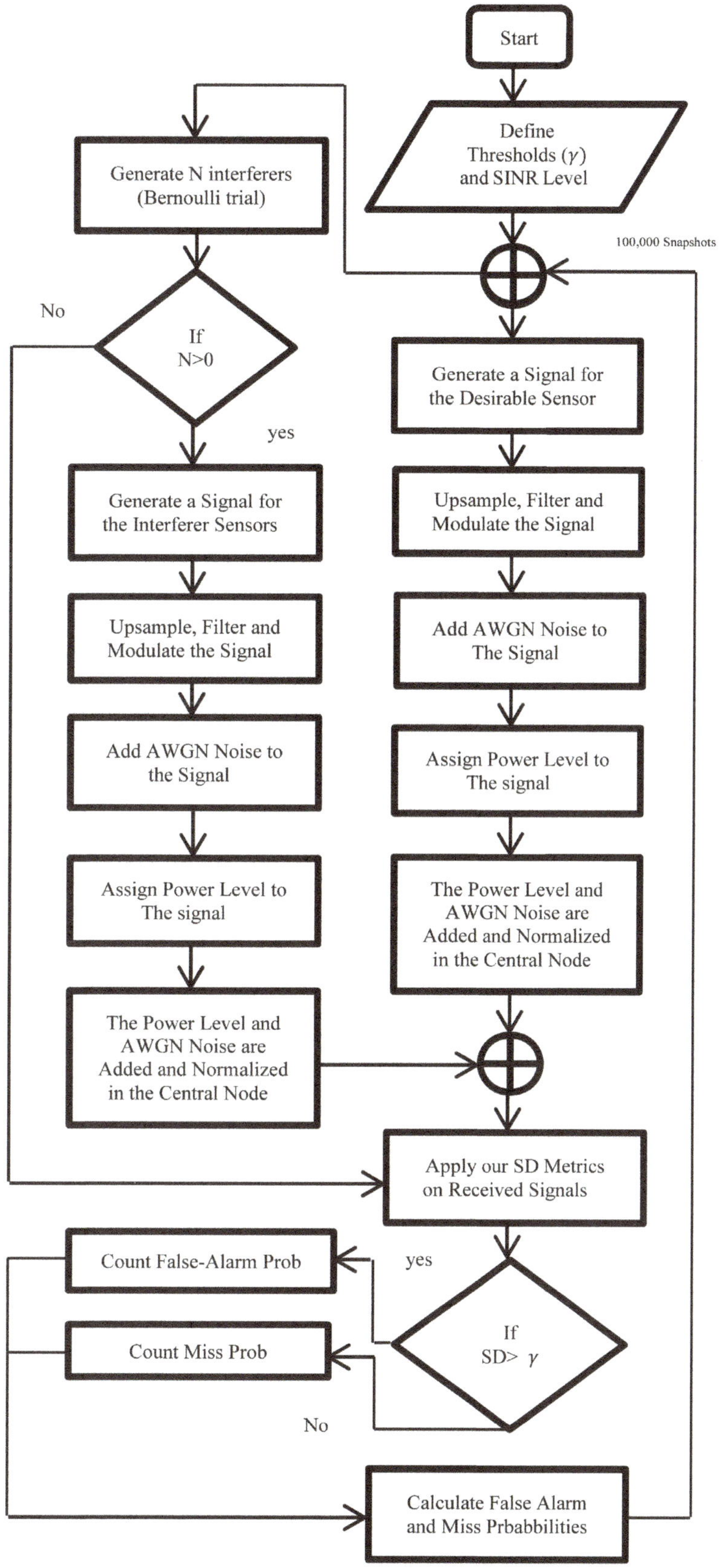

Figure 22. Flowchart for the simulation setup.

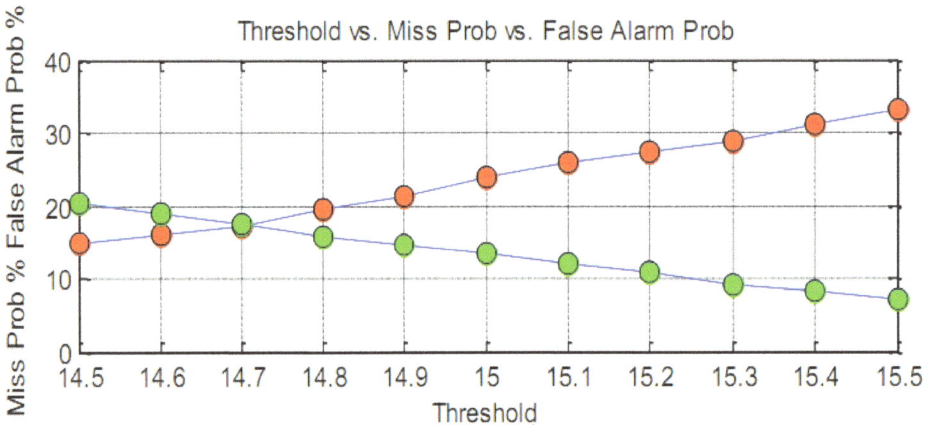

Figure 23. Miss probability = 17.33% (red) vs. False-Alarm probability = 18.36% (green) vs. threshold = 14.7, $\Delta_{SINR} = \pm 1$ dB, $SINR_{cut-off} = 5$ dB, logarithmic metric, QPSK, measurement period = 50 bits.

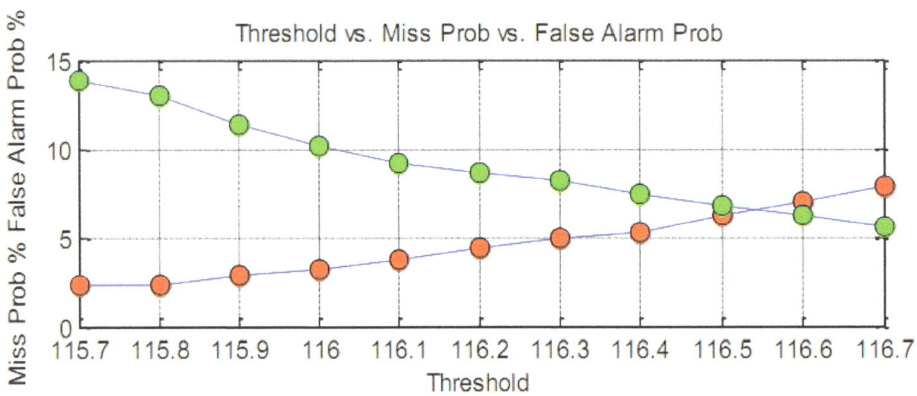

Figure 24. Miss probability = 6.85% (red) vs. False-Alarm probability = 7.27% (green) vs. threshold = 116.5, $\Delta_{SINR} = \pm 1$ dB, $SINR_{cut-off} = 5$ dB, 3rd moment metric, 8 PSK, measurement period = 200 bits.

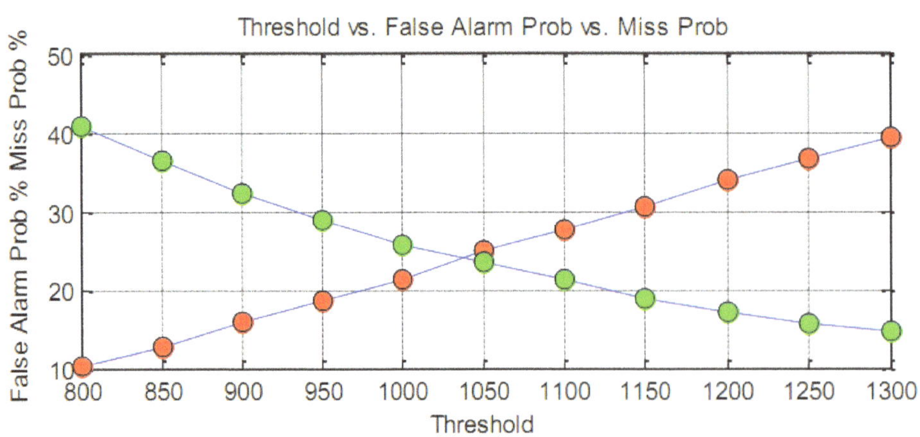

Figure 25. Miss probability = 24.22% (red) vs. False-Alarm probability = 23.07% (green) vs. threshold = 1050, $\Delta_{SINR} = \pm 1$ dB, $SINR_{cut-off} = 5$ dB, MAX-MIN based metric, 16 PSK, measurement period = 200 bits.

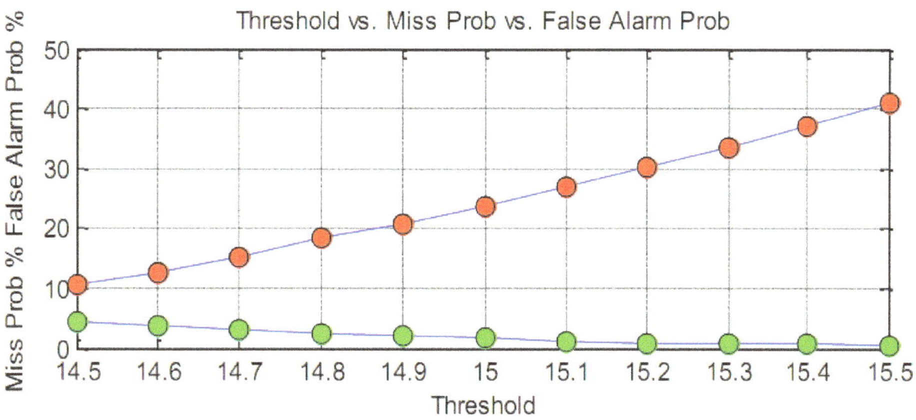

Figure 26. Improving the system throughput at threshold point 15.2, logarithmic metric, QPSK, measurement period = 200 bits.

analyze its performance. We propose three simple statistical discrimination metrics which are applied directly at the receiver's IQ ADC output to determine if the received signal represents a valid collision-free packet. Hence, saving a significant amount of processing power and collision detection processing time delay compares with conventional full-decoding mechanisms, which also requires going through the entire complex receiver and modem processing. We also analyze and demonstrate the amount of power saving achieved by our scheme compared with the conventional full-decoding scheme and provide a mathematical empirical characterization of the statistics of various quantities encountered in our scheme. As demonstrated by the numerical results and performance analysis, our novel scheme offers much lower computational complexity and shorter measurement period compared with a full-decoding scheme, and minimal impact on throughput, which can also be arbitrarily minimized per the system designer's choice of parameter setting and trade-offs.

References

[1] Latif, K., Ahmad, A., Javaid, N., Khan, Z.A. and Alrajeh, N. (2013) Divide-and-Rule Scheme for Energy Efficient Routing in Wireless Sensor Networks. *Procedia Computer Science*, **19**, 340-347. http://dx.doi.org/10.1016/j.procs.2013.06.047

[2] Ahmad, A., Latif, K., Javaid, N., Khan, A. and Qasim, U. (2013) Density Controlled Divide-and-Rule Scheme for Energy Efficient Routing in Wireless Sensor Networks. *26th Annual IEEE Canadian Conference on Electrical and Computer Engineering (CCECE)*, Regina, 5-8 May 2013, 1-4.

[3] Rasheed, M.B., Javaid, N., Khan, Z.A., Qasim, U. and Ishfaq, M. (2013) E-HORM: An Energy Efficient Hole Removing Mechanism in Wireless Sensor Networks. *26th IEEE Canadian Conference on Electrical and Computer Engineering (CCECE)*, Regina, 5-8 May 2013, 1-4.

[4] Rahim, A., Javaid, N., Aslam, M., Qasim, U. and Khan, Z.A. (2012) Adaptive-Reliable Medium Access Control Protocol for Wireless Body Area Networks. *Poster Session of 9th Annual IEEE Communications Society Conference on Sensor, Mesh and Ad Hoc Communications and Networks (SECON)*, Seoul, 18-21 June 2012, 56-58.

[5] Alvi, A.N., Naqvi, S.S., Bouk, S.H., Javaid, N., Qasim, U. and Khan, Z.A. (2012) Evaluation of Slotted CSMA/CA of IEEE 802.15.4. *7th International Conference on Broadband, Wireless Computing, Communication and Applications (BWCCA)*, Victoria, 12-14 November 2012, 391-396.

[6] Hayat, S., Javaid, N., Khan, Z.A., Shareef, A., Mahmood, A. and Bouk, S.H (2012) Energy Efficient MAC Protocols in Wireless Body Area Sensor Networks. *5th International Symposium on Advances of High Performance Computing and Networking (AHPCN-2012) in Conjunction with 14th IEEE International Conference on High Performance Computing and Communications (HPCC-2012)*, Liverpool, 25-27 June, 1185-1192.

[7] Biroli, A., Martina, M. and Masera, G. (2012) An LDPC Decoder Architecture for Wireless Sensor Network Applications. *Sensors*, **12**, 1529-1543.

[8] Grover, P., Woyach, K. and Sahai, A. (2011) Towards a Communication-Theoretic Understanding of System-Level Power Consumption. *IEEE Journal on Selected Areas in Communications*, **29**, 1744-1755.

[9] Pellenz, M., Souza, R. and Fonseca, M. (2010) Error Control Coding in Wireless Sensor Networks. *Telecommunication Systems*, **44**, 61-68. http://dx.doi.org/10.1007/s11235-009-9222-5

[10] Kashani, Z. and Shiva, M. (2007) Power Optimised Channel Coding in Wireless Sensor Networks Using Low-Density Parity-Check Codes. *IET Communications*, **1**, 1256-1262.

[11] Abouei, J., Brown, J., Plataniotis, K. and Pasupathy, S. (2011) Energy efficiency and Reliability in Wireless Biomedical Implant Systems. *IEEE Transactions on Information Technology in Biomedicine*, **15**, 456-466.

[12] Cai, Z.H., Hao, J. and Wang, L. (2008) An Efficient Early Stopping Scheme for LDPC Decoding Based on Check-Node Messages. 11*th IEEE Singapore International Conference on Communication Systems*, Guangzhou, 19-21 November 2008, 1325-1329.

[13] Ji, X.Y., He, Y., Wang, J.L., Dong, W., Wu, X.P. and Liu, Y.H. (2014) Walking down the STAIRS: Efficient Collision Resolution for Wireless Sensor Networks. *Proceedings of IEEE INFOCOM* 2014, Toronto, April 27-May 2014, 961-969.

[14] Peng, J., Cheng, L. and Sikdar, B. (2007) A Wireless MAC Protocol with Collision Detection. *IEEE Transactions on Mobile Computing*, **6**, 1357-1369.

[15] Kheirandish, D., Safari, A. and Kong, Y.N. (2014) A Novel Approach for Improving Error Detection and Correction in WSN. 27*th Canadian Conference on Electrical and Computer Engineering* (*CCECE*), Toronto, 4-7 May 2014, 1-4.

[16] Kim, W.T., Bae, S.J., Kang, S.G. and Joo, E.K. (2000) Reduction of Computational Complexity in Two-Step SOVA Decoder for Turbo Code. *IEEE Global Telecommunications Conference*, (*GLOBECOM*), San Francisco, 1887-1891.

[17] Robertson, P., Villebrun, E. and Hoeher, P. (1995) A Comparison of Optimal and Sub-Optimal MAP Decoding Algorithms Operating in the Log Domain. *IEEE International Conference on Communications* "*Gateway to Globalization*", Vol. 2, Seattle, 18-22 June 1995, 1009-1013.

[18] Dey, C. and Kundu, S. (2012) Hybrid Forwarding for General Cooperative Wireless Relaying in m-Nakagami Fading Channel. *International Journal of Future Generation Communication and Networking*, **5**, 29-42.

[19] LeBlance, D.C. (2004) Statistics Concepts and Applications for Science. 2nd Edition, Jones & Bartlett Publishers, Sudbury.

Appendix A (Tables for Simulation Results)

In this appendix, we provide more detailed performance results for our proposed scheme where $\alpha = 0.3$. In the following tables, *R* is the measurement period in bits, *B* is the number of quantization levels for the received signal envelop, *Z* is the oversampling rate, Δ_{SINR} is the tolerance level for the *SINR* (e.g. $\Delta_{SINR} = \pm 1.5$ dB means the *SINR* = 6.5 dB for calculating False-Alarm probabilities and the *SINR* = 3.5 dB for calculating Miss probabilities when the cut-off SNR is 5 dB), γ is the threshold level (in section II we explained how to select the threshold level. In logarithmic, 3rd moment and MAX-MIN based metrics we choose the intersection point between the False-Alarm and Miss probabilities curves in order to have a balanced consideration for such probabilities) and V is number of samples per measurement period. For example, let's assume our choice of parameters is as follows:

- Measurement period (*R*) = 25 bits
- Oversampling rate (*Z*) = 2
- Modulation scheme is 8PSK (*i.e.*, *M* = 3).

In order to show how many samples per measurement period (*V*) is used we use the following formula:

$$V = Z \times \left\lfloor \frac{R}{M} \right\rfloor = 2 \times \left\lfloor \frac{25}{3} \right\rfloor = 16 \text{ samples}$$

Table 1. Qpsk-Logarithmic based metric.

R	B	Z	Δ_{SINR}	P_{FA}	P_{miss}	γ	V
25	4	2	±1	33.33%	32.88%	14.7	24
25	6	4	±1	29.44%	29.92%	14.9	48
25	8	6	±1	26.90%	27.11%	14.9	72
25	10	8	±1	24.98%	26.02%	15.0	96
25	4	2	±1.5	26.47%	25.79%	14.6	24
25	6	4	±1.5	21.11%	20.37%	14.8	48
25	8	6	±1.5	18.41%	18.48%	14.9	72
25	10	8	±1.5	16.27%	17.11%	15.0	96
50	4	2	±1	24.72%	25.60%	14.6	50
50	6	4	±1	19.71%	20.97%	14.7	100
50	8	6	±1	18.36%	17.33%	14.7	150
50	10	8	±1	15.97%	16.23%	14.7	200
50	4	2	±1.5	16.24%	15.76%	14.4	50
50	6	4	±1.5	9.99%	10.64%	14.6	100
50	8	6	±1.5	8.12%	7.97%	14.6	150
50	10	8	±1.5	6.84%	7.62%	14.6	200
200	4	2	±1	9.10%	8.44%	14.2	200
200	6	4	±1	4.61%	3.96%	14.2	400
200	8	6	±1	3.03%	3.24%	14.3	600
200	10	8	±1	2.58%	2.06%	`14.3	800
200	4	2	±1.5	1.93%	2.02%	14.1	200
200	6	4	±1.5	0.47%	0.56%	14.2	400
200	8	6	±1.5	0.18%	0.26%	14.3	600
200	10	8	±1.5	0.16%	0.14%	14.2	800
500	4	2	±1	1.96%	1.99%	14.1	500
500	6	4	±1	0.56%	0.41%	14.1	1000
500	8	6	±1	0.35%	0.26%	14.2	1500
500	10	8	±1	0.20%	0.15%	14.2	2000
500	4	2	±1.5	0.11%	0.09%	14.1	500
500	6	4	±1.5	0.00%	0.00%	14.2	1000
500	8	6	±1.5	0.00%	0.00%	14.1	1500
500	10	8	±1.5	0.00%	0.00%	14.2	2000
1000	4	2	±1	0.25%	0.52%	14.1	1000
1000	6	4	±1	0.00%	0.00%	14.1	2000
1000	8	6	±1	0.00%	0.00%	14.1	3000
1000	10	8	±1	0.00%	0.00%	14.1	4000
1000	4	2	±1.5	0.00%	0.00%	14.1	1000
1000	6	4	±1.5	0.00%	0.00%	14.1	2000
1000	8	6	±1.5	0.00%	0.00%	14.1	3000
1000	10	8	±1.5	0.00%	0.00%	14.1	4000

Table 2. 8 PSK-logarithmic based metric.

R	B	Z	Δ_{SNR}	P_{FA}	P_{miss}	γ	V
25	4	2	±1	36.77%	36.71%	14.5	16
25	6	4	±1	33.62%	33.49%	14.8	32
25	8	6	±1	31.94%	30.72%	14.8	48
25	10	8	±1	29.56%	29.81%	14.9	64
25	4	2	±1.5	30.90%	30.08%	14.3	16
25	6	4	±1.5	25.62%	25.60%	14.7	32
25	8	6	±1.5	23.53%	23.17%	14.8	48
25	10	8	±1.5	22.52%	21.62%	14.9	64
50	4	2	±1	28.72%	29.63%	14.7	32
50	6	4	±1	24.87%	25.49%	14.9	64
50	8	6	±1	23.38%	23.64%	14.9	96
50	10	8	±1	22.25%	22.28%	14.9	128
50	4	2	±1.5	22.23%	22.46%	14.6	32
50	6	4	±1.5	16.39%	16.67%	14.8	64
50	8	6	±1.5	13.37%	13.66%	14.8	96
50	10	8	±1.5	12.38%	12.08%	14.8	128
200	4	2	±1	12.60%	13.01%	14.3	132
200	6	4	±1	7.62%	7.59%	14.3	264
200	8	6	±1	6.18%	5.55%	14.3	396
200	10	8	±1	4.50%	5.65%	`14.4	627
200	4	2	±1.5	4.85%	4.54%	14.1	132
200	6	4	±1.5	1.76 %	1.67 %	14.2	264
200	8	6	±1.5	0.82 %	0.97%	14.3	396
200	10	8	±1.5	0.77%	0.63%	14.3	627
500	4	2	±1	4.00%	3.87%	14.1	332
500	6	4	±1	1.04%	1.63%	14.2	664
500	8	6	±1	0.69%	0.84%	14.2	996
500	10	8	±1	0.53%	0.53%	14.2	1328
500	4	2	±1.5	0.46%	0.36%	13.9	332
500	6	4	±1.5	0.05%	0.05%	14.0	664
500	8	6	±1.5	0.00%	0.00%	14.2	996
500	10	8	±1.5	0.00%	0.00%	14.2	1328
1000	4	2	±1	0.74%	0.94%	14.1	666
1000	6	4	±1	0.00%	0.00%	14.1	1332
1000	8	6	±1	0.00%	0.00%	14.1	1998
1000	10	8	±1	0.00%	0.00%	14.1	2664
1000	4	2	±1.5	0.00%	0.00%	14.0	666
1000	6	4	±1.5	0.00%	0.00%	14.0	1332
1000	8	6	±1.5	0.00%	0.00%	14.0	1998
1000	10	8	±1.5	0.00%	0.00%	14.0	2664

Table 3. 16 PSK-logarithmic based metric.

R	B	Z	Δ_{SINR}	P_{FA}	P_{miss}	γ	V
25	4	2	±1	38.61%	39.45%	13.9	12
25	6	4	±1	35.49%	35.49%	14.3	24
25	8	6	±1	33.86%	33.84%	14.4	36
25	10	8	±1	31.33%	31.33%	14.4	48
25	4	2	±1.5	33.24%	32.76%	13.8	12
25	6	4	±1.5	28.70 %	29.25%	14.3	24
25	8	6	±1.5	23.04%	27.34%	14.4	36
25	10	8	±1.5	24.58 %	24.94%	14.5	48
50	4	2	±1	33.14%	33.04%	14.7	24
50	6	4	±1	28.83%	30.06%	15.0	48
50	8	6	±1	26..34%	27.51%	15.0	72
50	10	8	±1	25.85%	24.54%	14.9	96
50	4	2	±1.5	26.21%	25.80%	14.6	24
50	6	4	±1.5	20.65%	20.65%	14.8	48
50	8	6	±1.5	18.29%	18.43%	14.9	72
50	10	8	±1.5	16.77 %	15.71 %	14.9	96
200	4	2	±1	16.92%	16.86%	14.3	100
200	6	4	±1	11.26%	10.11%	14.3	200
200	8	6	±1	8.10%	8.44%	14.4	300
200	10	8	±1	7.68%	6.67%	`14.4	400
200	4	2	±1.5	7.10%	6.92%	14.2	100
200	6	4	±1.5	3.35%	2.85%	14.2	200
200	8	6	±1.5	2.00%	1.86%	14.3	300
200	10	8	±1.5	1.43	1.75%	14.4	400
500	4	2	±1	5.27%	6.48%	14.2	250
500	6	4	±1	2.25%	2.73%	14.2	500
500	8	6	±1	1.71%	1.56%	14.2	750
500	10	8	±1	0.99%	1.39%	14.3	1000
500	4	2	±1.5	1.00%	1.22%	14.0	250
500	6	4	±1.5	0.15%	0.24%	14.1	500
500	8	6	±1.5	0.07%	0.11%	14.2	750
500	10	8	±1.5	0.05%	0.03%	14.2	1000
1000	4	2	±1	1.44%	2.03%	14.1	500
1000	6	4	±1	0.38%	0.30%	14.1	1000
1000	8	6	±1	0.00%	0.00%	14.1	1500
1000	10	8	±1	0.00%	0.00%	14.1	2000
1000	4	2	±1.5	0.00%	0.00%	14.0	500
1000	6	4	±1.5	0.00%	0.00%	14.0	1000
1000	8	6	±1.5	0.00%	0.00%	14.0	1500
1000	10	8	±1.5	0.00%	0.00%	14.0	2000

Table 4. QPSK-3rd moment based metric.

R	B	Z	Δ_{SINR}	P_{FA}	P_{miss}	γ	V
25	4	2	±1	33.55%	34.10%	116.4	24
25	6	4	±1	30.36%	30.01%	116.9	48
25	8	6	±1	29.26%	28.04%	117.0	72
25	10	8	±1	27.31%	26.86%	117.0	96
25	4	2	±1.5	26.88%	26.99%	116.2	24
25	6	4	±1.5	22.31%	22.62%	116.8	48
25	8	6	±1.5	20.05%	20.94%	117.0	72
25	10	8	±1.5	18.80%	18.86%	117.1	96
50	4	2	±1	27.87%	27.77%	116.7	50
50	6	4	±1	23.11%	24.08%	117.0	100
50	8	6	±1	23.52%	22.67%	116.9	150
50	10	8	±1	21.20%	20.97%	117.0	200
50	4	2	±1.5	19.05%	19.16%	116.5	50
50	6	4	±1.5	14.50%	14.35%	116.9	100
50	8	6	±1.5	11.73%	12.16%	117.0	150
50	10	8	±1.5	11.35%	10.97%	117.1	200
200	4	2	±1	10.54%	10.91%	116.0	200
200	6	4	±1	6.77%	6.88%	116.1	400
200	8	6	±1	6.06%	5.50%	116.1	600
200	10	8	±1	5.15%	5.29%	116.2	800
200	4	2	±1.5	3.57%	3.57%	115.8	200
200	6	4	±1.5	1.59%	1.83%	116.0	400
200	8	6	±1.5	1.45%	1.35%	115.9	600
200	10	8	±1.5	1.44%	1.36%	116.0	800
500	4	2	±1	3.33%	3.33%	115.6	500
500	6	4	±1	1.47%	1.57%	115.6	1000
500	8	6	±1	1.17%	1.24%	115.6	1500
500	10	8	±1	1.12%	1.22%	115.5	2000
500	4	2	±1.5	0.48%	0.43%	115.2	500
500	6	4	±1.5	0.16%	0.15%	115.3	1000
500	8	6	±1.5	0.15%	0,17%	115.2	1500
500	10	8	±1.5	0.14%	0.16%	115.1	2000
1000	4	2	±1	0.78%	0.84%	115.4	1000
1000	6	4	±1	0.07%	0.08%	115.4	2000
1000	8	6	±1	0.00%	0.00%	115.4	3000
1000	10	8	±1	0.00%	0.00%	115.4	4000
1000	4	2	±1.5	0.00%	0.00%	115.2	1000
1000	6	4	±1.5	0.00%	0.00%	115.2	2000
1000	8	6	±1.5	0.00%	0.00%	115.2	3000
1000	10	8	±1.5	0.00%	0.00%	115.2	4000

Table 5. 8 PSK-3rd moment based metric.

R	B	Z	Δ_{SINR}	P_{FA}	P_{miss}	γ	V
25	4	2	±1	37.16%	37.16%	115.4	16
25	6	4	±1	33.49%	33.00%	116.2	32
25	8	6	±1	30.89%	30.81%	116.4	48
25	10	8	±1	29.80%	30.31%	116.5	64
25	4	2	±1.5	30.30%	30.65%	115.4	16
25	6	4	±1.5	26.42%	25.37%	116.0	32
25	8	6	±1.5	22.60%	22.97%	116.3	48
25	10	8	±1.5	22.37%	21.80%	116.4	64
50	4	2	±1	31.76%	31.05%	116.7	32
50	6	4	±1	27.97%	27.88%	117.1	64
50	8	6	±1	25.81%	26.53%	117.2	96
50	10	8	±1	24.91%	24.43%	117.2	128
50	4	2	±1.5	24.06%	24.56%	116.6	32
50	6	4	±1.5	19.39%	18.88%	117.0	64
50	8	6	±1.5	17.20%	17.29%	117.2	96
50	10	8	±1.5	15.74%	15.39%	117.2	128
200	4	2	±1	16.30%	15.56%	116.2	132
200	6	4	±1	11.25%	11.30%	116.4	264
200	8	6	±1	9.22%	9.46%	116.4	396
200	10	8	±1	7.27%	6.85%	116.5	627
200	4	2	±1.5	6.10%	6.19%	116.1	132
200	6	4	±1.5	4.01%	3.84%	116.3	264
200	8	6	±1.5	3.48%	3.27%	116.3	396
200	10	8	±1.5	3.04%	2.79%	116.4	627
500	4	2	±1	5.57%	5.75%	115.8	332
500	6	4	±1	3.26%	3.07%	115.8	664
500	8	6	±1	2.44%	2.34%	115.8	996
500	10	8	±1	2.14%	2.27%	115.8	1328
500	4	2	±1.5	0.95%	1.15%	115.6	332
500	6	4	±1.5	0.56%	0.53%	115.5	664
500	8	6	±1.5	0.43%	0.45%	115.6	996
500	10	8	±1.5	0.41%	0.36%	115.4	1328
1000	4	2	±1	1.69%	1.69%	115.5	666
1000	6	4	±1	0.79%	0.86%	115.4	1332
1000	8	6	±1	0.64 %	0.75%	115.4	1998
1000	10	8	±1	0.56%	0.56%	115.4	2664
1000	4	2	±1.5	0.19%	0.17%	115.2	666
1000	6	4	±1.5	0.05%	0.02%	115.0	1332
1000	8	6	±1.5	0.00%	0.00%	115.2	1998
1000	10	8	±1.5	0.00%	0.00%	115.2	2664

Table 6. 16 PSK-3rd moment based metric.

R	B	Z	Δ_{SNR}	P_{FA}	P_{miss}	γ	V
25	4	2	±1	38.37%	38.89%	114.1	12
25	6	4	±1	35.08%	34.92	115.0	24
25	8	6	±1	34.09%	33.37%	115.3	36
25	10	8	±1	31.57%	31.92%	115.6	48
25	4	2	±1.5	33.06%	33.38%	114.0	12
25	6	4	±1.5	28.26%	28.58%	114.9	24
25	8	6	±1.5	25.93%	25.75%	115.3	36
25	10	8	±1.5	24.22%	24.44%	115.4	48
50	4	2	±1	33.71%	34.13%	116.6	24
50	6	4	±1	31.16%	30.80%	117.0	48
50	8	6	±1	28.37%	28.07%	117.0	72
50	10	8	±1	27.59%	26.89%	117.1	96
50	4	2	±1.5	26.99%	27.58%	116.3	24
50	6	4	±1.5	21.89%	21.99%	116.9	48
50	8	6	±1.5	19.57%	20.01%	117.1	72
50	10	8	±1.5	18.90%	19.28%	117.2	96
200	4	2	±1	18.92%	19.78%	116.4	100
200	6	4	±1	13.99%	14.55%	116.5	200
200	8	6	±1	13.53%	12.73%	116.6	300
200	10	8	±1	11.30%	12.14%	116.7	400
200	4	2	±1.5	9.59%	9.59%	116.3	100
200	6	4	±1.5	6.40%	6.40%	116.5	200
200	8	6	±1.5	5.29%	4.91%	116.5	300
200	10	8	±1.5	4.33%	4.22%	116.6	400
500	4	2	±1	7.79%	8.51%	115.9	250
500	6	4	±1	4.84%	5.12%	116.0	500
500	8	6	±1	3.71%	4.30%	116.0	750
500	10	8	±1	3.48%	3.29%	116.0	1000
500	4	2	±1.5	1.98%	1.98%	115.7	250
500	6	4	±1.5	1.01%	1.09%	115.8	500
500	8	6	±1.5	0.94%	0.81%	115.8	750
500	10	8	±1.5	0.81%	0.78%	115.8	1000
1000	4	2	±1	3.00%	3.08%	115.6	500
1000	6	4	±1	1.50%	1.50%	115.6	1000
1000	8	6	±1	1.33%	1.10%	115.5	1500
1000	10	8	±1	1.07%	1.17%	115.5	2000
1000	4	2	±1.5	0.37%	0.40%	115.4	500
1000	6	4	±1.5	0.24%	0.20%	115.2	1000
1000	8	6	±1.5	0.21%	0.18%	115.1	1500
1000	10	8	±1.5	0.11%	0.13%	115.2	2000

Table 7. QPSK-Maximum to minimum based metric.

R	B	Z	Δ_{SINR}	P_{FA}	P_{miss}	γ	V
25	4	2	±1	36.33%	35.23%	550	24
25	6	4	±1	34.44%	33.92%	750	48
25	8	6	±1	29.90%	30.11%	950	72
25	10	8	±1	24.98%	26.02%	1100	96
25	4	2	±1.5	31.47%	30.79%	550	24
25	6	4	±1.5	27.11%	24.82%	750	48
25	8	6	±1.5	25.41%	23.90%	900	72
25	10	8	±1.5	21.37%	22.11%	1050	96
50	4	2	±1	33.52%	34.60%	800	50
50	6	4	±1	29.30%	31.97%	1100	100
50	8	6	±1	28.36%	30.33%	1300	150
50	10	8	±1	26.97%	27.23%	1550	200
50	4	2	±1.5	27.24%	28.76%	1350	50
50	6	4	±1.5	24.74%	23.64%	1300	100
50	8	6	±1.5	23.12%	22.65%	1100	150
50	10	8	±1.5	20.14%	21.01%	1050	200
200	4	2	±1	28.10%	27.44%	1450	200
200	6	4	±1	25.34%	23.21%	1500	400
200	8	6	±1	24.81%	22.01%	1500	600
200	10	8	±1	18.34%	18.93%	`1550	800
200	4	2	±1.5	17.34%	19.84%	1450	200
200	6	4	±1.5	15.73%	17.83 %	1500	400
200	8	6	±1.5	13.23%	15.34%	1500	600
200	10	8	±1.5	11.54%	12.34%	1500	800
500	4	2	±1	22.21%	23.43%	1550	500
500	6	4	±1	18.24%	19.34%	1550	1000
500	8	6	±1	16.26%	15.34%	1600	1500
500	10	8	±1	12.20%	12.15%	1600	2000
500	4	2	±1.5	14.65%	15.09%	1550	500
500	6	4	±1.5	12.00%	12.30%	1550	1000
500	8	6	±1.5	10.40%	10.60%	1650	1500
500	10	8	±1.5	8.34%	8.90%	1650	2000
1000	4	2	±1	20.65%	19.32%	1700	1000
1000	6	4	±1	18.34%	17.89%	1750	2000
1000	8	6	±1	15.55%	15.34%	1750	3000
1000	10	8	±1	11.78%	11.34%	1750	4000
1000	4	2	±1.5	12.32%	13.34%	1700	1000
1000	6	4	±1.5	9.90%	9.10%	1800	2000
1000	8	6	±1.5	7.45%	7.23%	1850	3000
1000	10	8	±1.5	5.93%	5.55%	1850	4000

Table 8. 8 PSK-Maximum to minimum based metric.

R	B	Z	Δ_{SINR}	P_{FA}	P_{miss}	γ	V
25	4	2	±1	39.07%	39.71%	450	16
25	6	4	±1	36.62%	35.49%	450	32
25	8	6	±1	33.94%	32.72%	450	48
25	10	8	±1	30.46%	30.81%	500	64
25	4	2	±1.5	32.70%	31.03%	450	16
25	6	4	±1.5	29.62%	29.60%	500	32
25	8	6	±1.5	27.53%	25.17%	550	48
25	10	8	±1.5	24.52%	23.62%	550	64
50	4	2	±1	34.72%	35.63%	650	32
50	6	4	±1	30.87%	33.45%	650	64
50	8	6	±1	29.38%	31.64%	650	96
50	10	8	±1	28.25%	28.28%	700	128
50	4	2	±1.5	29.33%	29.49%	600	32
50	6	4	±1.5	25.39%	24.67%	600	64
50	8	6	±1.5	24.37%	23.66%	600	96
50	10	8	±1.5	23.34%	22.03%	700	128
200	4	2	±1	31.20%	29.01%	1200	132
200	6	4	±1	30.01%	28.12%	1200	264
200	8	6	±1	28.91%	26.12%	1200	396
200	10	8	±1	26.22%	24.66%	`1250	627
200	4	2	±1.5	21.01%	22.22%	1200	132
200	6	4	±1.5	19.17%	19.10%	1200	264
200	8	6	±1.5	17.33%	17.12%	1350	396
200	10	8	±1.5	15.23%	15.74%	1350	627
500	4	2	±1	25.37%	26.34%	1900	332
500	6	4	±1	23.33%	24.15%	1900	664
500	8	6	±1	21.91%	22.45%	1950	996
500	10	8	±1	20.17%	21.84%	1950	1328
500	4	2	±1.5	21.45%	19.12%	1800	332
500	6	4	±1.5	18.26%	17.21%	1800	664
500	8	6	±1.5	16.14%	14.22%	1800	996
500	10	8	±1.5	14.87%	11.67%	1850	1328
1000	4	2	±1	27.23%	26.33%	2750	666
1000	6	4	±1	24.45 %	23.32%	2700	1332
1000	8	6	±1	20.12%	22.34%	2700	1998
1000	10	8	±1	19.65%	20.32%	2700	2664
1000	4	2	±1.5	21.00%	21.00%	2750	666
1000	6	4	±1.5	18.35%	15.45%	2750	1332
1000	8	6	±1.5	14.76%	12.35%	2800	1998
1000	10	8	±1.5	10.67%	9.38%	2800	2664

Table 9. 16 PSK-Maximum to minimum based metric.

R	B	Z	Δ_{SINR}	P_{FA}	P_{miss}	γ	V
25	4	2	±1	41.07%	40.73%	400	12
25	6	4	±1	39.62%	37.44%	400	24
25	8	6	±1	35.92%	33.76%	400	36
25	10	8	±1	31.49%	31.86%	500	48
25	4	2	±1.5	35.71%	33.07%	400	12
25	6	4	±1.5	30.62%	30.65%	400	24
25	8	6	±1.5	29.03%	27.37%	550	36
25	10	8	±1.5	26.50%	25.62%	550	48
50	4	2	±1	36.02%	37.73%	650	24
50	6	4	±1	33.82%	35.45%	650	48
50	8	6	±1	32.36%	33.64%	750	72
50	10	8	±1	31.21%	30.83%	750	96
50	4	2	±1.5	30.36%	29.49%	800	24
50	6	4	±1.5	27.29%	25.64%	900	48
50	8	6	±1.5	26.37%	24.66%	1000	72
50	10	8	±1.5	25.34%	23.03%	1000	96
200	4	2	±1	33.10%	30.01%	1000	100
200	6	4	±1	31.01%	29.12%	1000	200
200	8	6	±1	29.91%	27.12%	1000	300
200	10	8	±1	27.22%	26.69%	`1000	400
200	4	2	±1.5	23.07%	24.22%	1050	100
200	6	4	±1.5	20.17%	20.30%	1150	200
200	8	6	±1.5	19.33%	18.12%	1300	300
200	10	8	±1.5	16.83%	16.79%	1350	400
500	4	2	±1	27.37%	27.34%	1900	250
500	6	4	±1	24.63%	26.85%	1950	500
500	8	6	±1	23.91%	24.35%	1950	750
500	10	8	±1	21.07%	22.14%	1950	1000
500	4	2	±1.5	22.95%	20.12%	1800	250
500	6	4	±1.5	19.20%	18.21%	1900	500
500	8	6	±1.5	17.24%	16.81%	1900	750
500	10	8	±1.5	15.85%	13.67%	1950	1000
1000	4	2	±1	28.20%	27.73%	2300	500
1000	6	4	±1	26.05%	25.31%	2300	1000
1000	8	6	±1	22.42%	23.24%	2300	1500
1000	10	8	±1	20.95%	21.32%	2350	2000
1000	4	2	±1.5	23.03%	23.00%	2300	500
1000	6	4	±1.5	19.38%	16.43%	2300	1000
1000	8	6	±1.5	15.76%	13.35%	2350	1500
1000	10	8	±1.5	11.61%	10.91%	2350	2000

7

Analysis of Five Typical Localization Algorithms for Wireless Sensor Networks

Shelei Li, Xueyong Ding, Tingting Yang

Polytechnical School, Sanya College, Sanya, China
Email: lishelei@126.com

Abstract

In this paper, the self-localization problem is studied. It is one of the key technologies in wireless sensor networks (WSNs). And five localization algorithms: Centroid algorithm, Amorphous algorithm, DV-hop algorithm, APIT algorithm and Bounding Box algorithm are discussed. Simulation of those five localization algorithms is done by MATLAB. The simulation results show that the positioning error of Amorphous algorithm is the minimum. Considering economy and localization accuracy, the Amorphous algorithm can achieve the best localization performance under certain conditions.

Keywords

Wireless Sensor Networks (WSNs), Localization Algorithm, Centroid, Amorphous, DV-Hop, APIT, Bounding Box

1. Introduction

Wireless Sensor Network (WSN) is one of the hottest direction in the research areas of IT, developed on sensor technology, micro-electromechanical technology and wireless communication technology [1]. WSN is widely used in national defense, military, environmental monitoring, traffic management, medical and health care, manufacturing, anti-terrorism disaster relief and other fields. For wireless sensor nodes, most applications need to collect data and perceive sensor location. Localization of sensor nodes and localization of events occurred are the basic functions of WSN. So the study on self-localization of sensor node is very important and meaningful [2].

According to the positioning mechanism, WSN localization algorithm can be divided into two classes: Range-based localization algorithm and Range-free (non-ranging based) localization algorithm. Range-based location algorithm is got by measuring the adjacent section. There isn't distance and angle information in Range-free localization algorithm, so classic algorithms are: time of arrival (TOA) ranging method, time difference location method (TDOA), the received signal strength indicator (RSSI) ranging method, etc. Range-free localization algorithm can realize node localization without distance and angle information, only according to information such as network connectivity, classic algorithms include: centroid, DV-Hop (distance vector-hop), Amorphous, and APIT algorithm. To compare with non-ranging based localization algorithm based on distance measurement for the static characteristic of most sensor network application, the former is curbed in both cost and node size, while the latter does not need additional hardware, there is an advantage in cost and power consumption. On the other hand, the former performance is strongly influenced by environmental factors, while the latter performs well in this respect, and positioning accuracy can meet the requirements of most sensor network applications. In comprehensive view, the latter is more suitable for the characteristics of WSN large-scale network. As a result, most scholars of recent research focus on the Range-free localization algorithm [3].

This paper has analyzed the five classical localization algorithms: Centroid algorithm, Amorphous algorithm, DV-hop algorithm, APIT algorithm and Bounding Box algorithm. And through the comparison among the simulation results of five kinds of algorithm performance, and the study of the influence on the localization performance by anchor node proportion and communication radius, it can provide reference for selecting localization algorithm and configuring parameters [3].

2. Algorithm Introductions

The problem of node localization in wireless sensor networks is to ascertain the location of unknown node according to some localization mechanisms and based on known node. Localization mechanism embodies in localization algorithm. The centroid localization algorithm, Amorphous, DV-hop, APIT and Bounding Box algorithm are introduced.

2.1. Centroid Localization Algorithm

Centroid algorithm is put forward by outdoor positioning algorithm of network connectivity in the University of Southern California. The core of the algorithm is that the anchor nodes broadcast a beacon signal to the network every once in a while. The signal itself and location information contained in "received from different when the unknown node anchor node beacon signal exceeds a preset threshold or received after a certain period of time, the node will determine its position for the anchor nodes of polygon centroid", "literature of centroid algorithm is improved, the density of the proposed adaptive method, the best plans in areas of low anchor node density increase of anchor nodes, in order to improve the positioning accuracy". Fully advantages of the algorithm are based on network connectivity, without coordination between the unknown node and anchor node. The implementation is relatively simple, where shortcoming is positioning accuracy which is not high. A disadvantage of the centroid localization algorithm is that the localization accuracy depends on the density of the anchor node excessively.

2.2. Amorphous Localization Algorithm

Radhika Nagpal has proposed [4] Amorphous algorithm, unlike DV-Hop, it assumes that the average network connectivity of local is known and uses Equation (1) to compute network average hop distance. Among them, r is the communication radius, nlocal is the average network connectivity. The average distance of each hop estimated distance multiplied with minimum hop count is calculated to be positioning of nodes, and anchor nodes also need to work out at least three distances, using maximum likelihood estimation to estimate its position or three sided measurements.

$$\text{HopeSize} = r\left(1+\mathrm{e}^{-nlocal} - \int_{-1}^{1} \mathrm{e}^{-\frac{nlocal}{\pi}\left(\operatorname{ar}\cos t - t\sqrt{1-t^2}\right)} \mathrm{d}t\right) \quad (1)$$

In essence, the Amorphous algorithm is an enhanced version of DV-Hop algorithm, the gradient correction method and the distance of each hop algorithm calculation, and the introduction of a number of anchor nodes estimate refinement. The algorithm needs to predict the average connectivity of network and need higher node densities.

2.3. DV-Hop Localization Algorithm

DV-hop algorithm based on hop count is a range free localization algorithm [5] [6] proposed by Nicolescu *et al.* Its basic principle is to use inter node hop distance to replace the actual distance between nodes.

$$\text{HopSize}_i = \frac{\sum_{i \neq j} \sqrt{(x_i - x_j)^2 + (y_i - y_j)^2}}{\sum_{i \neq j} h_j} \tag{2}$$

(x_i, y_i) and (x_j, y_j) are the coordinates of anchor node i and j, respectively. h_j is the jump section number between anchor node i and j $(i \neq j)$, we can calculate the coordinates of the nodes to be located. The DV-hop algorithm is simple, less communication overhead, computation amount is less, the cost is low; but to achieve good positioning accuracy, it needs to decorate the large number of anchor nodes.

2.4. APIT Localization Algorithm [7]

In this section, we describe our novel area-based Range-free localization scheme, which is called APIT. APIT requires a heterogeneous network of sensing devices where a small percentage of these devices (percentages vary depending on network and node density) are equipped with high-powered transmitters and location information obtained via GPS.

The first step of APIT algorithmic is to determine a plurality of triangular region that contains unknown node, the intersection of these triangular areas is a polygon; then to determine the smaller area including unknown node and calculating the position of the centroid of the polygon, that is, the position of unknown node.

The APIT algorithm can be broken down into four steps: 1) gather information: information of the anchor node near the unknown node, such as the position, identification number, signal strength degree; 2) APIT testing: testing whether the unknown node is internal of the triangle that combined with different anchor nodes; 3) calculate overlapping areas: calculate the overlapping area of the all triangles that contain unknown nodes; 4) computing unknown node position: calculate the centroid of the overlapping area and the position of the unknown node. These steps are performed at individual nodes in a purely distributed fashion. Before providing a detailed description of each of these steps, we first present the basic pseudo code for our algorithm.

2.5. Bounding Box Algorithm

Semic *et al*. proposed [8] Bounding Box algorithm. This algorithm uses discrete communication model, and uses a square as communication area, which take r as communication radius and $2r$ as side length, as shown in **Figure 1**.

P is unknown node in **Figure 1**. Considering P as the center, within the area of r radius, there are A, B and C anchor nodes. To make these squares which take point of A, B, C. as the center respectively and $2r$ as side leng. then take the centroid of rectangular region of three squares' intersections as the estimated location of node P. If the P as the center, within the area of r radius, there are k as anchor nodes, the rectangular region centroid of k squares' intersection can be the position of the unknown node. The size and position of intersection region can be obtained by Equation (3).

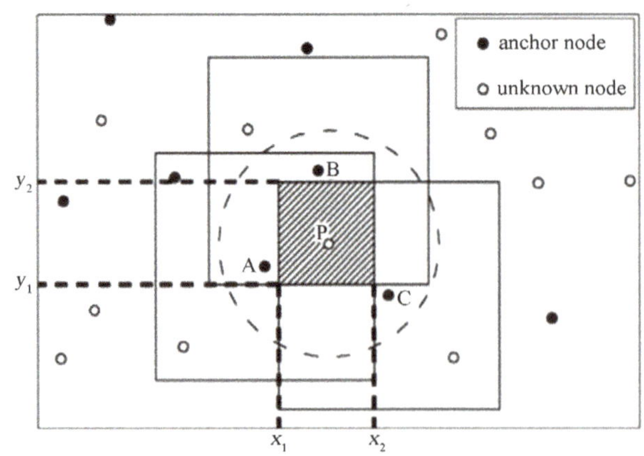

Figure 1. Positioning principle of bounding box.

$$\begin{aligned} &y_1 \quad \max \quad y_i - r; \\ &y_2 \quad \min \quad y_i + r; \\ &x_1 \quad \max \quad x_i - r; \\ &x_2 \quad \min \quad x_i + r; \end{aligned} \tag{3}$$

that (x_i, y_i) is the coordinates of the *i*th anchor nodes, $i = 1, 2 \cdots, k$.

3. Simulation and Analysis of Algorithm

To place 200 nodes in an area of 800 m * 800 m randomly, 40 nodes there into are anchor nodes, and the communication radius is 200 m. Each time postions of simulation node are generated randomly. **Figure 2** is a random simulation result. It can be obtained from nodes distribution that the average network connectivity is 29.65, the average number of network neighbor anchor nodes is 5.76. By calling the five positioning algorithm procedures, positioning error and positioning in the node distribution coverage were obtained, in which the positioning error of Centroid was 0.28498, the coverage rate of positioning was 100%; the positioning error of Amorphous was 0.21156, the coverage rate of positioning was 100%; the positioning error of DV-hop was 0.30731, the coverage rate of positioning was 100%; the positioning error of APIT was 0.31168, the coverage rate of positioning was 90%; the positioning error of Bounding Box was 0.29328, the coverage rate of positioning was 100%. As in **Figure 2**, red * represents anchor node, blue O represents an unknown node, black O represents an unknown node that con't be located, blue — connects the estimates location and the true location, so as to represent localization error.

Due to the random time simulation results cannot explain the problem, 100 times simulation was done under each of the different conditions, and calculated the average localization error to research five algorithms' performance.

In **Figure 3** the x-coordinate is the proportion of anchor nodes number to total nodes number, the Y-coordinate is the average location error. **Figure 3** shows the relation between anchor nodes proportion and localization error of centroid, Amorphous, DV-hop, APT and Bounding Box algorithm, while placing 200 nodes in an areas of m * 800 m. It can be seen from **Figure 3** with the increasing of the proportion of anchor node, the average localization error of five algorithms of node showed a trend of decline. The down trend of Bounding Box algorithm is the most obvious, because the increase of anchor nodes proportion inevitably leads to the increase of anchor nodes number in each unknown node's communication radius, and leads to the increase of the number of squares with side length of 2 times the communication radius and taking those anchor nodes as center, then the intersection region of those squares become smaller, so the estimated location of the unknown node becomes more accurate, and the localization coverage rate becomes higher. In the two kinds of algorithms: DV-hop and Amorphous, the main localization error comes from the calculation process of the average distance of each hop, so the decreased trend of localization error is relatively slow while the anchor node proportion increases. The Amorphous algorithm can always achieve fine localization accuracy in different anchor node proportion, taking into account economic factors, the localization performance is fine while the anchor node proportion is 15%.

4. Conclusion

Simulation experiments were carried out to compare the performance of five algorithms and we studied the performance change of five algorithms with anchor ratio and communication radius. The simulation results show that the localization error of Amorphous algorithm is minimum. Considering the economic benefits and positioning accuracy, the best localization performance of Amorphous algorithm was achieved when anchor ratio was 20% and communication radius was 200 m in the condition of distributing 200 nodes randomly in 800 m * 800 m square area. The next step will focus on the positioning precision and how to reduce the power consumption of nodes and prolong the network lifetime.

Acknowledgements

This work was supported by the project "Research on the scheme of fire off the coast of Sanya sea route optimization" and funded under Sanya College and local cooperation project, the grant number is 2011YD35. The authors wish to thank Sanya Scientific & Information Bureau.

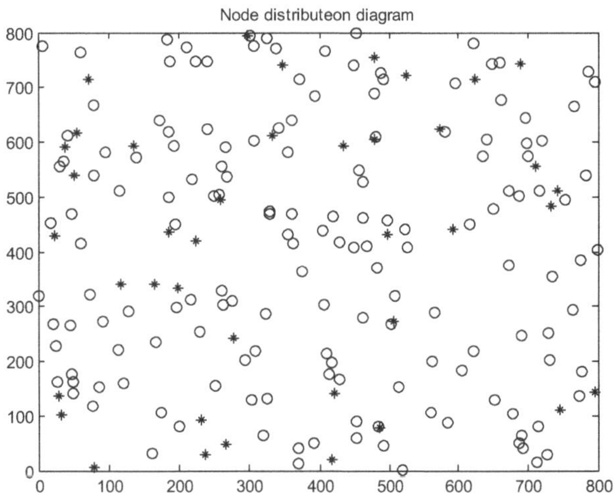
Node distributeon diagram
0
100
200
300
400
500
600
700
800

(a)

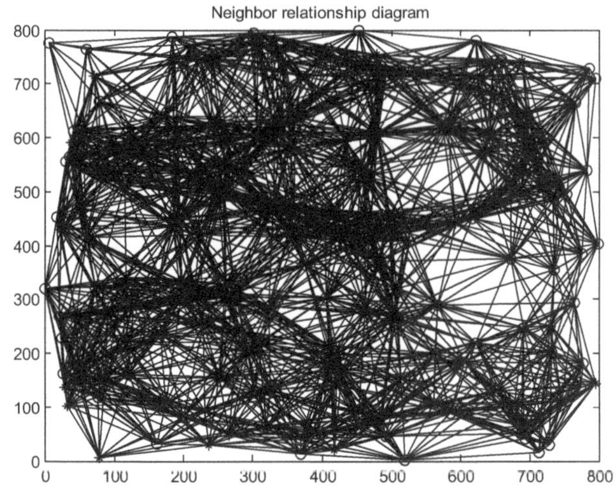
Neighbor relationship diagram
0
100
200
300
400
500
600
700
800

(b)

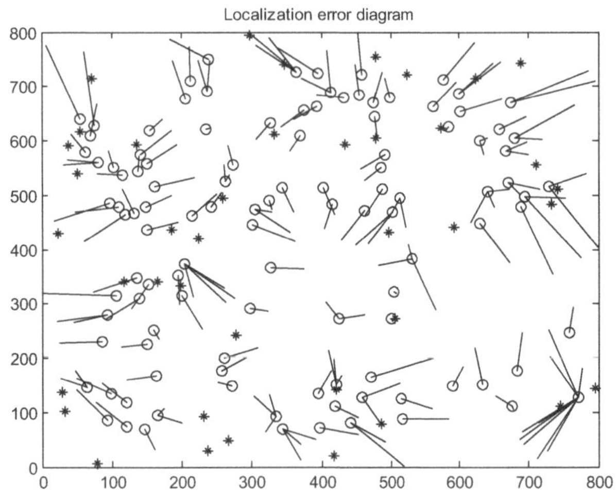
Localization error diagram
0
100
200
300
400
500
600
700
800

(c)

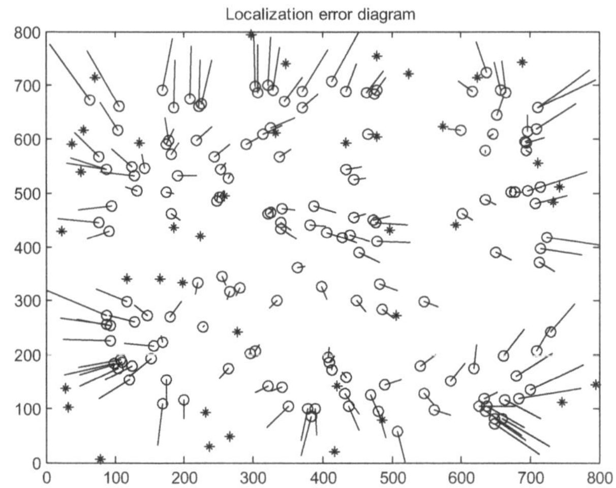
Localization error diagram
0
100
200
300
400
500
600
700
800

(d)

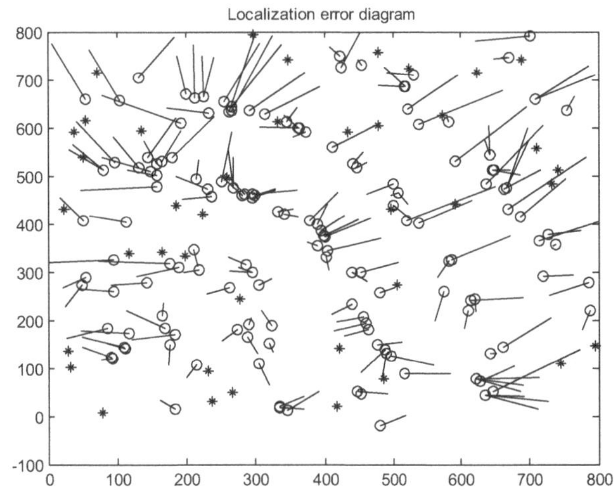
Localization error diagram
-100
0
100
200
300
400
500
600
700
800

(e)

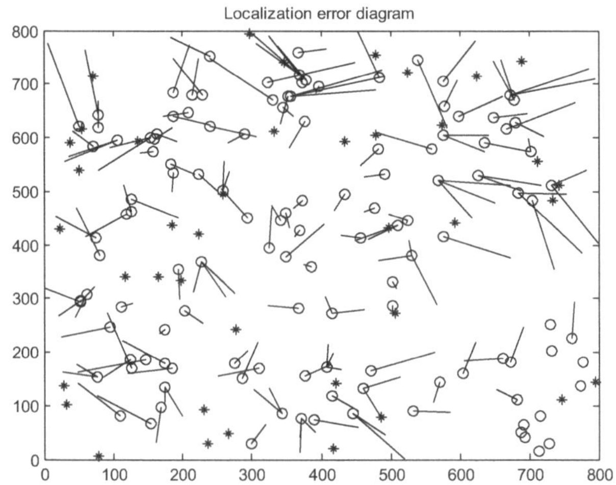
Localization error diagram
0
100
200
300
400
500
600
700
800

(f)

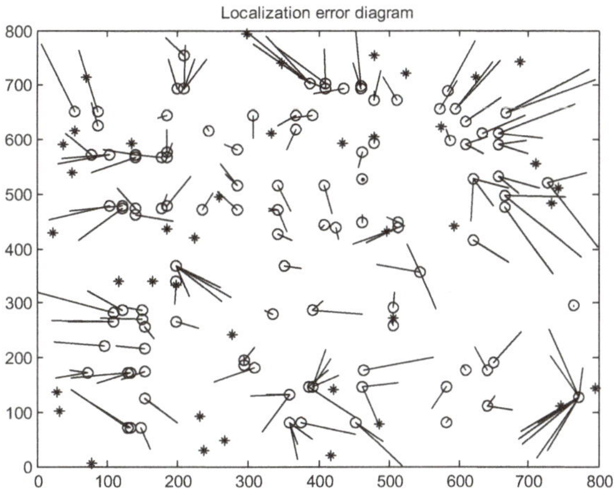

(g)

Figure 2. A random simulateion result. (a) Node distributeon diagram. (b) Neighbor relationship diagram. (c) Localization error diagram of Centroid algorithm. (d) Amorphous Localization algorithm. (e) DV-hop Localization algorithm. (f) APIT Localization algorithm. (g) Bounding Localization Box algorithm.

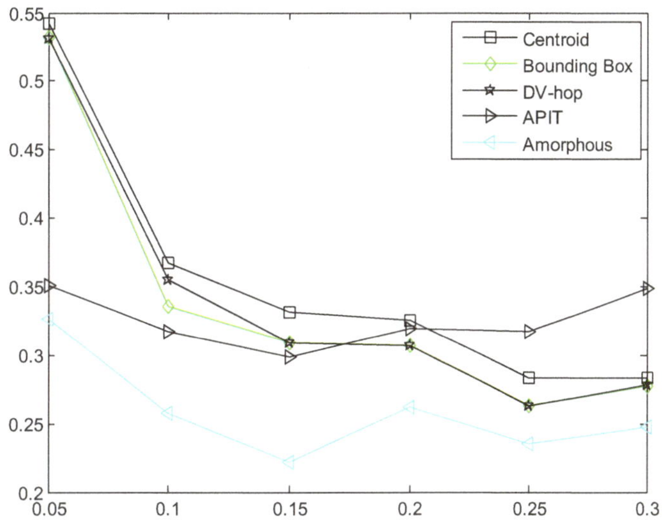

Figure 3. Localization error curve with the change of anchor radio.

This work was supported by the project "the Project Supported by the Scientific Research Fund of the Provincial Natural Science Foundation of Hainan", the grant number is 614252

Conflict of Interests

The authors declare that there is no conflict of interests regarding the publication of this article.

References

[1] Callaway, E.H. (2004) Wireless Sensor Networks: Architectures and Protocols. CRC Press, Boca Raton, 1-40.

[2] Priyantha, N.B., Balakrishnam, H., Demaine and Teller, S. (2003) Anchor-Free Distributed Localization in Sensor Networks. *Proceedings of the 1st International Conference on Embedded Networked Sensor System*, Los Angeles, 5-7 November 2003, 340-341.

[3] Zhang, Z., Hong, S. and Ren, T.J. (2010) Discussion on Localization Algorithm of Wireless Sensor Networks. *Journal of Guizhou University* (*Natural Science*), **27**, 99-102.

[4] Nagpal, R., Shrobe, H. and Bachrach, J. (2003) Organizing a Global Coordinate System from Local Information on an

Ad Hoc Sensor Network. *Information Processing in Sensor Networks Lecture Notes in Computer Science*, **2634**, 333-348.

[5] Nicolescu, D. and Nath, B. (2001) Ad-Hoc positioning systems (APS). *Proceedings of the* 2001 *IEEE Global Telecommunications Conference*, San Antonio, 25-29 November 2001, 2926-2931. http://dx.doi.org/10.1109/GLOCOM.2001.965964

[6] Dai, Y., Wang, J.P. and Zhang, C.W. (2010) Research and Improvement of Localization Algorithms for Wireless Sensor Network. *Journal of Transduction Technology*, **23**, 567-570.

[7] He, T., Huang, D.C., Blum, B.M., Stanjovic, J.A. and Abdelzaher, T. (2003) Range-Free Localization Schemes for Large Scale Sensor Networks. *Proceedings of the* 9*th Annual International Conference on Mobile Computing and Networking* (*MobiCom*'03), San Diego, 14-19 September 2003, 81-95.

[8] Shi, W.R., Xu, L. and Xu, Y.S. (2007) Localizations Algorithm with Mobile Anchor in Static Wireless Sensor Network. *Chinese Journal of Scientific Instrument*, **28**, 385-393.

8

Design of Crossbar Mixer at 94 GHz

Sanjeev Kumar Shah[1], Gyanendra Kumar Gupta[2]

[1]Shri Venkateshwara University, Gajraula, India
[2]Graphic Era University, Dehradun, India
Email: sanjeevkshah@yahoo.co.in

Abstract

This paper presents the design and performance of a 94 GHz crossbar mixer using chip diodes in crossbar configuration. The mixer has been designed in suspended stripline configuration on 2 mil fused quartz substrate. Co-simulation between Agilent Advance Design System (ADS) and CST Microwave studio is used to simulate and optimize the mixer circuit. The conversion loss of this crossbar mixer with ~10 dBm of LO power over RF bandwidth (89 to 93.8 GHz) with LO frequency at 94 GHz has been measured. Measurement results show minimum conversion loss of the order of 8 dB and achieved LO to RF isolation was greater than 26 dB.

Keywords

Crossbar Mixer, Waveguide, Suspended Stripline

1. Introduction

Mixer is an essential component of almost all receivers used in communication, radar and radio applications. The mixer generates an output spectrum, consisting of sum, difference and harmonics of the individual frequencies present at its input. The desired frequency can be filtered at output.

Millimeter wave mixers are generally designed in waveguide, suspended stripline, fin-line or combination of these lines. After going through many literatures, it has been found that suspended stripline has been successfully used to design mixer at 94 GHz [1]-[3]. Crossbar configuration of the mixer is well proven technology at this frequency.

Many researchers have designed crossbar mixer at 94 GHz with excellent results. Although cross-bar mixers have been described in many literatures, actual cross bar mixer modeling, simulation and optimization using commercial CAD software are not available. The aim of this paper is to describe the simulation technique which simplifies the design and reduces the time as well as cost for production.

2. Mixer Description

The configuration of cross bar mixer is shown in **Figure 1**. The mixer is realized in suspended stripline. The

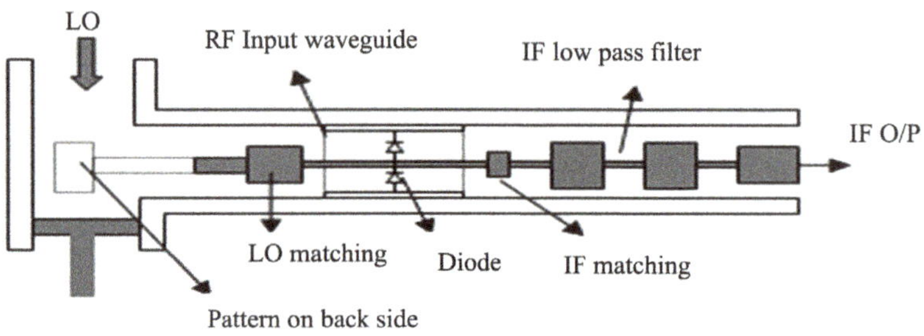

Figure 1. Configuration of crossbar mixer.

stripline circuit consists of two schottky barrier chip diodes, a suspended stripline to waveguide transition, an IF low pass filter and LO, IF & RF matching sections [4].

The LO power is coupled to the diode through a broadside coupler. The RF signal is incident perpendicular to the circuit.

Two diodes with opposite polarity are mounted in series across broad wall of the waveguide WR-10.

3. Analysis and Design

Cut-off frequency of the channel is also function of width of transmission line; hence channel dimension should ensure that cut-off frequency is beyond operating frequency band for the widest dimension of the transmission line used in the design. Cut-off frequency for the stripline channel can be calculated from standard formula [5] but they are not accurate at such a high frequency of operation. Hence complete port structure is simulated in CST microwave studio and port solution is obtained for different widths of transmission lines.

Analysis and design of crossbar mixer is carried out with the help of CST microwave studio and Agilent's ADS circuit simulator. In this method the basic structure of crossbar mixer without any matching, coupling and filter circuits is created and multiport (5 × 5) scattering parameter data is obtained. The LO port is labeled as port 1, IF as port 2, RF as port 3 port 4. Port 4 and 5 are used as internal ports where diode are placed in ADS circuit simulation. CST simulation of the above structure results in 5 × 5 S-matrix file. This S-matrix is renormalized with respect to distributed waveguide (WR-6) impedance and imported into the ADS circuit simulator in "Touchstone" (.SnP) format. The 2 × 2 S-matrix of LO coupler (along with suspended stripline to waveguide transition) and low pass filter is simulated separately and exported to ADS. All S-Parameters blocks are integrated in the ADS circuit simulator and harmonic balance analysis is done. The block diagram of the circuit used in ADS circuit simulation is shown in **Figure 2**.

The schematic circuit in ADS circuit simulator for harmonic balance simulation and optimization of the mixer is shown in **Figure 3**.

4. Simulation Steps

Mixer simulation has been done under following steps.

1) Basic crossbar mixer unit which consisting RF and LO waveguide and a small transmission line in LO-IF circuit is simulated in CST microwave studio and 5 × 5 scattering matrix file is exported to ADS circuit simulator.

2) Design and simulation of LO side broadside coupler is done in CST microwave studio and 2 × 2 scattering matrix file is exported to ADS circuit simulator. Broadside coupler can be replaced by band pass filter.

3) A 7 element Chebyshev low pass filter with 0.2 dB ripples in pass band is designed and simulated and resulting 2 × 2 S-matrix file is exported to ADS circuit simulator.

4) Complete harmonic balance simulation of crossbar mixer is done in ADS circuit simulator by integrating all S-Parameter blocks.

5) Finally equivalent matching circuit dimension is used from CST, which takes into account electromagnetic simulation of matching elements.

Harmonic balance is a frequency-domain analysis technique for simulating nonlinear circuits and systems. It is well-suited for simulating analog RF and microwave circuits.

Harmonic Balance Simulation calculates the magnitude and phase of voltages or currents in a potentially non-

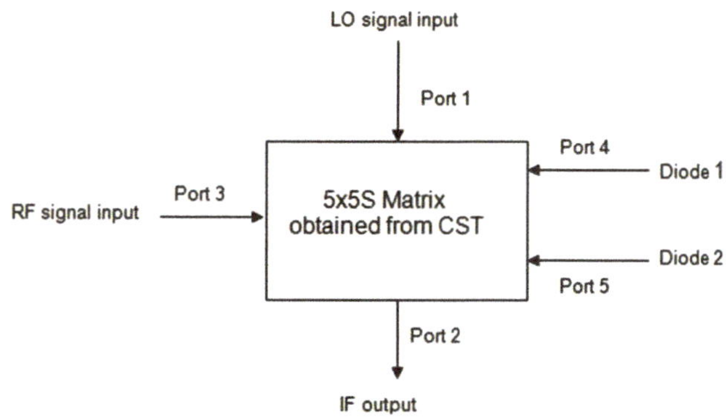

Figure 2. Block diagram of mixer circuit used in ADS.

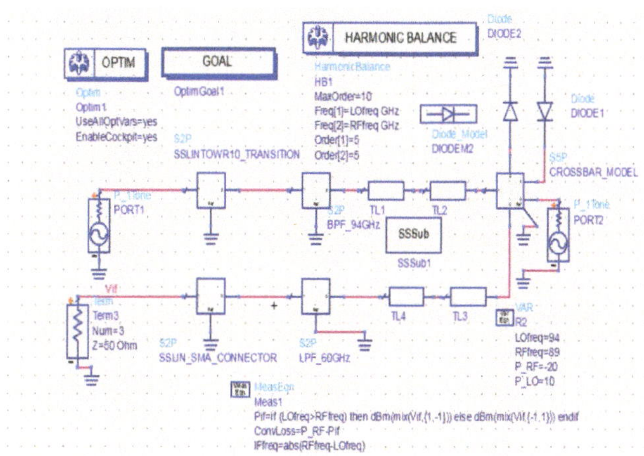

Figure 3. Schematic of the mixer in ADS.

linear circuit. In contrast, S-Parameter or AC simulation modes do not provide any information on nonlinearities of circuits.

Harmonic balance simulation makes possible the simulation of circuits with multiple input frequencies. This includes intermediation frequencies, harmonics, and frequency conversion between harmonics.

The HB method depends on calculating currents and voltages at many harmonically related frequencies for each fundamental signal under consideration.

5. Simulation Results

A) *Waveguide to SSLIN Transition*

The suspended stripline to waveguide transition consists of an electric probe inserted into the broad wall of the waveguide. The probe is an extension of a suspended stripline into the waveguide. When a probe is inserted into the waveguide it radiates and in certain position desired mode is excited or vice versa fields set up in the waveguide can be picked up. At one end of waveguide input is given and at other end tuning short is provided to provide optimum matching. Most critical parameters for the transition are probe shape and insertion distance of suspended stripline into the waveguide.

The A dimension of the channel is chosen as 1.27 mm to facilitate the waveguide (WR-10) transition. For WR-10 the dimension is a = 2.54 mm and b = 1.27 mm. The electric probe transition was chosen because it can be fabricated as an integral part of the stripline and difficulty of making variable contacts is avoided. The transition model and its simulated response are shown **Figure 4**.

B) *Band Pass Filter Simulation*

The end coupled suspended stripline band pass filters are most promising due to its compact size, lightweight, low cost and ease of fabrication. In E plane circuit supporting dielectric (e.g. Fin line) causes additional losses [6] [7].

The 94 GHz band pass filter exhibits an insertion loss of 1 dB with 3 dB relative bandwidth at a center frequency of 94 GHz and the return loss is better than −18 dB at a center frequency. The designed and fabricated 94 GHz band pass filter shows the good performance for planar integrated millimeter-wave circuits. The 3D view and simulated results of band pass filter are shown **Figure 5**.

C) *Low Pass Filter Simulation*

A 7 element Chebyshev low pass prototype has been designed using standard low pass filter design formulae given in Matthaei [8] for a cut-off frequency of 60 GHz and minimum of 30 dB rejection at 94 GHz. The 3D view and simulated results of low pass filter are shown **Figure 6**.

D) *Harmonic Balance Simulation Results*

Harmonic balance simulation results of the mixer after optimization are shown **Figure 7**.

6. Hardware and Results

The complete assembly of realized Quartz MIC card and mixer hardware is shown below in **Figure 8**. K-Type Connector is used to Extract IF output. To test the performance of the mixer, the conversion loss over the band is shown in **Figure 9**. With RF Frequency of 92 GHz, LO Frequency of 94 GHz and 10 dBm of LO power, LO leakage at RF port was found to be –18.3 dBm. Hence LO to RF isolation was 28.3 dB. For optimum conversion loss a LO power of 10 dBm is required. A conversion loss of the order of 8 dB was achieved.

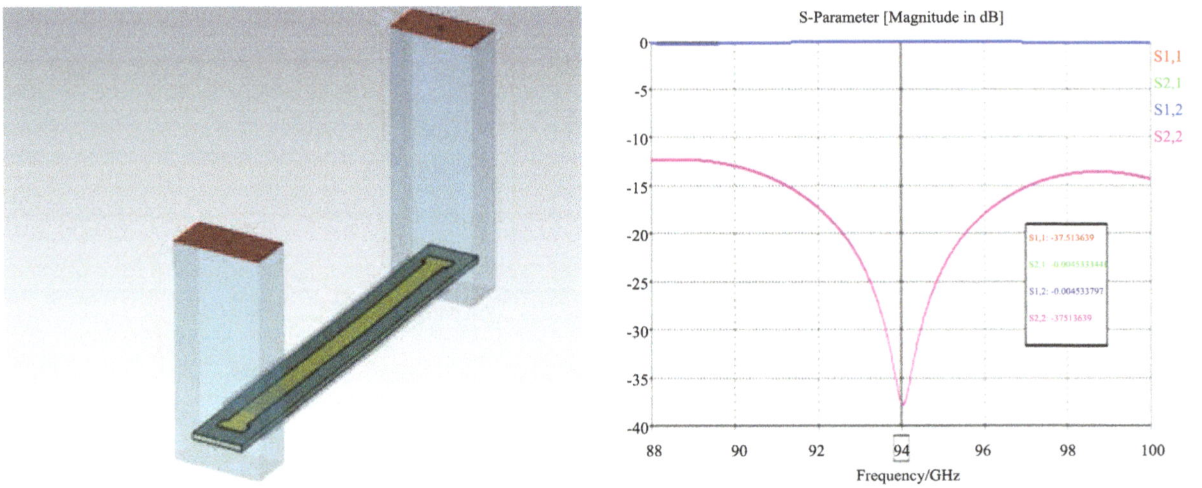

Figure 4. Waveguide to SSLIN Transition and simulation results.

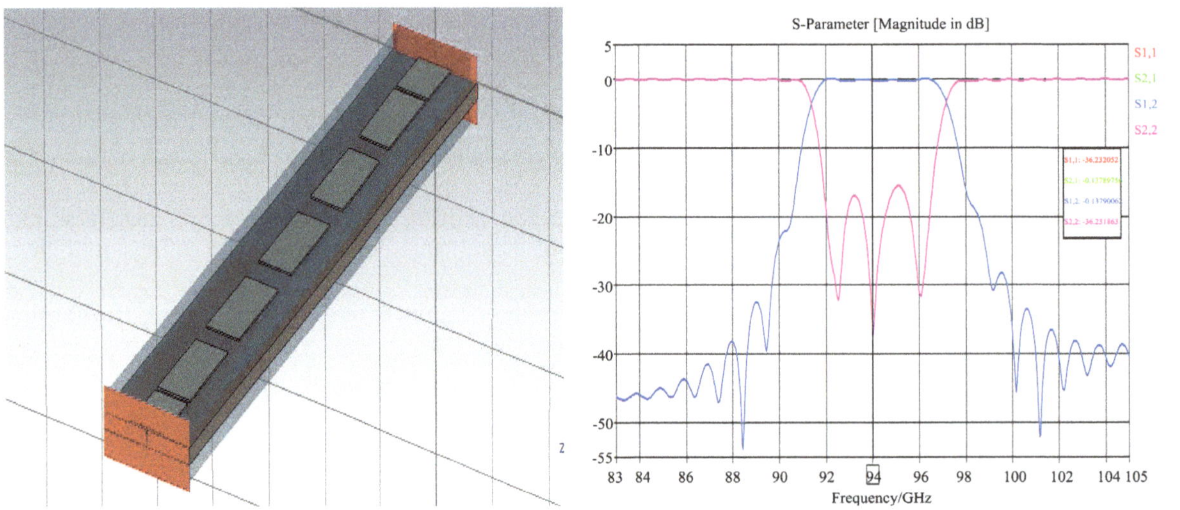

Figure 5. Band pass filter 3D view and simulation results.

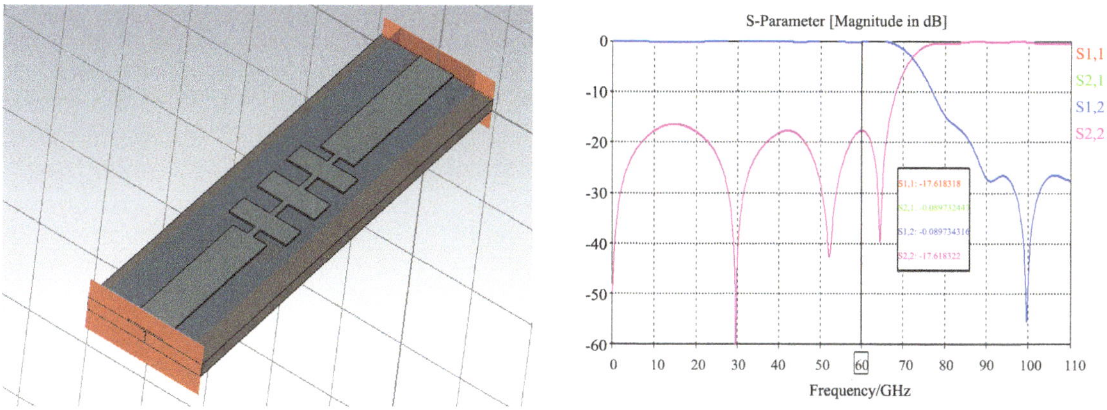

Figure 6. Low pass filter 3D view and simulation results.

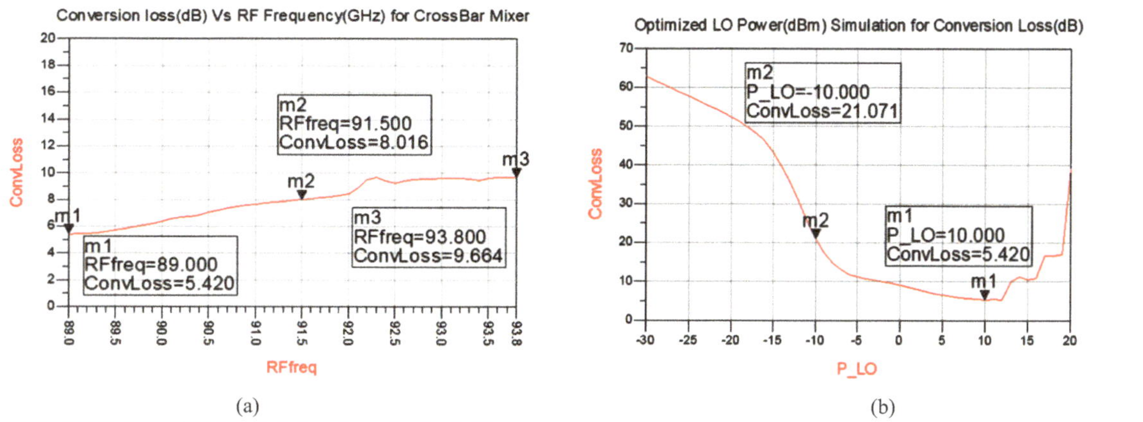

Figure 7. (a) Simulated results for conversion loss (dB) vs RF Frequency (GHz) LO Frequency = 94 GHz, LO power = 10 dBm; (b) Simulated results for conversion loss (dB) vs LO power (dBm) RF Frequency = 92 GHz, LO Frequency = 94 GHz.

Figure 8. MIC card of mixer on 2 mil quartz substrate.

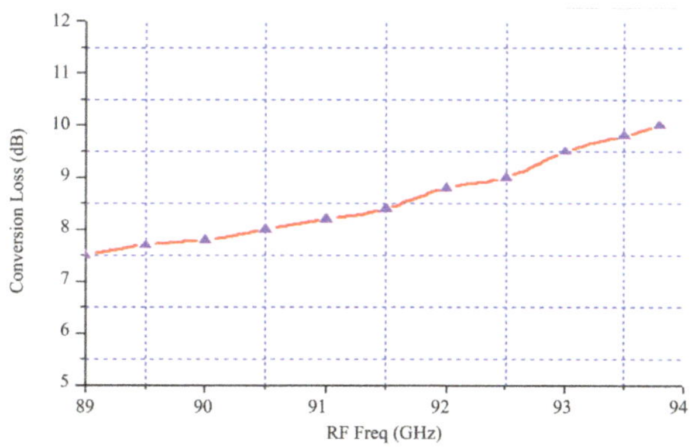

Figure 9. Conversion loss of realized mixer.

7. Conclusion

A simplified low conversion loss mixer at 94 GHz was developed using co-simulation techniques and optimization procedure. Deviation between simulation and measured results was mainly attributed to mechanical tolerances in the mixer cavity and bonding techniques. The realized mixer using chip diode meets the design objective and compares well at other crossbar mixer operating at W-Band.

References

[1] Paul, J., Yuan, L. and Yen, P. (1982) Beam Lead Dielectric Crossbar Mixers from 60 to 140 GHz. 1982 *IEEE MTT-S International Microwave Symposium Digest*, Dallas, 15-17 June 1982, 372-373.

[2] Nguyen, C. and Chang, K. (1984) 140 GHz Broadband Crossbar Stripline Mixer with over 20 GHz Instantaneous RF Bandwidth. *Electronics Letters*, **20**, 441.

[3] Bui, L.Q., Ton, N. and Ball, D. (1984) A D-Band Millimeter-Wave Crossbar Mixer. 1984 *IEEE MTT-S International Microwave Symposium Digest*, San Francisco, 30 May-1 June 1984, 555-556.

[4] Lioutas, N. and Fourikis, N. (1991) The Development of Simplified Crossbar Mixer Operating at Ka-Band. *International Journal of Infrared and Millimeter Waves*, **12**, 845-855.

[5] Lehtovuori and Costa, L. (1998) Model for Shielded Suspended Substrate Microstrip Line. Circuit Theory Laboratory Report Series, No. CT-38, Espoo, 12 p.

[6] Maas, S.A. (1986) Microwave Mixers. Artech House, Boston.

[7] Tahim, R.S., Chan, J.C. and Dutt, B. (1983) Millimeter Wave Image Enhanced Mixer. *4th International Symposium on Recent Advances in Microwave Technology*.

[8] Matthaei, G., Young, L. and Jones, E.M.T. (1980) Microwave Filters, Impedance Matching Networks, and Coupling Structures. Artech House, Boston.

9

Multiple Node Placement Strategy for Efficient Routing in Wireless Sensor Networks

Kirankumar Y. Bendigeri[1], Jayashree D. Mallapur[2]

[1]Department of Electronics, Jain University Bangalore, Karnataka, India
[2]Department of Electronics & Communication Engineering, Basaveshwar Engineering College, Bagalkot, Karnataka, India
Email: kiranbendigeri@gmail.com, bdmallapur@yahoo.co.in

Abstract

The advances in recent technology have lead to the development of wireless sensor nodes forming a wireless network, which over the years is used from military application to industry, household, medical etc. The deployment pattern of sensor nodes in Wireless Sensor Network (WSN) is always random for most of the applications. Such technique will lead to ineffective utilization of the network; for example fewer nodes are located at far distance and dense nodes are located at some reason and part of the region may be without the surveillance of any node, where the networks do consume additional energy or even may not transfer the data. The proposed work is intended to develop the optimized network by effective placement of nodes in circular and grid pattern, which we call as uniformity of nodes to be compared with random placement of nodes. Each of the nodes is in optimized positions at uniform distance with neighbors, followed by running a energy efficient routing algorithm that saves an additional energy further to provide connectivity management by connecting all the nodes. Simulation results are compared with the random placement of nodes, the residual energy of a network, lifetime of a network, energy consumption of a network shows a definite improvement for uniform network as that of with the random network.

Keywords

Energy Consumption, Network Lifetime, Node Deployment, Connectivity Management

1. Introduction

WSN is developed for small scale level applications that use the sensors for communication. It has found its importance in the current trends because of various advantages as compared with other networks. Sensors are capable of sensing the data, processing and transmitting with wireless communication. The data can be fragmented into sub units so as to be transmitted from multiple sensors that consider the load balancing. The recent advances in MEMS [1], nano technology, various processors, and wireless communication have lead to the advancement of WSN which are self configuring, flexible and scalable. Its intervention is most useful when human reach ability is too difficult. The deployment of sensors can be static or a random but yet a fast response. For example in a military application, it is difficult to reach the specified areas so as to practically place the sensors, in that case sensors are randomly thrown from an area above the ground; another case would be the scenario of an agricultural application which requires structured monitoring, where sensors are uniformly distributed to receive the data at regular intervals. WSN can be treated as distribution of nodes for a particular task. Sensor nodes are battery powered and are expected to operate without attendance for a relatively long period of time. Power is consumed for sensing, communicating and data processing. Sensors are also able to renew their energy from solar sources, temperature differences, or vibration. An analog signal produced by the sensors is converted to digital by an ADC and sent to controllers for further processing. Usually sensor node is independent, small in size and adaptive to the environment, and should consume extremely low energy. Sensors are responsible for collecting the data [1], where in ADC sends the type of data sensed by sensor to the processor and requests, what to be done by sensor. WSN can establish the network on its own, which usually consists of low cost, low power, low energy features, where the data transfer through these is done efficiently by various techniques in order to increase the network lifetime. It is always referred by most of the researchers that WSN uses maximum communication energy for communication, which intently can be considered as, sensor node can be made to use minimum energy to reach the neighbors thus trying to save the energy consumed. This technique can be very effective utilization to save the energy during sleep, idle and receive mode, and there is always a chance of death of node and hence the network should sustain itself to the current load or sometimes entry of new node into the network where WSN should withstand the changes for the newer nodes providing scalability. While guaranteeing good performance overall, the sensor nodes are battery powered optimized positions along with the routing protocols to be designed for sensors so as to consume minimal energy extending the network lifetime. As the placement of node is an important phenomenon, if done efficiently followed by energy routing protocol, such network can greatly enhance the network lifetime.

The prime objective of sensor node is to sense the data and send it to base station or the cluster heads, and these can be considered as a communication range of several orders by which the data moment is done; it is always preferred that data can always be transferred by hop to hop that is source and destination involves certain neighbors for the data moment. Energy is an important issue for WSN to prolong the network lifetime; hence effective utilization of the battery should be done to do so. It is considered that placement strategy is going to be an important strategy for WSN since effective utilization of the network can be done by having controlled placement of nodes, where the node positions are known which when compared with the random placement technique, where sometimes nodes are placed in a such a way that it is nearly far from the region to reach the neighboring nodes, thus leading to un effective utilization of such nodes in the network. The proposed work is intended to concentrate on the different placement of nodes like random, circular and grid based scenario of a network that is mainly worked out to save the energy consumed by the network on par with sensor nodes and to increase the network lifetime.

2. Related Work

The paper [2] concentrates on routing techniques involved in WSN that considers both static and mobile sensor nodes. Routing protocols which might vary based on the implementation and network architecture, but also on future directions of research on development of routing protocols have been discussed here for WSN. The author in paper [1] proposes a method to prolong the lifetime of the sensor nodes, designing efficient routing protocols is critical. Even though sensor networks are primarily designed for monitoring and reporting events, since they are application dependent, a single routing protocol cannot be efficient for sensor networks across all applications. In this paper, the author analyzes the design issues of sensor networks and presents a classification and comparison of routing protocols. This comparison reveals the important features that need to be taken into consideration while designing and evaluating new routing protocols for sensor networks. Routing algorithms in paper [3] for WSNs are responsible for selecting and maintaining the routes in the network and ensure reliable and effective communication in limited periods. The energy constraint of WSNs makes energy saving the most important objective of various routing algorithms. In this paper the author presents survey of routing protocols and algorithms used in WSN's with energy efficiency as the main goal. In WSN, [4] the sensor nodes have a limited transmission range, and their processing and storage capabilities as well as their energy resources are also limited. Routing protocols for wireless sensor networks are responsible for maintaining the routes in the network and have to ensure reliable multi-hop communication under these conditions. In this paper, author presents a survey of routing protocols for WSN and compares their strengths and limitations. In this paper [5] author presents a multi-objective optimization methodology for node placement in wireless sensor network design. Emerging computational intelligence leads to optimization of NP-hard problem in a simple way, differential evolution approach is used as a tool for optimization of most important parameters in node placement process of wireless sensor network. Optimal operational modes of the nodes in order to minimize the energy consumption and meet application-specific requirements have been investigated and also other optimizations have been done on clustering and communication range of sensors.

In this paper, [6] author presents how to place SNs by use of a minimal number to maximize the coverage area when the communication radius of the SN is not less than the sensing radius, which results in the application of regular topology to WSNs deployment. With nodes placed at an equal distance and equipped with an equal power supply, the energy imbalance problem and the mathematical formulation for maximizing network lifetime in grid-based WSNs are given. Author also generalizes the maximizing network lifetime problem to the randomly-deployed WSNs which shows the significance of mathematical formulation for this crucial problem. In this paper, [7] author formulates a constrained multivariable nonlinear programming problem to determine both the locations of the sensor nodes and data transmission pattern. The two objectives studied in the paper are to maximize the network lifetime and to minimize the application-specific total cost, given a fixed number of sensor nodes in a region with certain coverage requirement. Through numerical results, author shows that the optimal node placement strategies provide significant benefit over a commonly used uniform placement scheme. The author presents [8] a novel algorithm for autonomous deployment of active sensor networks. The algorithm aims to enhance the sensing coverage based on an initial placement of sensor nodes. The sensing regions are modeled as circular discs of variable sensing range limits. Based on the fact that a unique circle packing exists satisfying any given set of combinatorics and boundary conditions of a sensor network, minimum sensing range required for every interior nodes to fulfill such packing conditions can be done. Based on a number of numerical simulations, we have verified that the proposed algorithm always yields sensor deployments of wide coverage and minimize the sensing ranges required for every interior sensing node to satisfy the packing and boundary conditions. In this paper, [9] the author proposes to deploy sensors either with variable battery capacities or with non-uniform densities in order to counterbalance the non-uniform energy drainage, thus achieving a longer network lifetime. Monitoring region is concentric ring areas and deployed nodes in these areas such that the highest battery resources are allocated to the ring where the highest energy drainage takes place. Results show 6 to 7 times longer lifetime values are attained without any increase in costs with this approach.

In this paper author [10] considers the design issue of sensor networks by placing a few powerful aggregate nodes into a dense sensor network such that the network lifetime is significantly prolonged when performing data gathering. Specifically, the problem is to place K aggregate nodes into a dense sensor network of n sensor nodes with $K < n$ such that the lifetime of the resulting network is maximized. In a WSN [11] nodes are battery powered, therefore, they are energy constrained. A number of routing schemes, energy efficient and energy-aware, have been proposed to maximize the network lifetime. In this paper, author considers the lifetime maximization problem, routing protocols that take into account the battery, residual energy in sensor nodes and the energy required for transmission along the path are considered. For each transmission round, only the nodes which have their remaining energies greater than a threshold can participate as relays for other nodes data, in addition to sensing the environment. This choice allows the distribution of energy load among any sensor nodes, thus extends network lifetime. Simulation results show that the network lifetime is increased largely comparable to Direct and MTE schemes. In this paper [12] author presents a new routing scheme for Wireless Sensor Networks, in which routing is based on Radio-aware Metric, which contains radio information at the Medium Access Control (MAC) layer. In this new routing scheme, data are forwarded to a designated neighbor, instead of flooding data into the whole network like with Directed Diffusion. Simulation results shows that compared to Directed Diffusion, proposed routing scheme is almost 4.3 times more efficient in terms of energy consumption and 2.6 times more reliable in terms of data delivery. This paper [13] presents a method which intelligently determines the optimum node coordination of sensor nodes topology. The topology is in linear array, aimed for beam-forming in WSN's. The array is constructed in random sensor node deployment. The selected nodes should align similar to a uniform linear array (ULA) to minimize the position errors which will improve the beam-forming performance (gain, transmission range and characteristics). Instead of utilizing random beam-forming which needs a large number of sensor nodes to interact with each other and form a narrow radiation beam, the proposed approach is emphasized to only a number of sensor nodes which can construct a linear array. Beam-forming technologies can increase the system performance, increase the transmission range and control the directionality of the reception or transmission of a signal. The proposed work utilizes beam-forming technique by using ULA to establish a communication link in a WSN. In this paper [14], author proposes a systematic framework for choosing the minimum sensor nodes from those which are originally distributed randomly in a sensor network. Integer linear programming model is used to describe the optimal placement. A greedy algorithm is proposed under quite general conditions. Simulation results and theoretical analysis with different grid density shows that proposed framework is computationally feasible and the resultant sensor node placement performs near-optimum.

In this paper [15], environmental monitoring applications are considered where data may be continuously reported with the possibility of urgent alarming if necessary. Hierarchical architecture of the network is assumed in order to overcome the problem of energy constrained sensors. Two algorithms are proposed with the purpose of network lifetime elongation and the maximization of the use of the available energy. The first algorithm is a modification for LEACH-C to enhance its performance. It results in a 25% longer lifetime. The second algorithm is an energy efficient method to ensure full coverage of the network as long as sensors are still working. This achieves 32% longer lifetime than LEACH-C. This paper [16] presents a method of optimizing sensor node co-ordination in order to reassign the nodes in linear arrangement to form a linear antenna array. This linear sensor node array (LSNA) is constructed within random sensor node placement. The LSNA should be optimized as closely comparable as a conventional uniform linear array (ULA) to minimize the beam-forming performance errors. Beam-forming has been introduced in wireless sensor network in order to increase the transmission range of individual sensor nodes. Performance of LSNA demonstrates an excellent agreement over ULA. The positioning of nodes [17] in a sensor network has an effect in its performance. In this paper, two predefined configurations are compared to the random distribution to gauge the magnitude of the effect on energy consumption of each type of sensor allocation. The experiments assume a flat, obstacle free, rectangular field, with Directed Diffusion used as routing protocol, and random different positions for the querying entity (sink) and the event location in the field. The results confirm that in an environment such as this, it is worth investing in the uniform positioning of sensors as they offer a significant performance gain.

In wireless sensor networks [18], the number of sensor nodes has direct relation to the cost of total wireless sensor networks. The research on the number of sensor nodes has significant meanings of theory and practice to design of wireless sensor networks. By computation and analysis, the sensor deployments in the form of equilateral triangle, as a rule, are better than those in the form of square, and the efficient coverage area ratios, decrease with increasing number of sensor nodes. Sometime information is incompletely monitored or undetected. This is coverage and connectivity problems. The coverage problem is also one of basic problem in wireless sensor networks. The paper analyzes several sensor deployments and computes their efficient coverage areas and their efficient coverage area ratios. In addition, the relation between the number of sensors and efficient coverage area ratio is discussed [18]. Wireless sensor network [19] is a revolution of information collection and perception field, which has been drawing much more application gradually. However, The WSN with characteristic of self-organization, multi-hop, dynamic topology and limited energy resources, which make the network to prolong it's lifetime become extremely difficult. This paper analyzed variety multicast protocols in existence and the current achievements in the research based on a comprehensive studying of a mess of multicast routing protocols. This document provided the appropriate application environment for types of multicast protocols and the specific improvements. It has an active significance for WSN to increase it's performance in the near future. In wireless sensor networks, the major source [20] of a sensor node failure is battery exhaustion and replacing this energy source in the field is usually not practical. Therefore, the use of energy efficient infrastructure, such as repositioning the base station in a clustered wireless sensor network, is able to prolong the lifetime of the network and improve the overall network data. In this paper, author proposes an energy-efficient protocol for the repositioning of mobile base station using particle swarm optimization (PSO) in wireless sensor networks. Simulation results demonstrate that the proposed protocol can improve the network lifetime, data delivery and energy consumption compared to existing energy-efficient protocols developed for this network.

3. Implementation

The implementation for node deployment and best routing analysis is done in this section. We have considered three different node deployments for our experimentation and results. The **Figure 1** shows the uneven distribution of nodes. Node placement is a technique to place the nodes effectively in a simulation area so as to consume the minimum energy from each node that is intended for transmission of packets or a data. The present research is on deployment of wireless sensor nodes that are mainly concentrated in the static and the dynamic manner. In static deployment, nodes are fixed, whereas in dynamic deployment of nodes, the nodes are mobile. For the discussion we have classified the static deployment into, non deterministic (random) and deterministic deployment.

Randomized sensor placement often becomes the only option for WSN. Consider an example, in applications of WSN's in reconnaissance missions during combat, disaster recovery and forest fire detection. It is widely expected that sensors will be dropped by helicopter, grenade launchers or clustered bombs. Such means of deployment lead to random spreading of sensors. Although the node density can be controlled to some extent.

As with most of the research that have been carried out, communication network of WSN is always considered deploying the nodes with randomness. The randomness is a function to deploy the nodes randomly leading uneven or unequal distribution of nodes. When such network with the nodes that are deployed randomly is considered for the simulation, there are three issues that mainly arise, some nodes are densely deployed at particular region while the other region is distributed with only fewer nodes located at farer distances and leaving region will not at all be with the single coverage of node. The drawbacks of such random node deployment is that nodes with dense location, where routing is to take place, additional hops between source and destination in a region of dense location of nodes take part in routing leading to additional utilization of energy from each node, where as in the case of nodes at far location, additional energy again is to be spent in transmitting the data to neighbors located at far distance as well to the destination which is located at quite far distances. On a similar way it is difficult to manage the routing in a region where no nodes are located leading in all the three case, an uneven distribution of energy source and node deployment. **Figure 1** show the probability of random deployment of sensor node network, the node pattern is set in a simulation environment. As shown in **Figure 1**, the random function distributes the nodes close to each other for example node ID, 25, 30, 85, 79, 55, 48 are all closely located and node 77, 86 are located at a such far distance that they may not be in a communication range leading to loss of data transfer or finding more additional neighbors or energy to complete the routing. Such drawback can be overcome by placing the nodes uniformly in a simulation area which we call it as deterministic node deployment.

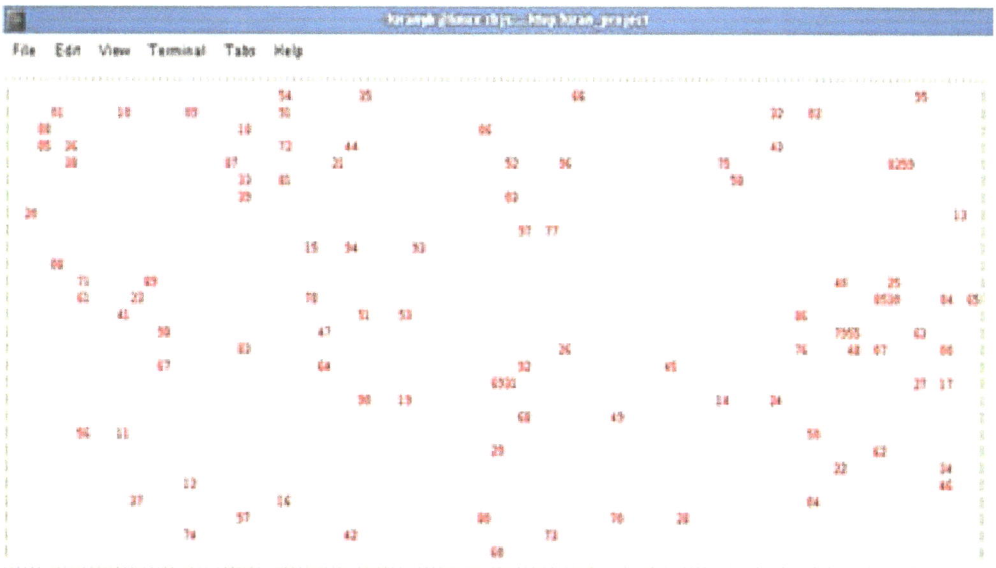

Figure 1. Random placement of sensor nodes.

In Deterministic Sensor Placement Schemes (DSPS), the nodes are placed in order to meet some of the desired performance objectives. For example, the coverage of the monitored region can be ensured through careful planning of node densities and fields of view and thus the network topology can be established at the setup time. DSPS are common in certain applications like room temperature monitoring, medical applications, underwater acoustics, imaging and video sensors among others. In our proposed work we have considered two different uniform models, *i.e.* grid and circular deployment of nodes.

Figure 2 shows the placement of nodes in grid manner, the same simulation area as that of random network is taken with 100 nodes and all the nodes are held at a uniform distance. Going through most of the research papers, where for any application nodes are just deployed randomly. The main intention of the work is to give optimized positions for the nodes stating that the nodes are intended to work from the stated position till the node dies, therefore the developed algorithm will initially fix the position of nodes then go for the routing. Improved placement can definitely leads to effective utilization of network like routing, energy, co-ordination to give increased network lifetime.

Figure 3 shows the circular deployment of nodes which provides mainly two objectives. First one is, careful node placement can be a very effective optimization means for achieving the desired design goals, and is classified as static approaches. On the other hand, some schemes have advocated dynamic adjustment of node location, since the optimality of the initial positions may become void during the operation of the network depending

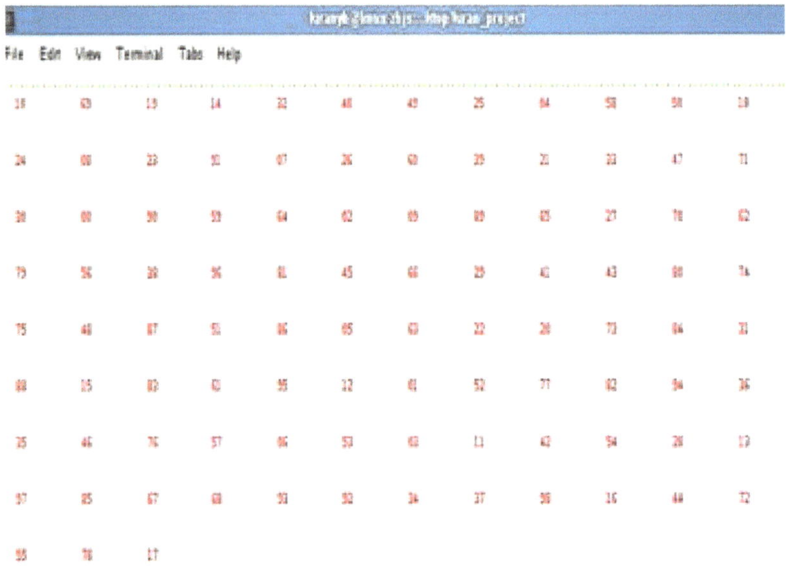

Figure 2. Sensor nodes placed in a grid pattern.

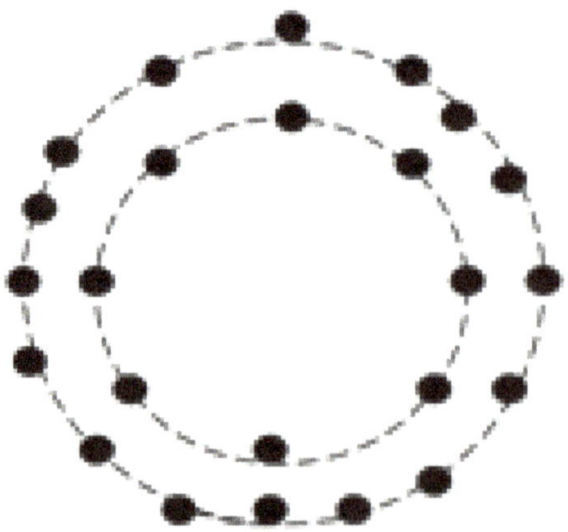

Figure 3. Circular deployment of sensor nodes.

on the network state and various external factors, We categorize the placement strategies into static and dynamic depending on whether the optimization is performed at the time of deployment or while the network is operational, this approach considers positions of node metrics that are independent of the network state or assume a fixed network operation pattern that stays unchanged throughout the lifetime of the network.

To do the routing be it for all the 3 different node topology, a simple shortest path from a set of multipath from source and destination is been considered, the area considered for simulation is 100 × 100 m, in **Figure 1** nodes are hardly distributed in a region of 50 × 50 m. The random deployment pattern indicates that nodes are densely deployed in one particular area, leaving behind most of the region uncovered. The proposed model considers a pattern of node distribution in grid and circular pattern in a way that for the grid it is with node at regular intervals and for the circular base station is located at the center surrounding a first layer of 8 sensor nodes covering equal area on a circle, followed by covering rest of the area on same pattern of deploying the nodes on a circular area so as to cover the complete simulation area. The advantage of optimized placement of node is that maximum area is covered; burden on nodes is reduced followed by increasing the network lifetime.

3.1. Mathematical Model

The WSN which uses a method of random deployment of nodes can typically be replaced by uniform distribution of nodes, be it in terms of grid or circular pattern of deployment of nodes. The work mainly concentrates on running a simulation using uniform node placement topology in a network, which is mainly worked for reducing the overall energy consumed by the network further increasing the network lifetime. Work is carried out considering three different network topology in which sensor nodes are placed at different positions, the routing is held at different instant in the network based on which the energy consumption can be found. The methodology is as follows, considering a random distribution of sensor network, as the name suggests randomness, the nodes are placed at un-equal distance at different positions as shown in **Figure 1**. It can be seen from the **Figure 1** that the nodes do not exhibit optimized location with randomness in x and y position, *i.e.* the two different nodes are never at equidistance. One of the important factor to consider is that, energy consumption depends on the distance the data is been transmitted to reach the destination. This is an important consideration for WSN. One of the examples for uniform network can be grid based network as shown in **Figure 2**, here the node topology is at uniform distance forming a grid pattern. The network is defined in (100 × 100) distance. Each node is placed at a uniform distance of 1 m, considering the total number of nodes to be 100. Noting that network is scalable to distribute the nodes at an equal distance, varying the nodes from 0 to 99. The neighbor of any node can be at a distance of only 1m, which we call it as a uniform pattern of nodes. Let "n" be total number of nodes in a network choosing "n" equal to 100, then the total number of grids "*k*" that can be required, can be found by the following equation,

$$k = \left(\sqrt{n-1}\right)^2 \tag{1}$$

where "*k*" be a number of grids

Let "*d*" be the distance between two nodes in a network and "*R*" be the radius within which, all the nodes are placed. The sensing range R_{sense} for a particular network can be given by,

$$R_{sense} = \sqrt{\frac{\pi R^2}{n}} \tag{2}$$

The concept is based on the fact that nodes are predetermined. This can result in many advantages, especially when the node is replicated of energy, a particular node can easily be found and replaced. Hence the node location "*NL*" of every node in a network is given by,

$$NL = \sqrt{\frac{R_{sense}}{\alpha \times n}} \tag{3}$$

α is in the range 0° to 180° with α varying 10° for every node.

Once the node location of every node is found, they are placed in a regular grid pattern, by this method a WSN in a grid pattern is created, which is called uniform distribution of nodes. As stated the node location, it is not just that nodes are simply placed as a grid pattern but an equation is been given mentioning the position of node which we call it as an optimized position of every node, which also holds good for circular placement of node topology. Another methodology for uniform node distribution can be a circular pattern of sensor nodes. The WSN in circular node placement topology is as shown in **Figure 3**. For simplicity circular deployment pattern of nodes is defined using two circles, here we have two types of communication. The data can be transmitted on a trajectory path or on circular path *i.e.* routing can go in both the paths depending on which the distance is shorter. Distance between the nodes for either of the methods is going to be varied. The trajectory method on trajectory path follows a distance of 1m between the neighbors, where the distance between the neighboring nodes are different for different circles. For the first circle it can be 0.39 m where 8 nodes are placed, for the second circle it is 1.56 m, by placing 16 nodes, and for the third circle it can be 3.51 m, for the fourth circle it can be 6.24 m. These distances can rather be called as "*sectorial separation*" which is given by,

$$\text{sector} = \left(\frac{1}{2} \times \theta \times \left(\frac{\pi}{180^\circ}\right) \times r^2\right) \tag{4}$$

r = a radius of a circle, θ be a angular difference between the nodes.

As with the uniform grid distribution pattern, node location is known for every node and it is not a random. Node locations are defined, where every node is placed on the on these locations, which is given by NL_o

$$NL_o = F(\alpha) r \tag{5}$$

$$F(\alpha) = \frac{\frac{2}{3}\sin\frac{\alpha}{2}}{\alpha} \tag{6}$$

α is in the range of 0° to 360°. $F(\alpha)$: Position for every node.

Once the nodes are deployed on a circle, the routing can be done by any means, that is routing can be on a trajectory path or it can be on a circular path or both. Definiteness can occur as the nodes are again at uniformity. Once the different types of patterns are set, each node can be assigned with energy, traffic, neighbor identity and their energy. *i.e.* each node is assigned with 50 J of initial energy "E_{ini}". The time it transmits the data, there is reduction in its total energy of a node that depends on the distance and traffic it carries on the path of its routing.

Once the network is set up, data in terms of packet can be sent from source to the destination. One general method that we follow for 3 different networks node deployment is that, source and destination are chosen on the edges and with same positions then the routing is done. For the routing simple technique that we adopted is multipath based routing, where multiple routes from source and destination have been chosen. Path based on distance, neighbor, and traffic have been selected for routing. Once the routing takes place, the energy consumed by the network nodes is found, that is related to the network lifetime, the overall energy consumed from a particular path between source and destination is given by "E_{total}"

$$E_{\text{total}} = \sum_{i=1}^{s} E_i \tag{7}$$

i = be the number of neighbors, s = total number of nodes, where E_i is given by,

$$E_i = \sum_{j=1}^{n'} \sum_{k=1}^{i_j} e_k \tag{8}$$

e_k is the energy consumption over the path of routing where data is been transmitted, n'—be the number of nodes in the path, i_j be number of hops, therefore e_k is given by "e_k"

$$e_k = e_{tx} + e_{rec} + E_{res} \tag{9}$$

e_{tx} and e_{rec} are transmission and reception energy constants for every hop, the mathematical model is been developed to find out the energy consumption (E_{res}) by the nodes during the routing using different deployment pattern of nodes.

All the nodes are initially assigned with 50 J of energy say E_{ini} which we call it as initial energy of a network (E_{ini}). A sensor consumes E_{sur} = 50 nJ for transmitting unit of data and k = 100 pJ/bit/m^2 for the transmitting amplifier. So the energy consumption of sensor node in order to be active in the network is given by "E_{sur}"

$$E_{sur} = E_{ene} \tag{10}$$

As the energy consumed depends on the number of bits transmitted and hence if sensor node transmits K bits, then energy consumed in order to be active in the network is given by "E_s(Actual)"

$$E_s\left(\text{Actual}\right) = K \times E_{ene} \tag{11}$$

At the destination nodes will receive packets, practically the received packets are less than actual packet sent. Hence energy consumed for receiving packets "E_r(Actual)" is given by, which is lesser than $E_s\left(\text{Actual}\right)$.

$$E_r\left(\text{Actual}\right) = \left(T_p - P_l\right) \tag{12}$$

where T_p is the energy due to transmission and P_l is the energy lost due to packet loss.

Energy consumption of sensor S_i in transmitting a unit of data packet to neighboring nodes "E_{nr}" is given by

$$E_{nr} = \sum_{n=0}^{n} E_{r(\text{Actual})} + n_n E_I d_{tr}^2 \tag{13}$$

where n_n is the number of neighoubrs, E_I is the energy of each node and d_{tr} is the distance of each neighbor from source node.

The neighboring node is always with the burden to deliver the data to the destination node, the energy consumption of it turns out to be an important factor, which is calculated as "E_r"

$$E_r = E_{r(\text{Actual})} + E_I\left(n_n - \left(n_n - 1\right)\right) d_{tr}^2 \tag{14}$$

Total energy consumed in network "E_t"

$$E_t = E_{nr} + E_r \tag{15}$$

If there are "n" no of sensor nodes in a coverage area which are active and transmitting data, then total energy is,

$$E_{tot} = nE_t \tag{16}$$

Residual energy for the sensor node is given by "E_{re}"

$$E_{re} = E_{ini} - E_t \tag{17}$$

3.2. Lifetime Definition

We assume each node has the same initial battery energy E_o, the lifetime T_i of node Si is defined as the expected time for the battery energy to be exhausted, given by "$T_{(n)}$"

$$T_{(n)} = \frac{e_k}{E_{ini}} \times 100 \tag{18}$$

where, n = number of sensor node. Taken with 100 to get the numerical in percentage.

Based on the energy model algorithm, simulation is been considered and comparative results are obtained for the pattern of nodes. Trajectory based node placement definitely saves the energy compared to random placement of nodes as shown in the below graphs.

4. Simulation

Simulations are done in OpenSuse platform using gcc C++. OpenSuse is a general purpose operating system built on top of the Linux kernel, developed by the community supported OpenSuse Project and sponsored by SUSE. In the simulation model we consider "n" number of nodes such that "n" can be in the range of 00 to 99, in the area of length "L" and breadth "B". The performance parameter, such as battery life of each node which is varied over an arbitrary range of values is considered. In the synthetic scenario of simulation "n" nodes are placed in L*B square meter field considering 3 different topologies (Random, circular and grid placement of nodes). Each node transmits a maximum of "P" packets at various times during simulation and their transmission range is "R" meters. For the simulation source and destination is chosen by user. All the nodes are initially assigned with same amount of energy. Depending upon the battery life and range of transmission we find out the neighbors of each node. Using neighbor and distance we find the shortest path. The transmission energy is calculated using the mathematical model. Thus we find out residual energy of the network after the routing takes place. The wireless sensor network is considered to be static, following inputs are considered for simulation. Number of nodes = 100, Initial energy of each node is 50 J, *Eele* = 50 nJ/bit/threshold distance, radio dissipation of transmitter and receiver *Eamp* = 100 pJ/bit/m^2, Communication range = 20 units. Simulation is to generate WSN topology, by applying AODV routing protocol algorithm to each topology, then calculating the energy, lifetime, hops that are performance parameter of the network.

The Performance parameters measured are as follows.

- *Intermediate Nodes*: It is the number of hops (nodes) required to transmit data from source to destination
- *Residual Energy*: The amount of energy that remains after the routing during data transfer
- *Network Lifetime*: It is the ratio indicating the utilization of total life of the network in transmitting the data

The result of simulation demonstrates that, initial deployment of node topology is considered by placing node on a uniform path, the energy consumption of each node and total network energy consumption is considered and compared with the random node placement pattern. The result that is shown proves that energy consumption is definitely less with grid deployment having distributed the node in whole of the simulation area, where as random placement tries to cover only half of the simulation area, following which the algorithm tries to save the energy consumed by the network, covering maximum area with increase in lifetime of network.

Figure 4 shows number of intermediate nodes or hops that are used to transmit the data between two ends, the results shows that it is based on the definiteness of the node position as the grid and circular network nodes use maximum communication radius for finding the next hop. The number of hops required to reach the destination in uniform network are definitely less as compared with random network as with the simulation results. The **Figure 5** gives the amount residual energy saved by the random, circular, grid based topologies that are calculated after the routing is been carried out. It is shown from **Figure 5** that random network consumes the highest amount energy for the routing. **Figure 6** gives the network lifetime that relates to usage of a network from the consumed energy. From the graph we can conclude that random based topology network lifetime is less compared to other two topologies. The simulations are done for a single routing, *i.e.* considering the part of data that is transmitted in one round of routing, when such 1000 rounds are considered for routing it is definite that energy as network parameter would definitely add to improve the performance of a network.

5. Conclusion

The work is carried to simulate the sensor network, to differentiate the behavior of network under different

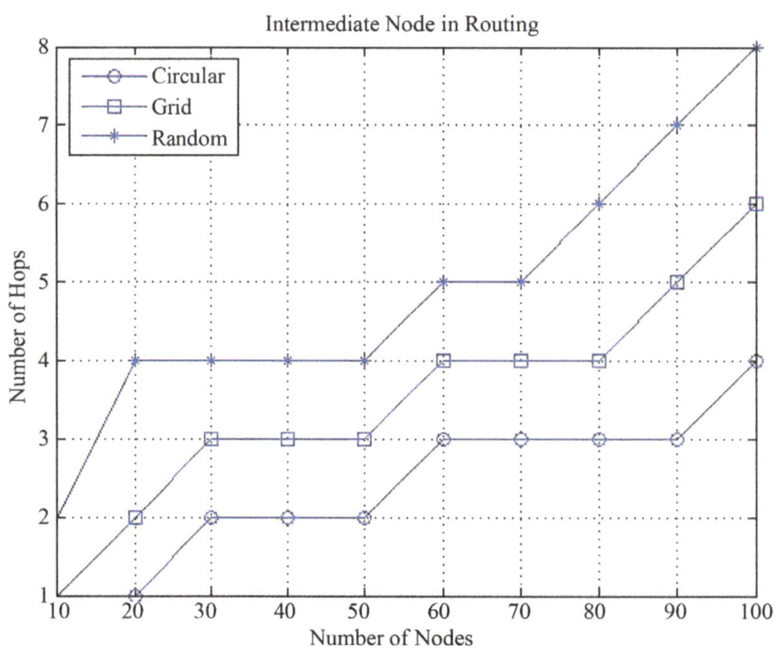

Figure 4. Graph showing number of hops vs number of nodes.

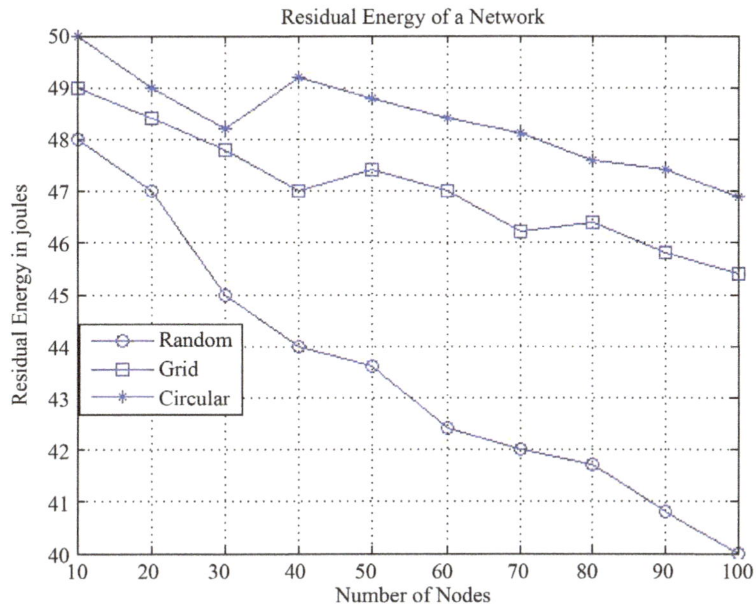

Figure 5. Graph showing residual energy in joules vs number of nodes.

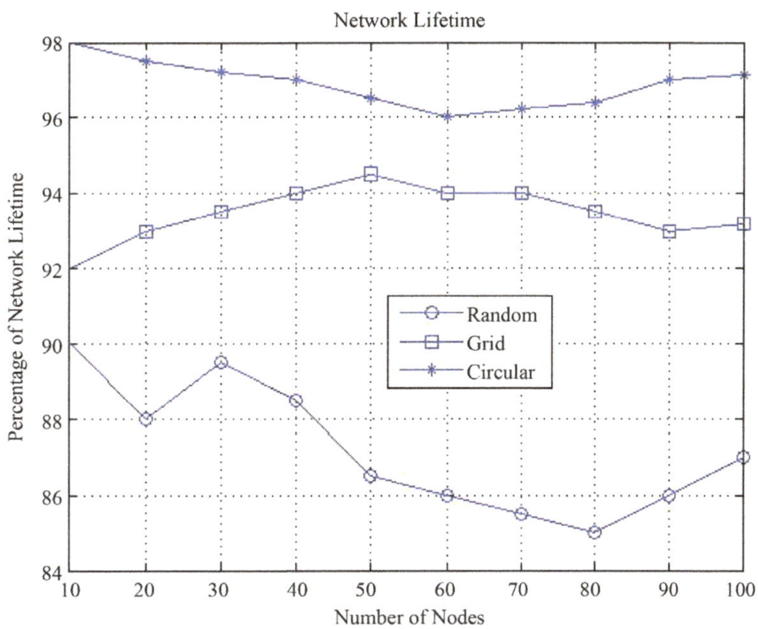

Figure 6. Showing graph of percentage of network lifetime vs number of nodes.

scenarios. Routing was run for random, grid and circular based network for different rounds. Based on the analytical and simulation results, it is proved that random deployment has smaller coverage area with more number of nodes deployed with un-equal distances leading to more energy utilization to decrease in overall network lifetime. Grid and circular deployment network have larger coverage area with less number of nodes deployed at equal distances. Hence energy utilization will be less with increase in network lifetime. The proposed model concludes that energy can be efficiently utilized in WSN's with the grid and circular node deployment. Work can be extended to run the algorithm till whole of the energy is exhausted. Different routing protocols can be considered to check the sustainability of the network.

References

[1] Biradar, R.V., Patil, V.C., Sawant, S.R. and Mudholkar, R.R. (2012) Classification and Comparision of Routing Protocol in Wireless Sensor Network. *Journal of Special Issue on Ubiquitous Computing Security Systems*, **4**, 704-711.

[2] Rupam, P.D. (2014) Analysis of Routing Protocols on Wireless Sensor Network: A Survey. *International Journal of Advanced Computational Engineering and Networking*, **2**.

[3] Roseline, R.A. and Sumathi, P. (2011) Energy Efficient Routing Protocols and Algorithms for Wireless Sensor Networks—A Survey. *Global Journal of Computer Science and Technology*, **11**.

[4] Singh, S.K., Singh, M.P. and Singh, D.K. (2010) Routing Protocols in Wireless Sensor Networks—A Survey. *International Journal of Computer Science & Engineering Survey* (*IJCSES*), **1**.

[5] Hojjatoleslami, S., Aghazarian, V. and Aliabadi, A. (2011) DE Based Node Placement Optimization for Wireless Sensor Networks. 3*rd International Workshop on Intelligent Systems and Applications* (*ISA*), 1-4.

[6] Tian, H., Shen, H. and Roughan, M. (2008) Maximizing Networking Lifetime in Wireless Sensor Networks with Regular Topologies. *Ninth International Conference on Parallel and Distributed Computing*, *Applications and Technologies*, Otago, 1-4 December 2008, 211-217. http://dx.doi.org/10.1109/pdcat.2008.29

[7] Cheng, P., Chuah, C.-N. and Liu, X. (2004) Energy-Aware Node Placement in Wireless Sensor Networks. *Global Telecommunications Conference*, 2004. *GLOBECOM* '04. *IEEE*, **5**, 3210-3214.

[8] Lam, M.-L. and Liu, Y.-H. (2006) Active Sensor Network Deployment and Coverage Enhancement Using Circle Packings. *IEEE International Conference on Robotics and Biomimetics*, 2006. *ROBIO* '06,*Kunming*, 17-20 December 2006, 520-525.

[9] Gun, M., Kosar, R. and Ersoy, C. (2007) Lifetime Optimization Using Variable Battery Capacities and Nonuniform Density Deployment in Wireless Sensor Networks. 22*nd international symposium on Computer and Information Sciences*, 2007. *ISCIS* 2007, 7-9 November 2007, 1-6.

[10] Liang, W.F., Ma, G.J., Xu, Y.L. and Shi, J.G. (2008) Aggregate Node Placements in Sensor Networks. 11*th IEEE Singapore International Conference on Communication Systems*, 2008. *ICCS* 2008, Guangzhou, 19-21 November 2008, 926-932. http://dx.doi.org/10.1109/iccs.2008.4737320

[11] Zhang, J., Ci, S., Sharif, H. and Alahmad, M. (2009) Lifetime Optimization for Wireless Sensor Networks Using the Nonlinear Battery Current Effect. *IEEE International Conference on Communications*, 2009. *ICC*' 09, Dresden, 14-18 June 2009, 1-6. http://dx.doi.org/10.1109/icc.2009.5199132

[12] Tariq, M., Kim, Y.-P., Kim, J.H., Park, Y.-J. and Jung, E.H. (2009) Energy Efficient and Reliable Routing Scheme for Wireless Sensor Networks. *International Conference on Communication Software and Networks*, 2009. *ICCSN*' 09, Macau, 27-28 February 2009, 181-185. http://dx.doi.org/10.1109/iccsn.2009.101

[13] Malik, N.N.N.A., Esa, M. and Yusof, S.K.S. (2009) Intelligent Optimization of Node Coordination in Wireless Sensor Network. *Innovative Technologies in Intelligent Systems and Industrial Applications*, 2009. *CITISIA* 2009, Monash, 25-26 July 2009, 328-331.

[14] Li, W. (2009) Wireless Sensor Network Placement Algorithm. 5*th International Conference on Wireless Communications*, *Networking and Mobile Computing*, 2009. *WiCom*' 09, Beijing, 24-26 September 2009, 1-4.

[15] Botros, S.S., Elsayed, H.M., Amer, H.H. and El-Soudani, M.S. (2009) Lifetime Optimization in Hierarchical Wireless Sensor Networks. *IEEE Conference on Emerging Technologies & Factory Automation*, 2009. *ETFA* 2009, Mallorca, 22-25 September 2009, 1-8. http://dx.doi.org/10.1109/etfa.2009.5347016

[16] Nik Abd Malik, N.N., Esa, M. and Yusof, S.K.S. (2009) Optimization of Adaptive Linear Sensor Node Array in Wireless Sensor Network. *Asia Pacific Microwave Conference*, 2009. *APMC* 2009, Singapore, 7-10 December 2009, 2336-2339. http://dx.doi.org/10.1109/apmc.2009.5385453

[17] Abbasy, M.B., Barrantes, G. and Marín, G. (2010) Performance Analysis of Sensor Placement Strategies on a Wireless Sensor Network. 2010 4*th International Conference on Sensor Technologies and Applications* (*SENSORCOMM*), Venice, 18-25 July 2010, 609-617. http://dx.doi.org/10.1109/SENSORCOMM.2010.96

[18] Akshay, N., Kumar, M.P., Harish, B. and Dhanorkar, S. (2010) An Efficient Approach for Sensor Deployments in Wireless Sensor Network. 2010 *International Conference on Emerging Trends in Robotics and Communication Technologies* (*INTERACT*), Chennai, 3-5 December 2010, 350-355. http://dx.doi.org/10.1109/interact.2010.5706178

[19] Li, X. and Qiu, S.B. (2011) Research on Multicast Routing Protocol in Wireless Sensor Network. *International Conference on Control*, *Automation and Systems Engineering* (*CASE*), Singapore, 30-31 July 2011, 1-4. http://dx.doi.org/10.1109/iccase.2011.5997642

[20] Abdul Latiff, N.A., Abdullatiff, N.M. and Ahmad, R.B. (2011) Extending Wireless Sensor Network Lifetime with Base Station Repositioning. 2011 *IEEE Symposium on Industrial Electronics and Applications* (*ISIEA*), Langkawi, 25-28 September 2011, 241-246. http://dx.doi.org/10.1109/ISIEA.2011.6108707

10

A Free Market Economy Model for Resource Management in Wireless Sensor Networks

Byeongwook Bae[1], Junyoung Park[2], Sunggu Lee[1]

[1]Department of Electrical Engineering, Pohang University of Science and Technology, Pohang, Korea
[2]System LSI Division, Samsung Electronics Co., Ltd., Hwasung, Korea
Email: lidwoogi@postech.ac.kr, zenojur@gmail.com, slee@postech.ac.kr

Abstract

This paper presents a free market economy model that can be used to facilitate fully distributed autonomous control of resources in massive heterogeneous wireless sensor networks (WSNs). In the future, it is expected that WSNs will exist as part of the global Internet of Things (IoT), and different WSNs can work together in a massive network of heterogeneous WSNs in order to solve common problems. Control of valuable processing, sensing and communication resources, determining which nodes will remain awake during specific time periods in order to provide sensing services, and determining which nodes will forward other nodes' packets are difficult problems that must be dealt with. It is proposed that just as the free market economy model enables the global human society to function reasonably well when individuals simply attempt to trade money and services in order to maximize their individual profits, and a similar model and mechanism should enable a massive network of heterogeneous WSNs to function well in a fully distributed autonomous manner. The main contributions of this paper are the introduction of the free market economy model for use with WSNs, the formal definition of a maximum profit price problem for multihop packet relaying, and the proposal of a distributed genetic algorithm for the solution of the maximum profit price problem. Simulation results show that the proposed distributed solution produces results that are 70% - 80% similar to a pareto optimal solution for this problem.

Keywords

Distributed Control, WSN Resources, Free Market Economy, Auction, Autonomous Control

1. Introduction

With the widespread use of mobile devices supporting communication, many modern day living areas are surrounded by various wireless and wired networks. The proliferation of these communication-capable devices creates the Internet of Things (IoT), in which nodes with unique IDs construct a network across diverse platforms and share data for their own purposes [1]. A central component of an IoT is a wireless sensor network (WSN) [2], of which various different types, composed of a variety of sensors, can be used to provide various different services to the users of the IoT. In the future, just as the Internet has spread across the globe and has become ubiquitous, the IoT is expected to grow in size to span the globe and become ubiquitous. This future IoT is expected to include many different wireless sensor networks (WSNs), composed of various different types of sensor nodes with widely varying sizes and capabilities.

In such massive heterogeneous WSNs, it will become extremely difficult to provide effective control of the available resources (sensors, processors, wireless radio modules, batteries, etc.) such that the system works well as a whole while ensuring long average lifetimes and high performance with low overall cost. As the number of WSN nodes increases to massive proportions, the only viable type of solution will be a fully distributed and autonomous control method.

Many of the "tasks" that a sensor node performs can be considered as "services" that benefit other nodes while incurring a "cost" to the node performing the service. For example, most of the area in a physical region that needs to be continuously monitored is covered by multiple sensors in order to provide extended sensing lifetimes by letting sets of nodes alternately enter sleep mode and to compensate for faulty nodes or external sabotage attempts. Thus, a node that stays active and provides processing and/or sensing services for a time period T incurs a "cost" (since valuable battery capacity must be utilized during T) while providing a "service" to neighboring nodes by permitting them to enter sleep mode during T. As another example, a node that relays packets between two other nodes in the network provides a "service" to those two other nodes while incurring a heavy "cost" since wireless radio transmission requires a high level of energy usage [3]. Clearly, it will benefit a node to not provide such sensing or packet relaying services. However, if all nodes behaved in such a selfish manner, then the WSN would become useless as no sensing or multihop packet deliveries would occur. Thus, for these types of distributed resource control problems, it is proposed that a free market economy model be used to provide fair distribution and use of resources and services in a fully distributed autonomous manner.

2. Free Market Economy Model

A free market economy model, based on an auction system, is proposed as a compensation scheme for the use of valuable sensor node resources for activities such as network processing, sensing and multihop packet relaying that can benefit other nodes in a massive heterogeneous WSN. For a given sensor node v, this compensation can take the form of credits that can then be used by v to request similar services from neighboring nodes. It can also take the form of actual money if a service provider uses such services to save costs by not having to introduce new sensor nodes to provide additional capabilities or by not building new wireless radio towers in sparsely populated areas. In this paper, "money" will be used to refer to both types of compensation instruments.

The target massive heterogeneous WSN can be modeled by an undirected graph $G = (V, E)$, where V and E are the sets of nodes (which could consist of sensor nodes, wireless access points, and gateway nodes) and wireless communication links, respectively, in the WSN. The nodes in V are time-synchronized using a mechanism such as that described in [4]. Time is partitioned into frames, but frame beacons to announce the beginnings of frames are not necessary since each node can determine the start time of a frame individually based on its globally time synchronized clock-the requisite time synchronization mechanism should operate in a completely distributed autonomous manner, such as in the method proposed in [4].

During each time frame, each node $v \in V$ determines prices for each service that it plans to offer and announces these prices to its neighboring nodes. The node v can also request services from neighboring nodes or go to sleep to conserve energy. However, for each service that node v requests, it must have the money available to pay for that service. If the node runs out of money, then it must set desirable service prices and wait to receive money from neighboring nodes for services rendered. If the node runs out of energy during this process, then it simply dies until its battery can be replaced. Thus, for each node $v \in V$, prices and requests for services should be determined in a manner that ensures the maximum "profit" for itself. Then, when operated properly, the free market model should ensure the longevity of the WSN while also ensuring a fair and efficiently operating WSN. As a first step towards showing the utility of the proposed free market economy model, the following sections will focus on one particular application, multihop packet relaying.

3. Problem Definition and Related Work

One WSN application that can benefit from the proposed free market economy model is the problem of multi-hop packet relaying. **Figure 1** illustrates the basic problem being addressed. At the start of a time frame, each node $v \in V$ needs to establish a price $p(v)$ that node v will charge in order to relay a fixed-size packet of data through itself. There are also a set of packet communication requests R. Each request $r \in R$ includes a source node $s(r) \in V$ and a destination node $d(r) \in V$. If $s(r)$ has a direct wired or wireless communication link to $d(r)$, then there is no cost for communication request r. However, if such a direct connection does not exist, then a multihop path to $d(r)$ is used provided such a path exists. In order to use this multipath path, node $s(r)$ must pay each intermediate node that it uses the relay price charged by that node. For example, in **Figure 1**, if a node v_1 decides to use the path $P = \{(v_1, v_2), (v_2, v_3), (v_3, v_4)\}$ to send a packet to v_4, $p(v_2)$ and $p(v_3)$ are paid to v_2 and v_3 respectively. If node $s(r)$ does not have sufficient money to pay all of the relay prices for a given path, then $s(r)$ cannot use that path. The prices of the packet relaying costs charged by each node are renewed during the beginning of each frame and the paths formed are valid until the end of each frame.

For each time frame, node $v \in V$ should set its price $p(v)$ in a manner that provides the most benefit for itself. A price $p(v)$ is established with the knowledge that other nodes will only use node v and pay the price$p(v)$ if the price is competitive with those charged on alternative routes for each communication request. For a given request $r \in R$, a distributed shortest-path (based on the sums of the prices of all intermediate nodes on a path) finding algorithm, such as a distributed version of Dijkstra's shortest path algorithm [5] or Ad-hoc On-demand Distance Vector routing (AODV), which floods the network with packets that store accumulated path information and then uses the information stored in return packets to find routing paths [6], needs to be used to determine the path that node $s(r)$ will select. Given that prices are established for each time frame and each node needs to establish pricing policies that ensure longevity and effective operation based on the specific needs of that node, each node $v \in V$ can establish conditions for its price $p(v)$ during a given time frame. For example, if its remaining energy level is extremely low or it has accumulated an extremely large amount of money, then $p(v)$ can be constrained to be greater than a very large constant. On the other hand, if v lacks sufficient money to request multihop packet deliveries that are urgently required, then it should utilize any available information about the prices expected to be established by neighboring nodes in order to specify that its price $p(v)$ should be less than the minimum of its neighbor's prices.

The *expected profit* $f(v)$ for a node $v \in V$ during a given time frame is defined to be the difference between the total reward and total cost for node v during that time frame, where the total reward is the sum of the prices paid by all requests $r \in R$ that are expected to utilize node v for multihop communication and the total cost is the sum of the prices paid by node v for all of its multihop communication requirements. Then the following optimization problem can be defined.

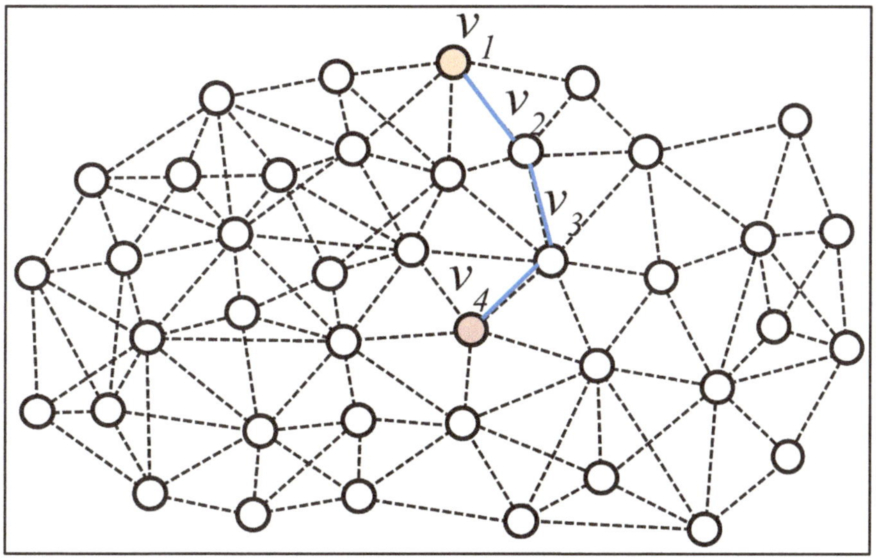

Figure 1. A multihop path from node v_1 to node v_4.

PROBLEM: Max Profit Price (MPP)

INPUT: Graph $G = (V, E)$, set of requests R with source $s(r) \in V$ and destination $d(r) \in V$ for each $r \in R$. Each node $v \in V$ can optionally have a set of conditions that it sets on the price $p(v)$.

OUTPUT: Prices $p(v)^*$ for all nodes $v \in V$, where the price $p(v)^*$ is established such that the expected profit $f(v)$ is maximized given that all requests in R use the shortest paths, where the length of a path P is the sum of the prices on path P. The prices are established in a fixed random order (e.g., based on a hash function of the clock values of individual nodes in V) determined in a distributed autonomous manner, with prices carried over from the previous time frame until new prices can be determined for the current time frame.

In Problem MPP, which can be considered to be a variant of shortest path routing, since each node can have its own objective function, there may not exist a single optimal solution. For instance, the price that maximizes the profit of v_1 may minimize the profit of v_2 and vice versa. To address multiobjective optimization such as in Problem MPP, a *pareto optimal* solution, which is a solution in which no change can be made to improve one objective without degrading another objective, can be used [7]. A formal definition for *pareto optimal* can be based on Pareto's original work [8] as shown below.

Definition 1: A multidimensional point $x_m^* \in X$, is *pareto optimal* if there does not exist another point $x_j \in X$ such that there exists $i \in \{1, \ldots, k\}$ with $F_i(x_j) > F_i(x_m^*)$, where $F_i(x_l)$, $i \in \{1, \ldots, k\}$, is an objective function on $x_l \in X$.

Since multiobjective optimization is an NP-hard [9] problem, a heuristic approach is necessary. Thus, in this paper, the authors have investigated the use of evolutionary algorithms [10] [11] for the solution of Problem MPP. Previous works that have used multiobjective problem modeling to solve problems related to WSNs include approaches for sensor deployment, sensor coverage control, network lifetime enhancement, and multihop routing [12]-[14].

4. Proposed Algorithm

The authors propose an implementation of the non-dominated sorting and crowding distance methods [15] in order to search for a pare to optimal solution for Problem MPP. The proposed method is based on a genetic algorithm (GA), which is a popular type of evolutionary algorithm. A GA models possible solution instances as chromosomes and uses multiple iterations involving crossover and mutation, based on behavior observed in nature, in order to find successively better solutions.

A direct implementation of a GA solution for Problem MPP requires a centralized approach. However, a centralized algorithm is impractical for the type of massive heterogeneous WSN targeted in this paper. Thus, a distributed algorithm that imitates the operations of a GA is proposed as a practical solution to Problem MPP. However, a centralized GA solution will still be used, but simply as an "ideal" solution that is used to evaluate the quality of the proposed algorithm.

An "ideal" solution to the MPP problem, used for comparison purposes only, is shown below as **Algorithm CGA** (Centralized GA). In this algorithm, chromosome vectors in the population consist of the prices of nodes for packet relaying, in a decimal number format, and offspring are generated using arithmetic crossover. As a characteristic of multiobjective optimization, the algorithm has multiple fitness functions as many objective functions and parents for offspring are selected by non-dominated sorting based on the profit vector. **Algorithm CGA** is a GA that operates in a centralized manner. Because of the distributed nature and scale of the types of massive heterogeneous WSNs targeted, **Algorithm CGA** cannot be directly implemented for the target WSN.

The proposed algorithm, **Algorithm DGA** (Distributed GA), also shown below, is a distributed algorithm that imitates a GA and uses genetic operations such as selection and crossover within 1-hop subnetworks. Since operations involving setting prices, broadcasting price information, and finding shortest paths are required to incrementally produce solutions with higher cumulative profits, only one offspring is inserted into the population per round. Also, because the distributed algorithm cannot directly adopt a non-dominated sorting approach, the algorithm attempts to find solutions that are close to a pareto optimal solution by iteratively attempting to maximize one objective at a time. In order to do this, rather than faithfully implementing all aspects of a GA, **Algorithm DGA** simply uses a "GA-type" approach to maximize one objective at a time. Pseudocode descriptions of **Algorithms CGA** and **Algorithms DGA** are shown below.

5. Experimental Result

The performance of **Algorithm DGA** was evaluated using simulations. Since this is a completely new model for

Algorithm CGA. A genetic algorithm for Problem MPP, used for comparison purposes only.

$C = \{c_0, c_1, \ldots, c_{n-1}\}$, $F = \{f_0, f_1, \ldots, f_{n-1}\}$
// c_i = (price vector of nodes $v \in V$)
// f_j = (profit vector of nodes $v \in V$), $|c_i| = |f_i| = |V| = N$

1) create initial population with $|C| = n$;
2) **for** ($i = 0$; $i < n$; i++) **begin**// initial phase
3) calculate profit vector $f_i \in F$ of $c_i \in C$;
4) **end for**
5) **while** (# of generation $< G$) **begin**// main phase
6) non-dominated sorting in C;
7) select c_i and c_j using tournament selection;
8) arithmetic (average) crossover from c_i, c_j to p;
9) mutate p to p_t;
10) find the lowest non-dominated rank solution c_k from C;
11) $C = C \cup \{p_t\} - \{c_k\}$;
12) **end while**

Algorithm DGA. The proposed distributed algorithm for Problem MPP.

$C = \{c_0, c_1, \ldots, c_{n-1}\}$, $F = \{f_0, f_1, \ldots, f_{n-1}\}$
// c_i = (price), f_j = (profit)

For all $v \in V$
1) create initial population $|C| = n$
2) **for** ($i = 0$; $i < n$; i++) **begin**// initial phase
3) $p(v) = c_i$ and notify $p(v)$;
4) find shortest path if requests exist from v;
5) calculate profit $f_i \in F$;
6) **end for**
7) **while** (# of rounds $< G$) **begin**// main phase
8) find index x with maximum profit from F;
9) find index y with minimum profit from F;
10) broadcast x to neighbor nodes;
11) receive j_m from neighbor node m;
12) broadcast y to neighbor nodes;
13) receive k_m from neighbor node m;
14) find the majority t_1 of $j_m^{\forall}$ and x;
15) find the second majority t_2 of $j_m^{\forall}$ and x;
16) $p = (c_{t1} + c_{t2})/2$; //generate offspring
17) mutate p to p_t;
18) find shortest path if requests exist from v;
19) calculate profit f;
20) r = (the majority of $k_m^{\forall}$ and y)
21) swap (c_0, c_r); swap (f_0, f_r);
22) $c_0 = p_t$ and $f_0 = f$;
23 **end while**

WSNs (based on all available open literature that the authors have been able to examine), other previous methods were not available for comparison. Thus, comparisons were made with a fixed price policy and a pareto optimal solution. *Angular similarity* to the pareto optimal solution, defined in [16], was used as the performance metric for profit vector comparisons.

The pareto optimal solution was obtained as the result of 500 generations of **Algorithm CGA**. **Algorithm DGA** also used 500 rounds in a distributed implementation. For each algorithm, the results, shown in **Figure 2**, were the averages of 10 different randomly connected networks. Overall, **Algorithm DGA** performed significantly better than a uniform fixed price policy. In the first few rounds, a few solutions produced by **Algorithm DGA** were below the results of the uniform fixed price policy. However, after 120 rounds, for almost every round, **Algorithm DGA** produced results that were closer to the pareto optimal solution (based on angular simi-

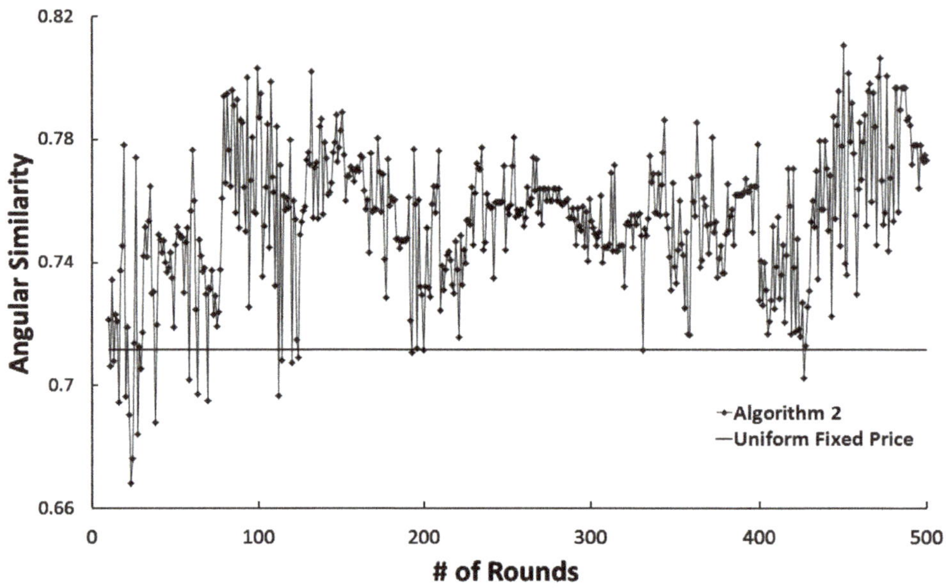

Figure 2. Angular similarity to the profit vector of a pareto optimal solution.

larity) than the uniform fixed price policy-this shows that **Algorithm DGA** derives better solutions as more rounds are used. As the number of rounds was increased, in certain cases, the similarity rapidly dropped before going back up because of the mutation nature of a GA-in a GA, mutations are used to avoid local minima solutions.

6. Conclusion

This paper has proposed the use of a free market economy model for the effective control of the resources in a massive heterogeneous WSN. The utility of this model is described in general times, and then a specific target application, packet relaying for multihop wireless networks, is examined. A maximum profit price problem is defined in a formal manner and a distributed algorithm solution, based on genetic algorithm principles, is proposed. The proposed algorithm is compared with a fixed price policy and a pareto optimal solution. Simulation results demonstrate that the proposed distributed algorithm performs similarly to the pareto optimal solution and performs successively better as more iteration rounds are used.

References

[1] Gubbi, J., Buyya, R., Marusic, S. and Palaniswami, M. (2013) Internet of Things (IoT): A Vision, Architectural Elements, and Future Directions. *Future Generation Computer Systems*, **29**, 1645-1660. http://dx.doi.org/10.1016/j.future.2013.01.010

[2] Yick, J., Mukherjee, B. and Ghosal, D. (2008) Wireless Sensor Network Survey. *Computer Networks*, ***52***, 2292-2330. http://dx.doi.org/10.1016/j.comnet.2008.04.002

[3] Jang, U., Lee, S. and Yoo, S. (2012) Optimal Wake-Up Scheduling of Data Gathering Trees for Wireless Sensor Networks. *Journal of Parallel and Distributed Computing*, **72**, 536-546. http://dx.doi.org/10.1016/j.jpdc.2012.01.008

[4] Marler, R.T. and Arora, J.S. (2004) Survey of Multi-Objective Optimization Methods for Engineering. *Structural and Multidisciplinary Optimization*, **26**, 369-395. http://dx.doi.org/10.1007/s00158-003-0368-6

[5] Dijkstra, E.W. (1959) A Note on Two Problems in Connexion with Graphs. *Numerische Mathematik*, **1**, 269-271. http://dx.doi.org/10.1007/BF01386390

[6] Perkins, C.E. and Royer, E.M. (1999) Ad-Hoc On-Demand Distance Vector Routing. *2nd IEEE Workshop on Mobile Computing Systems and Applications*, New Orleans, 25-26 February 1999, 90-100. http://dx.doi.org/10.1109/mcsa.1999.749281

[7] Zitzler, E. and Thiele, L. (1999) Multiobjective Evolutionary Algorithms: A Comparative Case Study and the Strength Pareto Approach. *IEEE Transactions on Evolutionary Computation*, **3**, 257-271. http://dx.doi.org/10.1109/4235.797969

[8] Pareto, V. (1906) Manuale di Economia Politica. Vol. 13, Societa Editrice.

[9] Garey, M.R. and Johnson, D.S. (1979) Computers and Intractability: A Guide to NP-Completeness. Freeman Press.

[10] Deb, K. (2001) Multi-Objective Optimization Using Evolutionary Algorithms. Vol. 16, John Wiley & Sons, Hoboken.

[11] Srinivas, N. and Deb, K. (1994) Muiltiobjective Optimization Using Non-Dominated Sorting in Genetic Algorithms. *Evolutionary Computation*, **2**, 221-248. http://dx.doi.org/10.1162/evco.1994.2.3.221

[12] Konstantinidis, A., Yang, K., Zhang, Q.F. and Zeinalipour-Yazti, D. (2010) A Multi-Objective Evolutionary Algorithm for the Deployment and Power Assignment Problem in Wireless Sensor Networks. *Computer Networks*, **54**, 960-976. http://dx.doi.org/10.1016/j.comnet.2009.08.010

[13] Jia, J., Chen, J., Chang, G. and Tan, Z. (2009) Energy Efficient Coverage Control in Wireless Sensor Networks Based on Multi-Objective Genetic Algorithm. *Computers and Mathematics with Applications*, **57**, 1756-1766. http://dx.doi.org/10.1016/j.camwa.2008.10.036

[14] Rajagopalan, R., Mohan, C.K., Varshney, P. and Mehrotra, K. (2005) Multi-Objective Mobile Agent Routing in Wireless Sensor Networks. *The* 2005 *IEEE Congress on Evolutionary Computation*, **2**, 1730-1737. http://dx.doi.org/10.1109/cec.2005.1554897

[15] Deb, K., Pratap, A., Agarwal, S. and Meyarivan, T. (2002) A Fast and Elitist Multiobjective Genetic Algorithm: NSGA-II. *IEEE Transactions on Evolutionary Computation*, **6**, 182-197. http://dx.doi.org/10.1109/4235.996017

[16] Apaydin, T. and Ferhatosmanoglu, H. (2006) Access Structures for Angular Similarity Queries. *IEEE Transactions on Knowledge and Data Engineering*, **18**, 1512-1525. http://dx.doi.org/10.1109/TKDE.2006.165

11

Wireless Sensor Networks for Water Quality Monitoring and Control within Lake Victoria Basin: Prototype Development

Anthony Faustine[1], Aloys N. Mvuma[1], Hector J. Mongi[1], Maria C. Gabriel[1], Albino J. Tenge[1], Samuel B. Kucel[2]

[1]The University of Dodoma, Dodoma, Tanzania
[2]Busitema University, Tororo, Uganda
Email: sambaiga@gmail.com, anmvuma@gmail.com, hjmongi@yahoo.com, magc5185@yahoo.co.uk, ajtenge@gmail.com, sbkucel@gmail.com

Abstract

The need for effective and efficient monitoring, evaluation and control of water quality in Lake Victoria Basin (LVB) has become more demanding in this era of urbanization, population growth and climate change and variability. Traditional methods that rely on collecting water samples, testing and analyses in water laboratories are not only costly but also lack capability for real-time data capture, analyses and fast dissemination of information to relevant stakeholders for making timely and informed decisions. In this paper, a Water Sensor Network (WSN) system prototype developed for water quality monitoring in LVB is presented. The development was preceded by evaluation of prevailing environment including availability of cellular network coverage at the site of operation. The system consists of an Arduino microcontroller, water quality sensors, and a wireless network connection module. It detects water temperature, dissolved oxygen, pH, and electrical conductivity in real-time and disseminates the information in graphical and tabular formats to relevant stakeholders through a web-based portal and mobile phone platforms. The experimental results show that the system has great prospect and can be used to operate in real world environment for optimum control and protection of water resources by providing key actors with relevant and timely information to facilitate quick action taking.

Keywords

Wireless Sensor Networks, Lake Victoria Basin, Gateway, Sensor Nodes

1. Introduction

Lake Victoria Basin (LVB) is an important ecosystem in the Eastern African Region. It is shared among five East Africa countries which are Burundi, Kenya, Rwanda, Tanzania and Uganda. The Basin is also rich in natural resources which include fresh water, fishery, minerals, forests and fertile land for agriculture. These resources have recently contributed to rapid urbanization and increasing human population. The negative impacts of climate change and variability are compounding the existing challenges. Consequently, increasing number of human activities has gradually speed-up contamination and damage to fresh water resources in the LVB [1]. Water quality monitoring and evaluation is therefore very critical for sustainable water resources management and utilization programs and has increasingly attracted a great deal of research and development attention. However, in most developing countries like Tanzania, water quality monitoring and evaluation is entirely a manual process. This traditional approach of water quality monitoring based on sampling and subsequent analyses in water laboratories is expensive, time-consuming and has a low resolution both in time and space.

Over recent years, wireless sensor networks (WSNs) have received considerable attention in environmental and industrial monitoring and control applications. Being one of the main building blocks in ambient intelligence, WSNs provide significant advantages both in cost as well as in distributed intelligence [2]. On the other hand, installation and maintenance expenses are reduced because of the use of cheap devices which do not require wiring. It is for this reason that WSN technology has been attracting rapidly growing attention both from academia and in industry. This technology can definitely be applied in the water quality monitoring and evaluation process by providing best approach to real time data acquisition, transmission and processing.

Chaamwe [3] stressed the fact that WSNs are suitable for water quality monitoring. In [4], Zennaro *et al.* proposed the design of WSN for water quality monitoring system. They used Sun SPOT motes and 90-FLT series E sensors by TPS to measure water pH, conductivity, total dissolved salt (TDS), dissolved oxygen (DO), turbidity and temperature. The authors reported that the proposed system cost around US $3400. However their solution is not suitable for long-term outdoor environments. Using ARM controller, Kumar *et al.* [5] presented WSN solar-based water quality monitoring system for measuring water pH, Oxygen level and turbidity. Furthermore, Aravinda *et al.* [6] presented a low-cost wireless water physic-chemistry sensing system using Arduino Mega 2560. They measured pH, DO, conductibility, oxidation reduction potential, light and temperature of water. The authors reported that the system cost around US $1040. However their approach used a computer as the gateway which is costly.

In this paper, a proposed WSN water quality monitoring system prototype is presented. The system consists of an Arduino microcontroller, a set of water quality sensors, and a network connection module. It measures water quality parameters including temperature, DO, pH and electrical conductivity (EC) values in real-time. The experimental results showed that the system has great prospect and can be used to operate in real world environment for optimum control of aquaculture environment. Considering the previously proposed systems, the proposed prototype provides software module that allow users to visualize the data from the WSN without installing specific software. Moreover, the prototype is suitable for long-term outdoor environments and implement low-cost gateway module.

The rest of the paper is organized as follows. Section 2 describes a development of the proposed WSN system prototype and outlines its key features together with details on how the proposed system was implemented and tested. Section 3 provides the preliminary results and discussion. Finally, Section 4 presents the concluding remarks of this work and introduces relevant open issues which have been identified.

2. Proposed WSN System Prototype

The proposed WSN system prototype was developed as a component of the Water Resource Governance System (WaGoSy); an innovative and integrated ICT solution which was designed to address challenges of water resources governance facing the LVB. The ultimate purpose of the proposed WSN system prototype is to automate the monitoring of water quality for supporting sustainable utilization and management of water resources in the basin. The proposed WSN system consists of three main building blocks namely, WSN sensor nodes, WSN gateway node and application software.

2.1. WSN Sensor Node

The WSN sensor node is the main building block of the developed WSN system prototype. It is equipped with sensor and microcontroller units, Global Positioning System (GPS) receiver, power supply and Radio Frequency (RF) transceiver. The sensor unit essentially consists of several sensors used for detection of water quality parameters. In this prototype, four sensors for measuring pH, EC, DO and temperature are employed to provide general characterization of water quality. However, the sensor node is scalable to allow more sensors depending on needs. **Table 1** summarizes the specifications for sensors used in the prototype.

The pH, EC and DO sensors are interfaced to the microcontroller using their respective circuits from Atlas Scientific with the exception of the temperature sensor which is directly connected to the microcontroller. All the sensors are powered by the power supply for their operation. The RF transceiver transmits sensed parameter values and GPS information to a gateway node. **Figure 1** shows the schematic diagram of sensor unit configuration used in the proposed system. To connect all the sensors to the microcontroller unit, a WaGoSy sensors carrier board as shown in **Figure 2(a)** was designed. The sensors were calibrated to ensure correct operation and accuracy in the resulting water quality parameter values. The calibration was done at the Lake Victoria water quality laboratory in Mwanza with the assistance from water quality laboratory technician as shown in **Figure 2(b)**.

The microcontroller unit consists of a microcontroller and a software programme that determines the behavior of the WSN sensor node. Arduino Mega 2560 microcontroller, an open-source electronics prototyping platform

Table 1. Summary of specifications for the sensors.

S/N	Sensor	Manufacture	Model	Range
1	pH	Atlas Scientific	ENV-40-pH	0 - 14 (Na^+ error at >12.3 pH)
2	EC	Atlas Scientific	ENV-40-EC-K0.1	(K = 0.1) 0.1 - 0.5 µs/cm to 50,000 µs/cm
3	DO	Atlas Scientific	ENV-40-DO	0 - 20 mg/L
4	Temperature	Maxim Integrated	DS18B20	−55˚C to +125˚C

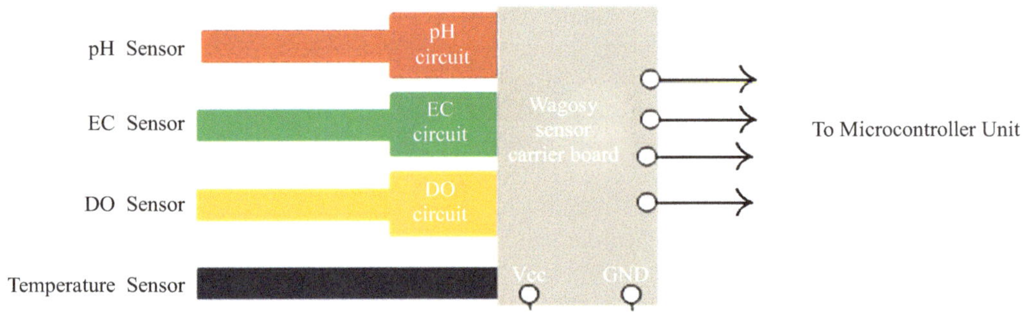

Figure 1. Schematic diagram of WSN sensor unit configuration.

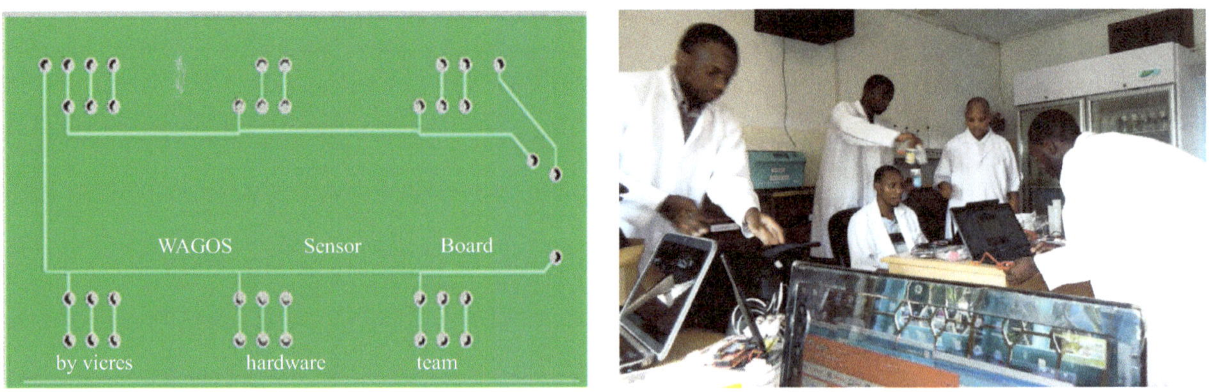

(a) (b)

Figure 2. (a) WaGoSy sensor carrier board; (b) Calibration of sensors.

based on flexible, easy-to-use hardware and software was used to acquire and process sensor data. Each sensor node is further equipped with EM-506 GPS receiver from USGlobalSat to provide location and time for the measured data. This is a complete GPS smart antenna receiver that includes an embedded antenna and GPS receiver circuits. The GPS module is also used to ensure that the sensor nodes are not tampered with for security reasons. However the GPS data is captured and sent to the gateway in preset time interval and not continuously to reduce power consumption. The architecture of the sensor node is shown in **Figure 3**.

2.2. WSN Gateway Node

The WSN system gateway is one of the most important and unique block in the proposed system. It collects all the information received from multiple sensor nodes. The developed gateway is equipped with microcontroller unit, General Packet Radio Service (GPRS) module, RF transceiver, memory card and power supply. Arduino Uno microcontroller is used to acquire and process received sensor data from WSN sensor nodes through the RF transceiver. The same XBee module as used in WSN sensor nodes is used to implement WSN gateway RF transceiver module. The GPRS module residing on top of the gateway node is used to communicate with the cellular network to forward the Short Message Services (SMS) data to a WaGoSy system. To implement the GPRS module, Arduino GPRS shield from Seed Studio was selected. This is quad-band low power consumption Global System for Mobile Communications (GSM)/GPRS module SIM900 with a compact PCB antenna and consumes 1.5 mA in sleep mode. A 1 GB compact memory card is used to store measurements data from WSN sensor nodes when the GSM connectivity is not available. **Figure 4** shows the architecture of the proposed gateway node.

2.3. WSN RF Transceiver

The WSN based on ZigBee was adopted because of lower cost and power consumption. It provides a data rate of 250 kbps at 2.4 GHz [7]. It thus allows small, low-cost devices to transmit quickly the small amount of data

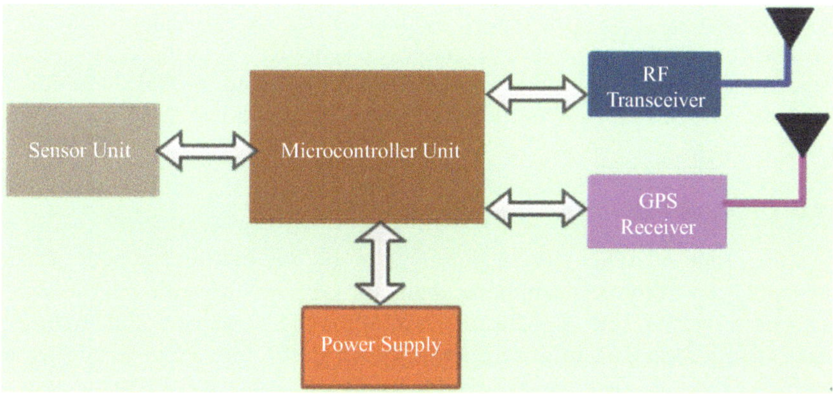

Figure 3. Sensor node architecture.

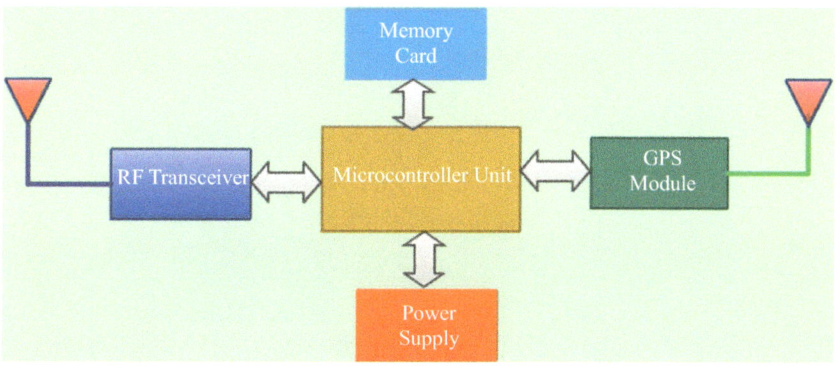

Figure 4. Proposed WSN gateway node architecture.

such as sensor readings. To provide wireless link between WSN sensor nodes and WSN gateway node, XBee Pro Series 2 module from Digi was used. The XBee product family is a series of modular products that make deploying wireless technology easy and cost-effective based on the IEEE 802.15.4/ZigBee. The module was configured to operate in a point to multipoint topology in sleeping-mode function because of its low latency communication between the remote WSN sensor nodes and the WSN gateway node. The XBee on sensor nodes was programmed to be an END DEVICE in AT mode while the XBee on the gateway node was set to be a COORDINATOR in AP mode. The configuration was done using X-CTU software developed by Digi for configuring and testing RF products modems [8].

2.4. WSN RF Power Supply

In a practical WSN system for water quality monitoring, where sensor nodes are distributed in remote sites, power supply is an extremely important issue to consider. The only viable method to power the sensor nodes in WSN application is to use battery. The advantages are evident, but batteries have limited lifespan and cannot stand for a long time. Replacing depleted batteries regularly is difficult. To avoid unnecessary work and make the system more flexible to deploy, different energy harvesting schemes like solar panel can also be used to recharge the battery. In the proposed prototype, a 3.7 V 6 AH rechargeable polymer lithium ion battery was used to power the WSN nodes. This battery has longer life and excellent long-term and self-discharge rates. However it provide less than 5 V required by most components on the WSN sensor nodes and gateway node. Therefore, a d.c power booster was used to boost the output from a battery voltage to 5V as depicted in **Figure 5**. This d.c booster solar charging control also provides the switching button for switching ON and OFF the WSN nodes. It further allows charging the lithium battery through USB or Solar Panel. A 10 W solar panel was used to provide continues battery charging in day time.

2.5. WSN Application Software

The WSN system software includes the WaGoSy system database used to store the received data and provides for visualization of information. The visualization module allows users to visualize the data from the WSN sensor nodes without installing specific software. It enables stakeholders to visualize, map and compare water quality parameters and locations in graphical representation. This software module was developed using MySQL, PHP, HTML and Java Scripts. **Figure 6** and **Figure 7** show the complete WSN node and WSN gateway modules of the system.

3. Results and Discussion

The main actor in the proposed WSN system prototype is the wireless sensor node. During measurement, the WSN sensor node wakes up from the sleep mode, waits for GPS receiver to fetch the latest geospatial information and latest data from sensors. It then passes the data to the WSN gateway node before going back into sleep mode. The process of acquiring data from sensors is shown in the flow chart in **Figure 8(a)**. A software programme was developed and uploaded into the microcontroller to allow the WSN sensor nodes to measure water quality parameters and GPS data at time intervals of 20 minutes. The rest of the time, the sensor nodes are in sleep mode to conserve battery power. The time interval can be changed depending on need, especially when

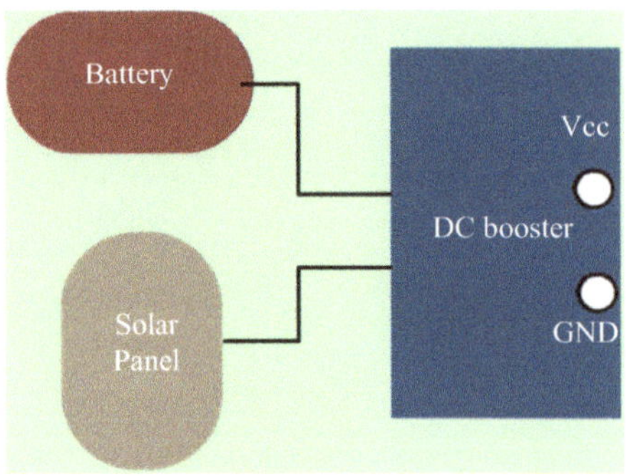

Figure 5. Power supply module.

(a) (b) (c) (d)

Figure 6. (a) WSN sensor nodes in lab testing; (b) WSN sensor nodes on WaGoSy sensor carrier board; (c) A complete WSN sensor node in an open cover; (d) A WSN sensor node in a closed cover.

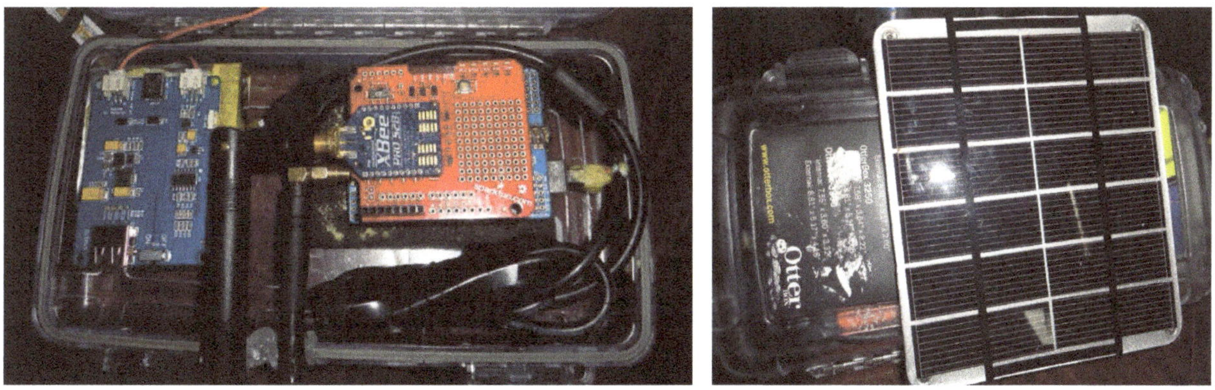

(a) (b)

Figure 7. (a) A gateway node in an open cover; (b) A gateway node in a closed cover.

power needs to be conserved for longer battery life. Once the measurements are done, the WSN sensor nodes relay the captured data through the XBee transceivers to the WSN gateway node. The gateway then aggregates these data and sends to the WaGoSy System as SMS for data analysis. **Figure 8(b)** shows the process of receiving data from sensors and sending to the database.

The system was tested in the field and it worked perfectly. Field testing of the system verified the functionalities of the proposed system and its practical application in actual environment as shown in **Figure 9**. The proposed solution gives water resources stakeholders' ability to test water quality parameters at selected sensitive sites continuously. Examples of such sites include domestic water sources, domestic water intake points and industrial water discharge points. The captured dataset is sent to the central database and can be accessed through the web-based portal or via mobile phone through SMS in graphical or tabular forms as shown in **Figure 10**. It also promotes more accountability and transparency as data collected are available to all stakeholders. The system cost, including sensors nodes and gateway stands at US \$1250. This is in stark contrast to commercial water quality monitoring systems US \$3400 [4], which is a threefold reduction in cost. From the results shown above, with the use of low-cost and open-source hardware, we can achieve acceptable sensor readings.

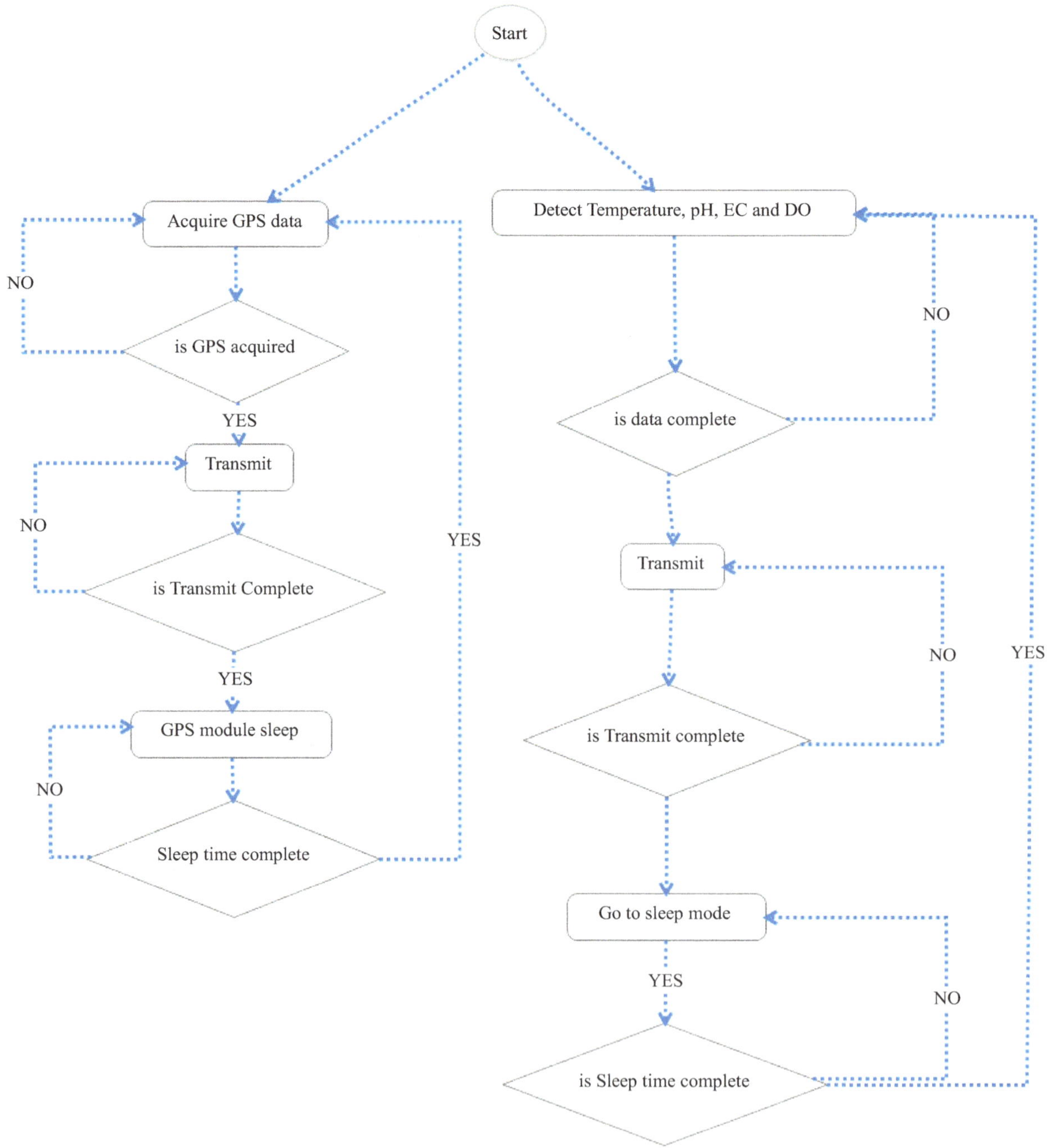

(a)

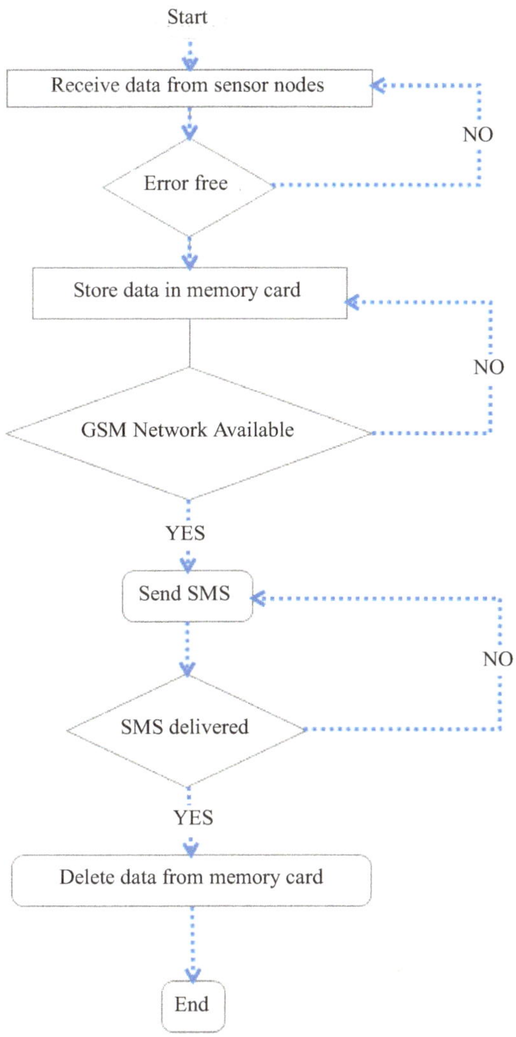

(b)

Figure 8. (a) Flowchart of acquiring data from sensors; (b) Flowchart of data processing at the gateway.

(a) (b)

Figure 9. Field testing of the WSN system (a) Field assistant installing the sensor nodes; (b) Sensor nodes taking measurement.

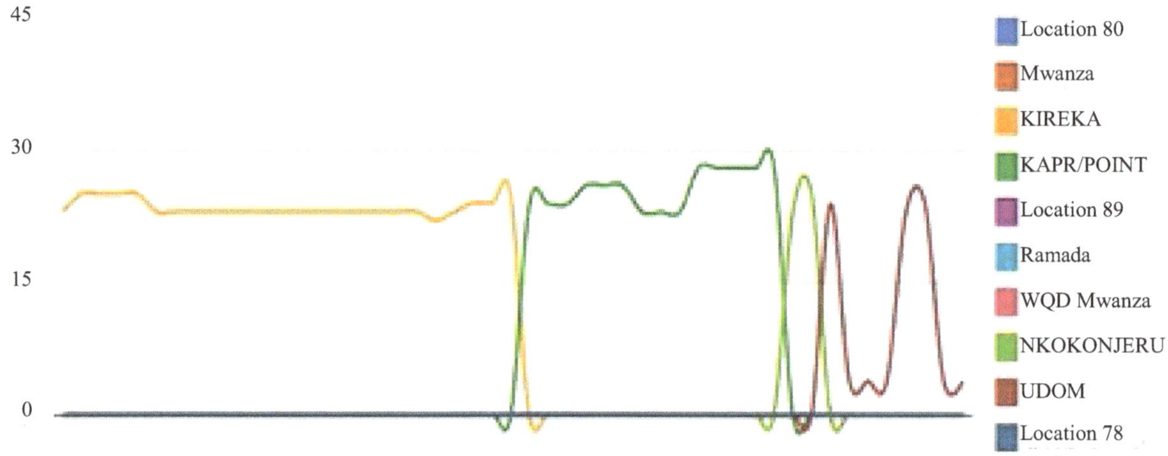

(a)

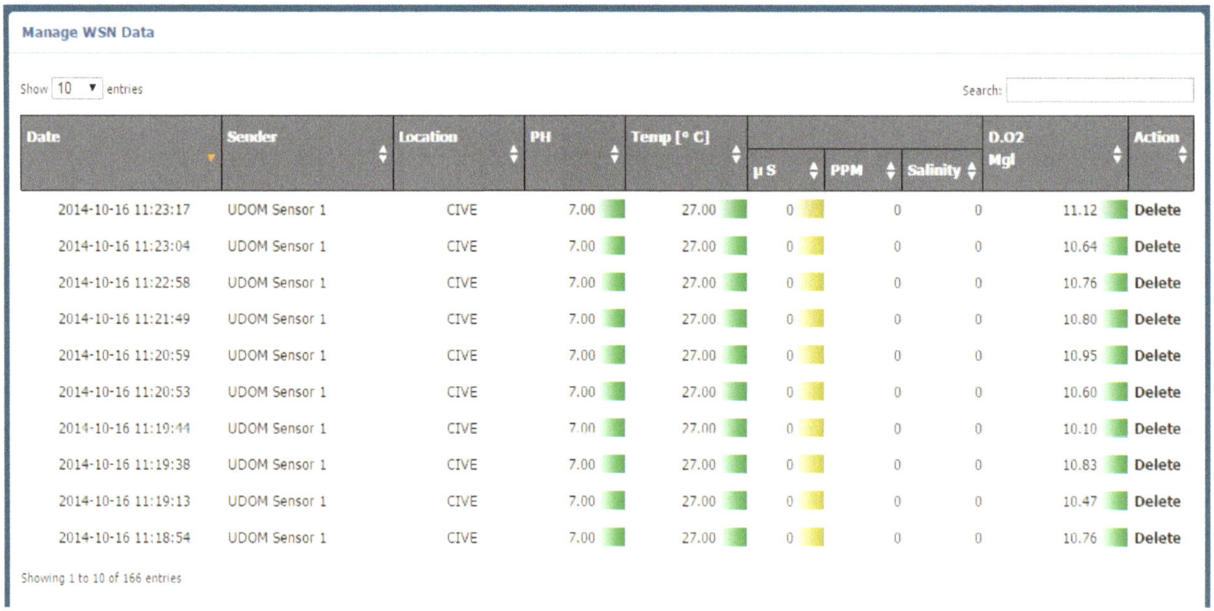
Manage WSN Data

Show 10 entries Search:

Date	Sender	Location	PH	Temp [° C]	µS	PPM	Salinity	D.O2 Mgl	Action
2014-10-16 11:23:17	UDOM Sensor 1	CIVE	7.00	27.00	0	0	0	11.12	Delete
2014-10-16 11:23:04	UDOM Sensor 1	CIVE	7.00	27.00	0	0	0	10.64	Delete
2014-10-16 11:22:58	UDOM Sensor 1	CIVE	7.00	27.00	0	0	0	10.76	Delete
2014-10-16 11:21:49	UDOM Sensor 1	CIVE	7.00	27.00	0	0	0	10.80	Delete
2014-10-16 11:20:59	UDOM Sensor 1	CIVE	7.00	27.00	0	0	0	10.95	Delete
2014-10-16 11:20:53	UDOM Sensor 1	CIVE	7.00	27.00	0	0	0	10.60	Delete
2014-10-16 11:19:44	UDOM Sensor 1	CIVE	7.00	27.00	0	0	0	10.10	Delete
2014-10-16 11:19:38	UDOM Sensor 1	CIVE	7.00	27.00	0	0	0	10.83	Delete
2014-10-16 11:19:13	UDOM Sensor 1	CIVE	7.00	27.00	0	0	0	10.47	Delete
2014-10-16 11:18:54	UDOM Sensor 1	CIVE	7.00	27.00	0	0	0	10.76	Delete

Showing 1 to 10 of 166 entries

(b)

Figure 10. Data Visualizations: (a) Variations patterns representation of WSN; (b) Tabular and graphical form representation of WSN dataset.

4. Conclusion

In this work, the development and demonstration of a low cost, continuous water quality monitoring WSN system prototype is described. The system uses low cost sensors and open source hardware aiming at providing continuous water quality measurements at substantially reduced cost. Preliminary field test results demonstrated that with appropriate calibration, the system has capability to continuously measure water quality parameters and transmit them to a database in real-time. With the use of web-based portal and mobile phones platforms, the values of measured parameters are displayed in easy-to-comprehend graphical and tabular formats anytime and anywhere. This capability has potential to enhance accountability, transparency, and participation which are attributes for good governance of water resources. More testing of the prototype is needed to identify areas that will require fixing.

Acknowledgements

Research team acknowledges Lake Victoria Research Initiative (VicRes), Inter-University Council for East Africa (IUCEA) and Swedish International Development Agency (SIDA) for supporting the research of which this paper is one of the outputs. The team also acknowledges the kind and material support received from stakeholders at different levels in the Lake Victoria Basin during the entire period of the research. The team wishes to thank the management and staff of Busitema University, Kigali Institute of Science and Technology and the University of Dodoma for their administrative support.

References

[1] Sridharan, S. (2014) Water Quality Monitoring System Using Wireless Sensor Network. *International Journal of Electronic Communications Engineering Advanced Research*, **3**, 399-402.

[2] Losilla, F., Garcia-Sanchez, A.-J., Garcia-Sanchez, F., Garcia-Haro, J. and Haas, Z.J. (2011) A Comprehensive Approach to WSN-Based ITS Applications. *Sensors*, **11**, 10220-10265. http://dx.doi.org/10.3390/s111110220

[3] Chaamwe, N., Liu, W. and Jiang, H. (2010) Wireless Sensor Networks in the Context of Zambia: A Developing Country. 2010 *Second International Conference on Information Technology and Computer Science* (*ITCS*), Kiev, 24-25 July 2010, 474-478. http://dx.doi.org/10.1109/ITCS.2010.122

[4] Zennaro, M., Floros, A., Dogan, G., Sun, T., Cao, Z., Huang, C., Bahader, M., Ntareme, H. and Bagula, A. (2009) On the Design of a Water Quality Wireless Sensor Network (WQWSN): An Application to Water Quality Monitoring in Malawi. 2009 *International Conference on Parallel Processing Workshops*, Vienna, 22-25 September 2009, 330-336. http://dx.doi.org/10.1109/ICPPW.2009.57

[5] Kumar, R.K., Mohan, M.C., Vengateshapandiyan, S., Kumar, M.M. and Eswaran, R. (2014) Solar Based Advanced Water Quality Monitoring System Using Wireless Sensor Network. *International Journal of Science*, *Engineering and Technology Research* (*IJSETR*), **3**, 385-389.

[6] Rao, A.S., Marshall, S., Gubbi, J., Palaniswami, M., Sinnott, R. and Pettigrove, V. (2013) Design of Low-Cost Autonomous Water Quality Monitoring System. 2013 *International Conference on Advances in Computing*, *Communications and Informatics* (*ICACCI*), Mysore, 22-25 August 2013, 14-19.

[7] Boonsong, W. and Ismail, W. (2014) Wireless Monitoring of Household Electrical Power Meter Using Embedded RFID with Wireless Sensor Network Platform. *International Journal of Distributed Sensor Networks*, **2014**, Article ID: 876914.

[8] XCTU Software. http://www.digi.com/products/wireless-wired-embedded-solutions/zigbee-rf-modules/xctu

12

Context Aware Wireless Sensor Network Suitable for Precision Agriculture

Nour Brinis, Leila Azouz Saidane

National School of Computer Science, University of Manouba, Manouba, Tunisia
Email: nour.brinis@ensi-uma.tn

Abstract

Using the Wireless Sensor Networks WSNs in a wide variety of applications is currently considered one of the most challenging solutions. For instance, this technology has evolved the agriculture field, with the precision agriculture challenge. In fact, the cost of sensors and communication infrastructure continuously trend down as long as the technological advances. So, more growers dare to implement WSN for their crops. This technology has drawn substantial interests by improving agriculture productivity. The idea consists of deploying a number of sensors in a given agricultural parcel in order to monitor the land and crop conditions. These readings help the farmer to make the right inputs at the right moment. In this paper, we propose a complete solution for gathering different type of data from variable fields of a large agricultural parcel. In fact, with the in-field variability, adopting a unique data gathering solution for all kinds of fields reveals an inconvenient approach. Besides, as a fault-tolerant application, precision agriculture does not require a high precision value of sensed data. So, our approach deals with a context aware data gathering strategy. In other words, depending on a defined context for the monitored field, the data collector will decide the data gathering strategy to follow. We prove that this approach improves considerably the lifetime of the application.

Keywords

Wireless Sensor Network, Precision Agriculture, Data Collector, Context Aware

1. Introduction

Water consumption in agriculture is the most part of water resources consumption in domestic, industrial and agricultural purposes. So, valuation of agricultural water productivity is important for improving the agricultural water resources utilization, especially in the arid areas [1]. For this reason, the science of precision agriculture has been involved, having the purpose to enhance efficiency in crop farming and water resources. This science consists of precise understanding, estimating and evaluating crops condition with the aim of different parameter readings. WSN (Wireless Sensor Networks) is one of the promising technologies for monitoring crops and doing the suitable readings. A number of studies have been published in the last few years on this theme [2]-[5]. We distinguish between two kinds of applications: reactive WSN applications and data gathering applications. Using a reactive WSN, even the person in the field is replaced by an automatic actuator to do some critical tasks like irrigation and fertilization when sensed parameters achieve a given value; this is called the Wireless Sensor Actuator Network WSAN. In the other case, the actuation task is not considered. So, the WSN has the limited purpose of monitoring the field. In case of reactive WSN, the precision farming system has the following parts:

1) Identification of the sensing agricultural parameters.
2) Identification of the sensing location and deployment of the sensor network.
3) Transferring data from crop field to control station for decision making.
4) Actuation and Control decision based on sensed data.

In the second kind of applications, the actuation part is not considered. These parts are detailed in the following paragraphs.

2. A WSN-Based Precision Farming System

2.1. Identification of the Sensing Agricultural Parameters

This task is performed by the agricultural engineer. It consists of choosing the agricultural parameters to sense in order to achieve the target. Some of the common and most critical measured factors in agricultural WSN are air temperature (T) and relative humidity (RH), soil moisture, salinity, etc. [6] [7]. Of course the decision of what factors to monitor depends on the crop type. In addition to these, there are other factors to measure such as leaf thickness and atmospheric pressure, which depends on the crop being monitored. The option of the sensor types such as Sensirion SHT75 or SHT71 for RH and T depends on factors related to sensor accuracy, resolution, range, power consumption, precision, cost, etc. **Figure 1** present different sensors used for agriculture field. We notice that some platforms like Mica2 support many kind of sensor boards in order to do different readings at the same time [8].

2.2. Sensor Network Deployment

Every deployment has its own needs imposed by the type of the monitored crop or plant or by other special application. The coverage of the sensor nodes in agricultural WSN must be dense, *i.e.* one sensor node every one square meter (1 m^2) [9] [10]. Dense deployment serves to capture all the necessary measurements in order to have complete and reliable knowledge of the monitored area. For instance, crops have always different climate, known as microclimate. The micrometeorological parameters outside the field are monitored with the installation of weather stations. Besides the aforementioned issues, precision agriculture has maintenance-safety issues. In fact, maintenance of the WSN system must be considered, because of the long-term nature of the deployment. The sensor nodes must be put into protective cases preventing them from moisture, mud, etc. These cases have ratings in form of IP00, which means no protection. The first digit means protection against solid objects, while the second one means protection against liquids and every level has its definition. For example, in IP67, the 6 digit means total protection against dust and the 7 digit means protection against the effects of temporal immersion till 1 m underwater [11].

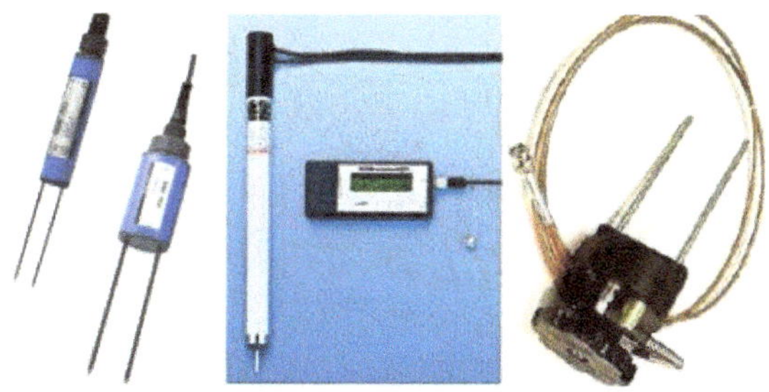

Figure 1. Agricultural sensors.

2.3. Transferring Data from Crop Field to Control Station

Many agricultural deployment strategies assume the sensor nodes sensing and sending data to be time-driven, such as in [12]-[14], in order to acquire a complete picture of the crop circumstances and act accordingly. There are other sensing strategies like event-driven, which are mostly used when monitoring phenomena such as volcanoes, earthquakes, etc. In these cases the monitored subject is the event itself. Finally, the sensing task can be done on-demand based on the user/application requirements. The sensing intervals vary from one minute to one hour.

2.4. Actuation and Control Decision

Actuators like sprinklers, foggers, and valve controlled irrigation system are available and are controlled by the base station.

As a delay tolerant application, precision agriculture may rely on a data gathering strategy better than a tree-routing technique. In particular, we are interested in designing data gathering protocols in WSNs to increase the lifetime of the system, making it more energy efficient. Many data gathering schemes have been proposed for wireless sensor networks. Our data gathering strategy inspires from the strategies presented in the following paragraphs.

3. Data Gathering Schemes

3.1. EDGM: A Single Hop Gathering Strategy Using Data Mules

The goal of this strategy [14] is to determine the optimal number of data mules required to optimize the covered network lifetime while meeting a given data Delivery Delay Deadline *DDD*. The goal of each sensor node is restricted to capture data and then transmit them to a one-hop node: directly to the sink, if this node is located in its transmission range, or to a Data Mule *DM*, when it is one hop away. In other words, sensor nodes do not relay the traffic generated by other nodes. At the proximity of each static node, a *DM* stops at a Break Point, denoted *BPoint*, for a defined Break Period, denoted *BPeriod*, and sends a *HELLO* message. When it arrives at the proximity of the sink and the energy terminal, a *DM* sends the collected data to the sink. Each sensor node *SN* is associated to a unique *BPoint*. The break point positions are defined as the tops of equilateral triangles of edge length $\sqrt{3}rc$ with *rc* is the communication range of sensor nodes [15]. Each *SN* remains awake only during its break period to send its captured data. Then, it turns off its communication unit. During the break period, since several sensor nodes are authorized to send data to the *DM*, collisions may occur. A deterministic medium access scheduling is set by the *DM*, in order to avoid the collision problem. More precisely, each node starts sending its sensed data when it receives an invitation delivered by a *DM* through an *INVIT* message which contains the medium access scheduling. During the first tour, each *SN* receives the information needed to synchronize itself with the *DM*s present in the considered area. Once it is synchronized, it keeps awake only during a *DM* break period in its associated *BPoint*. During this time, it is authorized to send its collected data through an *INVIT* message. **Figure 2** presents the data collector trajectory and break points.

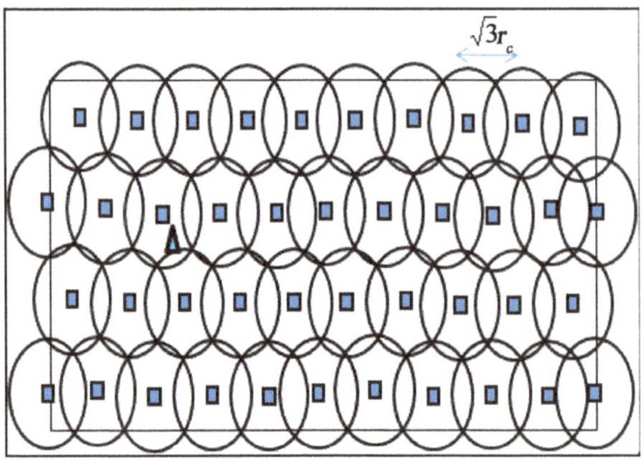

Figure 2. EDGM strategy.

3.2. ADA-DG: Adaptative Delay Aware Data Gathering Strategy

The key idea of ADA-DG [16] consists of finding the suitable trajectory of the Data Collector DC for which we guarantee sending data to the sink before a required data delivery delay deadline. For this purpose, each static sensor node sends their collected data to the DC with incorporating the tolerated data delivery delay deadline DDD. Knowing the DDD, the DC selects the suitable trajectory that allows carrying data to the sink at the appropriate time while minimizing as much as possible the static nodes energy consumption. The suitable trajectory is selected according to a suitable selection of a relay hops bound, denoted Hmax. Therefore, a subset of the whole break points is selected and it is colored Red, denoted *RBPs*. Only the *RBPs* must be visited by the DC. The remaining nodes attached to the remaining break points, colored green, will be affiliated to one of the *RBPs*. So, ADA-DG strategy aims at coloring the sensor network area into red-green disks. This problem is resolved by two steps. The _first step consists of determining the red breakpoints, at which the *DC* stops to gather data. This sub-problem is converted into a logical satisfiability problem. The propositional variable are the set of breakpoints {*BP*1, *BP*2, ..., *BPn*} where *Bpi* is set *true* means that the corresponding breakpoint is red. Knowing, the network graph, we set *n* logical clauses corresponding to the possibilities of each breakpoint to be colored red. For instance, let *Ci* be the clause corresponding to the break point *BPi*:

Ci: BPj or (BPk and BPl)

this constraint means that:

"*BPi* is Green"→ "*BPj* is red" or "*BPk* and *BPl* are red".

Our goal consists of satisfying all the clauses *Ci*. By resolving this problem, the different truth statement defines the different possibilities to construct red-green network. The goal of ADA-DG strategy is to find the trajectory visiting the minimum number of red break points; this means that the required solutions is that having the minimum number of truth values in their truth statement. We may have many solutions. So, the second criterion for selecting the *RBP* consists of minimizing the cost of the selected trajectory, in terms of number of relay hops. Therefore, [16] computes this cost for each truth solution. Then, the word offering the minimal cost is chosen to select the *RBPs*.

At the second step, the DC computes the shortest possible route that visits each *RBP* exactly once and turns back to the *RBP* containing the sink. This route is deducted using an heuristic for the Traveling Salesman Problem TSP.

Figure 3 presents an example of sensor network coloring using the ADA-DG strategy.

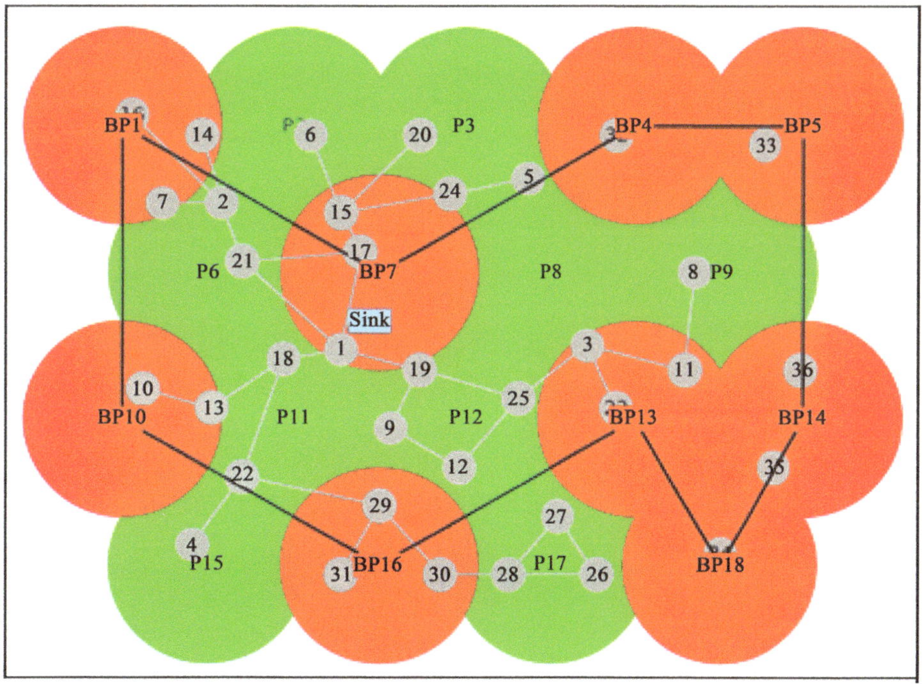

Figure 3. ADA-DG strategy.

Our contribution in this paper consists of including context awareness in data gathering schemes for considering the in-field variability agricultural parcel. In fact, context awareness would incorporate a degree of intelligence in the sensor nodes so that they can decide the action to do, depending on a local context. Since we are dealing with agriculture, two types of contexts may be considered: the first one consists of the sensed data value such as temperature or humidity (current context) and the second one is an implicit input defined for different land conditions such as crop types, location or time of the day.

4. Context-Aware Data Gathering Strategy

In this paper, we deal about the topology illustrated in **Figure 4**. An agricultural parcel is divided into a number of fields, each of them are planted with a type of crop. We suppose that the data collector makes a smart decision about its trajectory depending on each field context. This idea tends to get over supporting a fixed kind of sensor topology for variable fields. In the two following section, we propose two data gathering approaches. Each of them uses a context definition in order to decide the data collector strategy.

4.1. Data Gathering Scheme: Based on ADA-DG Strategy

We now explain how we incorporate context awareness in choosing the data gathering strategy. Our approach is based on the *ADA-DG* strategy. Besides, we suppose that a Field Master *FM* is selected in each field. His role consists of saving the last context of the field and sending it to the Parcel Master *PM* using *WIFI* technology. The field master saves the worst value of captured data, adds the date of capture and sends them to the *P*M. We also note that the data gathering may be done by a data collector robot or any other instrument that visits the field to make insecticide, to harvest crops or any other reason. The decision of the strategy is made only by the *PM* after collecting the n last contexts. The *PM* will visit k fields ($k \leq n$) depending on the time of the last data gathering of each field, the last context and the required delay. Concerning the context model, we inspire from the pyramid model proposed in [16]. The agriculture information space is represented as a concept pyramid with the vertices representing the various contexts of information retrieval (temperature, humidity, land type, crop type). In our methodology, we consider a space level query for the whole agricultural parcel (crop, land type, temperature) and an edge level query for each field (land type, temperature). The *FM* saves the worst context which is defined as the worst values of sensed data (temperature). We denote *DDDj* the required Data Delivery Delay of field j from the k fields to visit. The fields are organized in croissant order of the *DDD*. These fields will be visited by the *DC* under control of the *PM*. The control is done by the *PM* depending on the last received context of each field and their *DDD*. We apply the *ADA-DG* strategy in each field. In other words, the DC goal consists of applying the pre-explained ADA-DG strategy for each part of the field. The DDD of each part is decided by the *PM* based on the last context of each part. **Algorithm 1** resumes the behavior of the DC when receiving the decision from the base station. **Figure 5** illustrates an example of this data gathering scheme application.

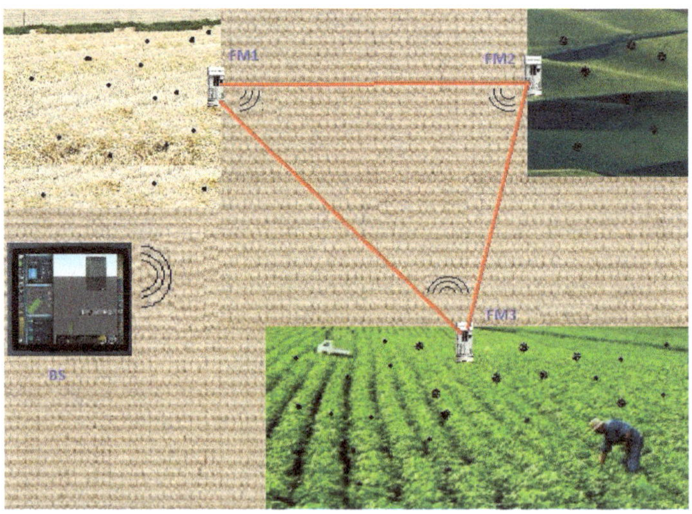

Figure 4. Parcel topology.

Algorithm 1: DC moving path INPUT: the network graph OUTPUT: the DC moving path
1: j: the counter of fields 2. When arriving to a *FM*, update *D*. *D* is the time elapsed from the *PM* decision make until reaching the concerning *FM* 3. $DDD_j = DDD_j - D$ 4. ADA-DG(DDD_j) 5. k is the counter of RBP 6. for each BP_k in RBP do 7. collect data from the associated nodes 8. end for 9. move to the next *FM*

Algorithm 1. DC moving path.

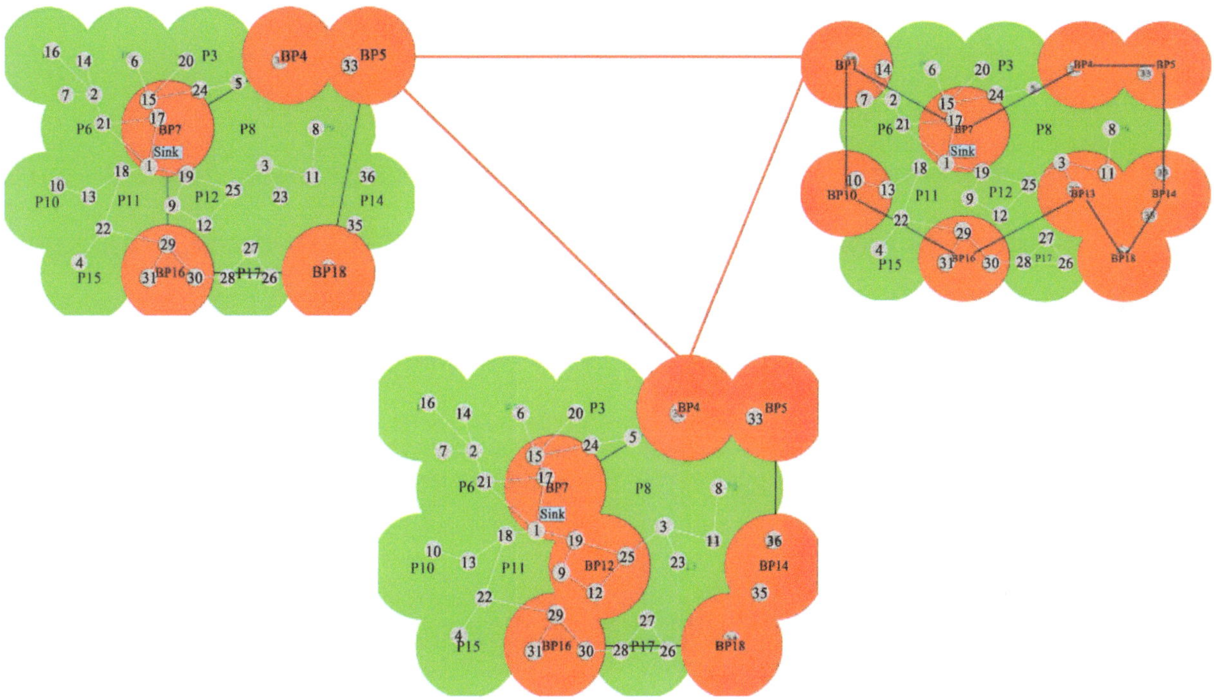

Figure 5. Example: data gathering based on AD-DG strategy.

4.2. Data Gathering Scheme: Based on Islands Creation

In the first approach, we suppose that all the sensors deployed in a field send their data separately to the data collector. However, as a fault-tolerant application, precision agriculture does not require a high precision value of sensed data. So, in order to improve the lifetime of the application, we propose to group the sensor nodes into islands with near sensed values. For instance, in a given land, we may have some permanent variabilities, for example, a hill or plain space, a shadow or light area for a given time-of the day... So, having the same permanent condition of a given part of the field may imply finding very near values of sensed data. Then, sensing data from only one node of a part of the field, having near conditions, may satisfy the precision agriculture application. For this purpose, we propose in this section to group the nodes into islands having near sensed values. In the following paragraph, we define the islands definition. Once the islands are defined, alternatively each node is visited in the same time-of the day. The other nodes sleep in order to preserve their energy. We notice that a trigger of new island creation is defined. In our case, a high difference between the actual and the last sensed data from the same island triggers new islands creation. Other triggers may be defined depending on the type of crops. In the following paragraphs, we explain the island creation algorithm and the corresponding data gathering approach.

4.2.1. Island Creation Algorithm

The goal of this algorithm consists of grouping the sensor nodes having usually near values of sensed data into the same island. For this purpose, deployed sensors exchange the sensed data and compare them with the received values of their neighbors. Neighbor nodes having sensed data, that values are in a given precision interval [*Vmin*, *Vmax*], are considered in the same island. Otherwise, a new island is defined when a neighbor node senses data that values are out of the precision interval. We suppose that network is connected, which means that each node can be reached. The neighboring nodes exchange the following messages:

- *CMsg* (*Check Message*): containing the dynamic sensed data (such as temperature T), the neighbor nodes check the similarity of T with their sensed data (belonging to a given [*Vmin*, *Vmax*] interval).
- *NIMsg* (*New Island Message*): Announce a new island definition.
- *EMsg* (*End Message*): allow the Island Master *IM* to know the node's identifiers of its island.

We denote:

- *Di*: the data sensed by node *i*.
- Ti: the instant when node *i* receives a *CMsg*.
- LTi: the last time when node *i* received a *CMsg*.
- Delay: the maximum delay to construct an island.
- Dbt: the minimum delay between two triggers to reconstruct islands.
- *i* is an end node if it has not descendant neighbors.
- CMsg (n, E, D): a CMsg sent by node n, passing by nodes in the set E. E is initially an empty set. D is the data captured by node n.

The flowchart of the proposed algorithm is shown in **Figure 6**. The first node triggered to launch the algo-

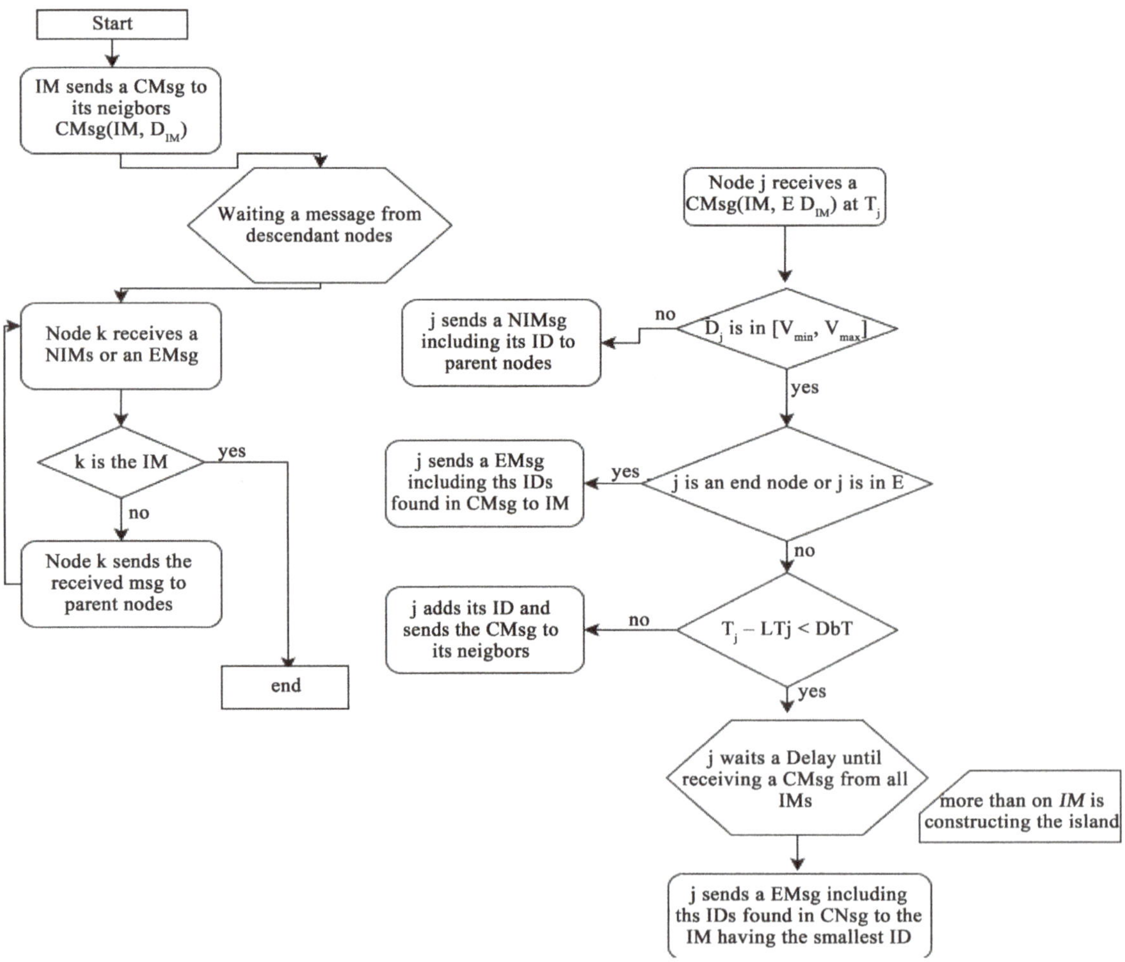

Figure 6. Flowchart of the island creation algorithm.

rithm, is declared *IM*. It sends a *CMsg* (*IM*, *TIM*) containing its *ID* and the sensed data to its neighbors. Each node *j*, receiving a *CMsg*, compares the received data with that sensed. In case of belonging to the precision interval, then if node *j* is neither an end node nor a node in a loop situation, than it sends the data to its neighbors after adding its *ID*. Otherwise, it sends an *EMsg* to the parent node (source node). In the other case, a *NIMsg* is sent to parent node.

An alternative circumstance could occur when more than one *IM* is declared in the process. This situation is marked by receiving more than one *CMsg* from different source nodes within a *Dbt* delay. In that case, the node waits until receiving all *CMsg*s, and then it send an *EMsg* to the *IM*, having the smallest *ID*, after including all nodes known from other received *CMsgs*.

We notice that a node *i* concludes that it has received a *CMsg* (*E*1, *T*) in a loop situation when the last received *CMsg* (*E*2, *T*) is such that $E1 \subset E2$. After performing the algorithm, each *IM* maintains the *IDs* of nodes in its island and the *IMs* of neighbor islands.

4.2.2. Island Creation Example

For the sake of clarity, an example of islands creation with 3 distinct temperature intervals: (27) (29) (30) (32) and (33) (35) is illustrated in **Figure 7**.

Initially, the node 1 sends a *CMsg* including its identifier and the sensed temperature (30) to its neighbors. Each node receiving *CMsg* checks if its sensed data is in the corresponding temperature interval. In that case, it adds its identifier and forwards the *CMsg* to its neighbors. Otherwise, it sends a *NIMsg* to its parent nodes. Nodes receiving a *NIMsg* forward it to the IM. In this case, node 9 senses a data that value is out of the interval (30) (33). So, it is declared *IM* and sends a *NIMsg* to the source node (node 3). This one forwards the message until the *IM* (node 1). Node 9 does not have descendant node. So, it forms an island containing only itself (node 9). Node 6 receives a *CMsg* in a loop situation, and then it will send an *EMsg* to IM (node 4).

4.2.3. Data Gathering Scheme

In this paper, we are interested to precision agriculture applications. In that case, network is not necessarily connected. However, knowing that the DC maintains the positions of all deployed nodes, we apply the island creation algorithm to each connected part of the network. Then, the DC will visit the different nodes that are declared IM. The first tour of the DC is a discovery tour. In the issue, the DC will maintain the real IM's list in an ascendant order of their identifiers and the identifiers of nodes composing each island.

The *DC* will visit only one sensor of each defined island *IM*. In order to balance the energy consumption between all nodes, *IM* is alternatively selected from island nodes and with the condition that at the same time of the day different *IM* are selected. Knowing the *IM* in each tour, a node can select its tour to be an *IM*. During the

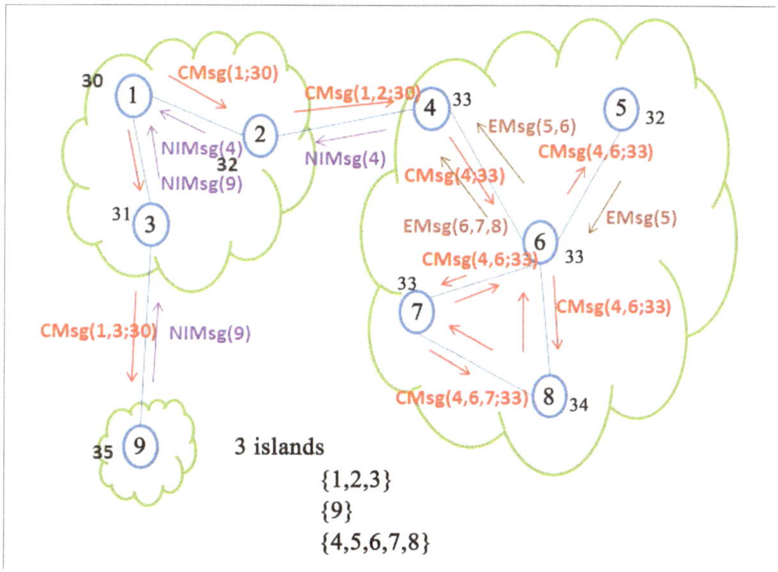

Figure 7. Example of island creation.

other tours, it remains in sleeping mode in order to conserve its energy. The reconstruction of islands is triggered by the *DC*.

5. Simulations and Results

In this section, we present simulation results to evaluate the efficiency of the proposed data gathering strategies. We carried out several simulations using NS2 simulator.

5.1. Simulation Parameters

Table 1 summarizes the simulation networking parameters.

5.2. Simulation Results

We analyze the strategies efficiency in the two following case:

- case 1: using the first approach for which we suppose that readings are done every 1 hour when the temperature goes on or above the threshold level, here it is 30 degree Celsius. Otherwise, readings are done every 2 hours.
- case 2: using the first approach, readings are done every 1 hour whatever the previous sensed temperature.

In **Figure 8**, we plot the average residual energy. We observe a minimum gain of 25% of energy in the first

Table 1. Simulation parameters.

Network configuration	Number of fields	3
	Fields area	(200 m × 200 m), (300, 300), (400, 400)
	Rc	50 m
	Bandwidth	2 Mbps
	Mac protocol	802.11
	DC speed	1 m/s
Energy model	Type	Battery
	Initial energy	100 Joules
	Transmit power	0.36 watt
	Receive power	0.24 watt
	Sensing power	0.015 watt

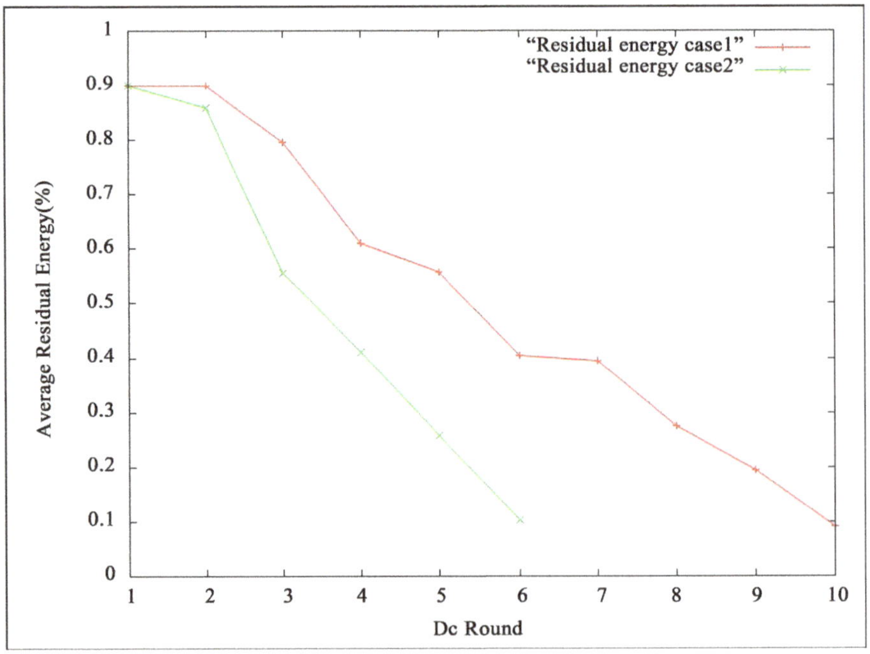

Figure 8. Average residual energy—first approach.

case. So, when introducing context awareness based on the dynamic parameter temperature, we can gain in terms of network lifetime. In the first case, the DC can do more than 10 tours before sensors failure; however in the second case only 6 tours are done.

Similarly, in **Figure 9**, we compare the first approach case 2 and the second approach using readings every 1 hour. We observe that the second approach is more relevant with making sleep some nodes of a given field. In fact, only one sensor of the island, IM, must sense and send the sensed data to the DC, the other island nodes sleep.

In this section, we evaluate the effect of the node islands creation on the efficiency of the solution. **Figure 10**

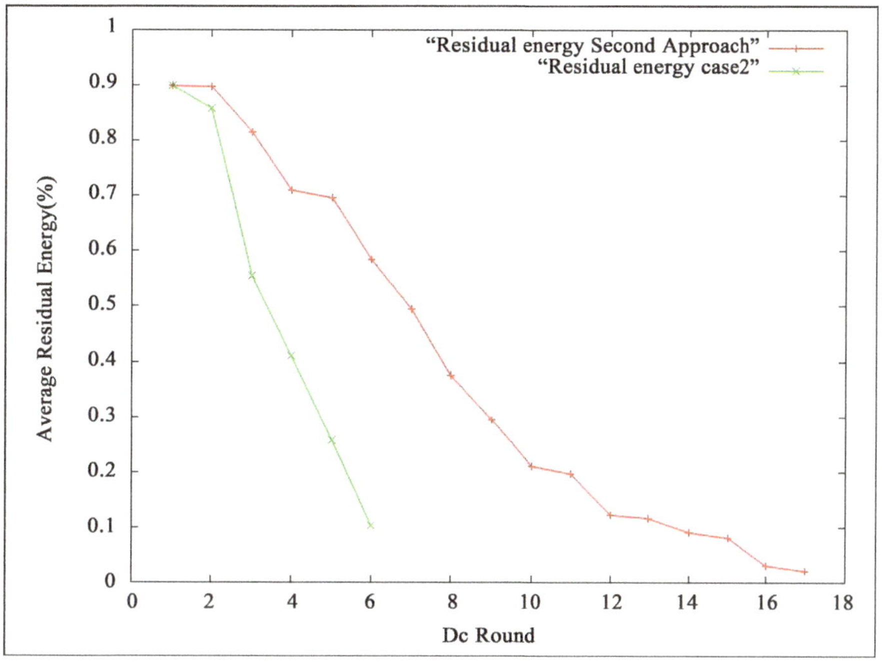

Figure 9. Average residual energy—first and second approaches.

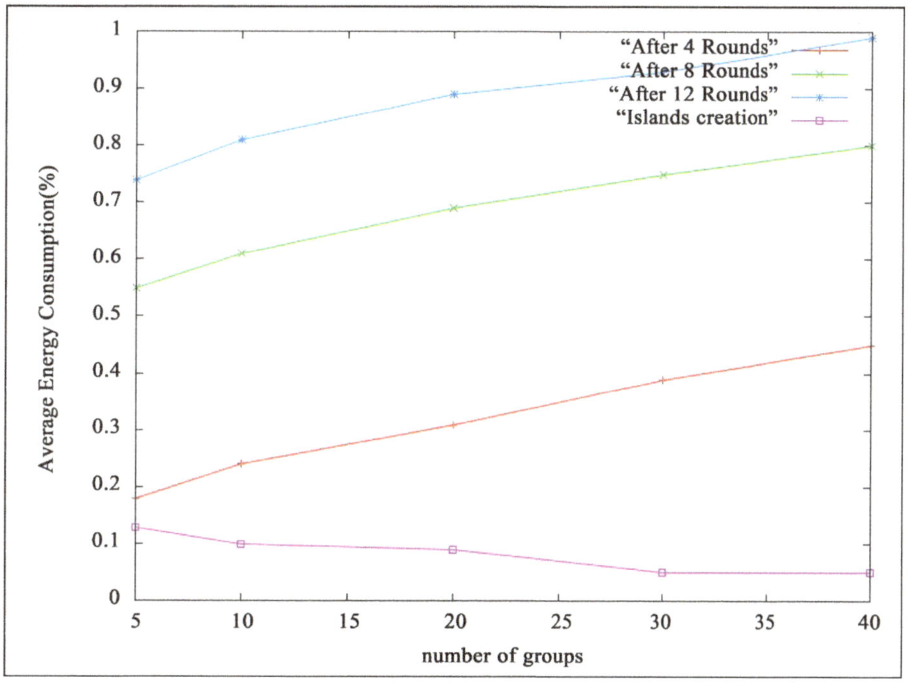

Figure 10. Average energy consumption.

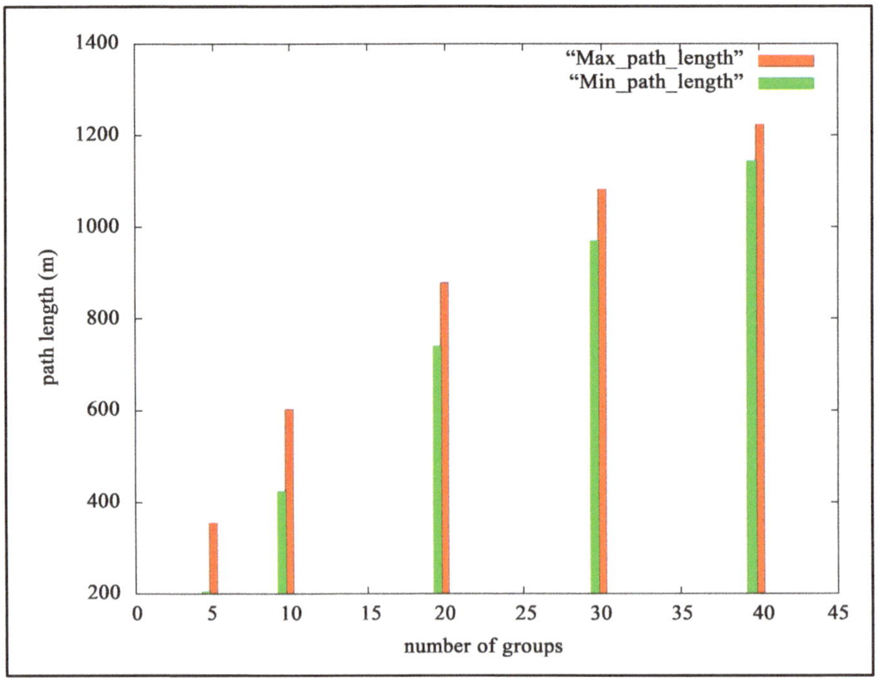

Figure 11. Maximal and minimal path length.

presents the energy consumed rate after 4, 8 and 12 tours. Besides, we present the energy consumed for island creation. Obviously, we found that the energy increase with the number of tours. Moreover, when the number of groups increases, the corresponding number of island masters increase. So, we observe that the energy consumed will increase with the number of active nodes representing the island master nodes.

Figure 11 depicts the maximal and minimal path length according to the number of groups. This result has an effect on the data delivery delay from sensed nodes to the base station. The path length increases with the number of islands. However the gap between minimal and maximal path length decrease when the number of islands increases. In fact, regrouping more nodes in the same island means that the distance between nodes increase, so, the island master selection will affect considerably the path length.

6. Conclusion

The aforementioned approaches have incorporated context awareness for gathering data from an agricultural parcel. Uniform and dynamic parameters are defined. Depending on these parameters, data gathering strategy is defined with optimum energy preservation. Our approaches have been evaluated based on some simulations scenarios but the feasibility of the system and their productivity cannot be proved without real system tests. In order to prove their feasibility, our future work consists of modeling the system with an adequate analytical approach.

References

[1] Martinelli, M., Ioriatti, L., Federico, V., Benedetti, M. and Massa, A. (2009) A WSN Based Solution for Precision Farm Purposes. IGARSS, 469-472.

[2] Zhang, W., Kantor, G. and Singh, S. (2004) Integrated Wireless Sensor/Actuator Networks in an Agricultural Application. *Proceedings of the 2nd Academic International Conference on Embedded Networked Sensor Systems* (*Sensys* 04), Acm Press, 317-318. http://dx.doi.org/10.1145/1031495.1031560

[3] Anurag, D., Roy, S. and Bandopadhyay, S. (2008) Agro-Sense: Precision Agriculture Using Sensor-Based Wireless Mesh Networks. *First ITU-T Kaleidioscope Academic Conference*.

[4] Burrell, J., Brooke, T. and Beckwith, R. (2004) Vineyard Computing: Sensor Networks in Agricultural Production. *Sensor and Actuator Networks*, **3**, 38-45.

[5] Lpez Riquelme, J.A., Soto, F., Suardaz, J., Snchez, P., Iborra, A. and Vera, J.A. (2009) Wireless Sensor Networks for

Precision Horticulture in Southern Spain. *Computers and Electronics in Agriculture*, **68**, 25-35. http://dx.doi.org/10.1016/j.compag.2009.04.006

[6] Garcia-Sanchez, A.-J., Garcia-Sanchez, F. and Garcia-Haro, J. (2011) Wireless Sensor Network Deployment for Integrating Video-Surveillance and Datamonitoring in Precision Agriculture over Distributed Crops. *Computers and Electronics in Agriculture*, **75**, 288-303. http://dx.doi.org/10.1016/j.compag.2010.12.005

[7] Liu, Y., Zhang, J. and Richards, M. (2009) Towards Continuous Surveillance of Fruit Flies Using Sensor Networks and Machine Vision. *Proceedings of 5th International Conference on Wireless Communications, Networking and Mobile Computing*, 1-5.

[8] Liu, H., Meng, Z. and Wang, M. (2009) A Wireless Sensor Network for Cropland Environmental Monitoring. *Proceedings of the IEEE International Conference on Networks Security, Wireless Communications and Trusted Computing* (*NSWCTC* 09), Wuhan, 25-26 April 2009. http://dx.doi.org/10.1109/nswctc.2009.306

[9] Goumopoulos, C., Kameas, A. and O'Flynn, B. (2007) Proactive Agriculture: An Integrated Framework for Developing Distributed Hybrid Systems. *Proceedings of the 4th International Conference on UIC*, 214-224. http://dx.doi.org/10.1007/978-3-540-73549-6_22

[10] Dursun, M. and Ozden, S. (2011) A Wireless Application of Drip Irrigation Automation Supported by Soil Moisture Sensors. *Scientific Research and Essays*, **6**, 1573-1582.

[11] Ayday, C. and Safak, S. (2009) Application of Wireless Sensor Networks with GIS on the Soil Moisture Distribution Mapping. *Proceedings of the GIS Ostrava Symbosium*.

[12] Cao, X., Chena, J., Zhang, Y. and Sun, Y. (2008) Development of an Integrated Wireless Sensor Network Micro-Environmental Monitoring System. *ISA Transactions*, **47**, 247-255. http://dx.doi.org/10.1016/j.isatra.2008.02.001

[13] Brinis, N., Azouz Saidane, L. and Minet, P. (2012) EDGM: Energy Efficient Data Gathering with Data Mules in Wireless Sensor Networks. *6th International Conference on Sensor Technologies and Applications*, *SENSORCOMM* 12, Rome, 19-24 August 2012.

[14] Brinis, N., Saidane, L.A. and Minet, P. Adaptative Delay Aware Data Gathering Strategy for Wireless Sensor Networks. *ICAIT*12.

[15] Commuri, S. and Watfa, M.K. (2006) Coverage Strategies in Wireless Sensor Networks. *International Journal of Distributed Sensor Networks*, **2**, 333-353. http://dx.doi.org/10.1080/15501320600719151

[16] Ingole, P., Pawar, J. and Vhatkar, H. (2013) Context Aware, Mobile Phone Based Sink Model of Wireless Sensor Network for Indian Agriculture. *Advances in Computing*, *Communication and Control Communications in Computer and Information Sciences*, **361**, 301-308. http://dx.doi.org/10.1007/978-3-642-36321-4_28

13

Congestion Control in Wireless Sensor Networks Based on Bioluminescent Firefly Behavior

Mukhdeep Singh Manshahia[1*], Mayank Dave[2], Satya Bir Singh[1]

[1]Department of Mathematics, Punjabi University, Patiala, India
[2]Department of Computer Engineering, National Institute of Technology, Kurukshetra, India
Email: *mukhdeep@gmail.com

Abstract

Congestion in Wireless Sensor Network (WSN) is an issue of concern for several researchers in recent years. The key challenge is to develop an algorithmic rule which may realize the optimased route on the idea of parameters like residual energy, range of retransmissions and the distance between source and destination. The Firefly Algorithmic rule is implemented in this paper that relies on the attractiveness issue of the firefly insect to control congestion in WSN at transport layer. The results additionally show that the projected approach is best as compared to Congestion Detection and Avoidance (CODA) and Particle Swarm Optimization (PSO) on network lifetime and throughput of the network.

Keywords

Congestion Control, WSN, Firefly Algorithm

1. Introduction

A Wireless Sensor Network is formed by spatially distributing low powered sensor nodes. Due to the recent developments in low powered tiny sensor technologies, the sensor nodes are used in wide range of applications in environmental monitoring [1]. During event detection, congestion can occur while transferring the data from source to sink. Congestion control in WSNs means to improve the performance when demand for the finite transmission capacity exceeds the supply. Wireless sensing nodes have bound constraints due to which the wireless implementation of sensing protocol is not easy. One of the main constraints among them is the energy

*Corresponding author.

of each node [2]. Energy is an important parameter as it decides the network lifetime of each node [3]. Once a sensing element node transmits knowledge onto the base station, the energy loss incurred may be quite intensive depending on the placement of the sensing element nodes relative to the base station. Energy constraints also make design of the WSN protocols very complex. Therefore, minimizing congestion and the energy consumption is the key demand among the sensor network protocols and algorithms.

The sensor networks are self-healing, so a node can be inserted or removed without having to restart the network. It is possible to have a high sampling rate over large spatial scales. This high sampling rate enables changes to be seen clearly. It is possible to have many users viewing the data simultaneously, and also manipulating it [4]. High sampling rate also leads to a congestion situation.

Wireless sensing element networks have lots of potential application areas like oceanographic knowledge assortment, home networking, disaster management, pollution watching, offshore exploration, aircraft control and health management systems [4], precision agriculture, traffic control, remote environment monitoring, target tracking, power monitoring, inventory location monitoring, marine environment monitoring, factory and process automation, crime investigation and military applications [5]-[7]. Wireless Sensor Network may be a new space of wireless sensing element network in underwater environments that has challenges to be overcome like long propagation delay, severely restricted information measure, and time varied multi-path propagation [8].

The main aim of the paper is to propose an approach to control congestion in wireless sensor networks at transport layer. Section 2 of the paper discusses about the brief literature study; Section 3 formulates the problem. Section 4 proposed methodology to solve the problem is shown. Section 5 explains the simulation environment, after that results and conclusion are discussed in Sections 6 and 7 respectively.

2. Literature Review

Fister *et al.* [9] proposed a comprehensive study of the living and evolving discipline of Swarm Intelligence, and their application in algorithmic programs. On the other hand, it encourages new researchers and algorithmic program developers to use this straightforward and nonetheless terribly economical algorithmic program for drawback determination. It typically guarantees that the obtained results can meet the expectations.

Sarma *et al.* [1] proposed energy economical cluster formation of sensor nodes using jumper firefly. Firefly Algorithms have been proposed to minimize the intra-cluster distance in order to optimize the energy consumption of the network. The performance of the network is then compared with the LEACH protocol.

Xu, Ming *et al.* [2] proposed energy-efficient algorithms to cut back energy consumption of Wireless Sensor Networks to some extent by neglecting the correlation existed between the native knowledge of nodes. A multi-population Firefly Algorithm is proposed in this work to correlate the knowledge in Wireless Sensor Network.

Schenato *et al.* [3] proposed a replacement of consensus-based protocol, remarked as Average TimeSynch, for synchronizing the clocks of a wireless device network. This algorithm is a combination of two algorithms whose main task is to average native data. The projected algorithmic program has the advantage of being wholly distributed, asynchronous, strong to packet drop and device node failure, and it's adaptation to time-varying clock drifts and changes the communication topology.

Sai Prakash *et al.* [10] proposed a firefly inspired tree formation in wireless sensor networks. In this cluster based energy aware technique the responsibility of cluster head is distributed among nodes to distribute the energy drain factor. The proposed approach has shown improvement in the network lifetime and minimizes the partitioning problem.

Kumar, E. Sandeep *et al.* [8] proposed Wireless sensing element networks encompass little nodes with sensing, computation, and wireless communications capabilities. In this work Firefly Algorithm is used in clustering of the sensor nodes in the network.

Sun, Yi *et al.* [11] proposed a clustering scheme for the network which employs Firefly Algorithm. The clustering of the network is considered on the basis of parameters considered together which includes energy and distance and a reach-back technique is employed for clustering of the sensor network.

3. Problem Formulation

The problem of congestion control in WSN is a very wide area of research. From the past few years many researchers use various algorithms to solve the same problem. Among the many reasons of congestion, Buffer overflow, Channel contention and packet collision are considered as the major focus areas. A fitness function is considered based on various parameters like throughput of the network, packet lost rate and residual energy. These parameters are:

- Residual Energy of the network: Residual Energy is an important factor in calculating the network lifetime of a node in the network. The energy exhausted by the node in communicating with other nodes or the in idle mode is used. Residual Energy is:

$$\text{Residual Energy} = \text{Total Energy} - \text{Energy Used}$$

- Distance between the nodes: Distance between two nodes communicating with each other at a particular time is calculated. The distance is an important factor as the energy consumed in transmitting a packet directly depends on the distance between the nodes. Distance is calculated using the Euclidean Formula which is given by:

$$\text{Distance} = \sqrt{(x_1 - x_2)^2 + (y_1 - y_2)^2}$$

- Normalized Packet Loss Rate: The packet lost while transferring packets from one node to another is also an important factor in detection of congestion. Queue length of the node and the packet collision in the network are two main reasons of packets loss in the network. The packet loss rate must be minimized for improving the performance of the congestion detection and control algorithm.
- Throughput of the network: Throughput is defined as the ratio of packet size and the delay. It is an important parameter which provides the information about the packet transmissions per unit time taken along a fixed path.

$$\text{Throughput} = \frac{\text{Packet Size}}{\text{Receive Time} - \text{Send Time}}$$

To improve the performance of the network and to reduce the congestion we need to optimize the fitness function.

$$\text{Fitness Function}\left(f_j\right) = \sum_{i=1}^{N}\left(w_1 * \tau_i + w_2 * \left(1 - P_{Li}\right) + w_3 * E_i + w_4 * d_{j,i}\right)$$

where, i is the iteration which ranges from 1 to N (total number of nodes).

w_1, w_2, w_3 and w_4 are the weights supplied to the algorithm.

τ is the throughput of the network.

P_L is the normalized Packet lost rate in the network.

$d_{j,i}$ is the distance between node i and j.

E is the residual energy of each node in the network.

4. Proposed Methodology

Reactive routing technique [12] is used in the proposed work. In reactive routing the route is computed on demand basis. When a node wishes to transmit data then it will generally generate RREQ message which is a route request message and is flooded in the network. It will result in a delay in initiating the communication from source to destination. In reactive routing the route is considered in which the RREQ message reaches earlier through either single hop or multiple hops.

A Firefly Algorithm based approach is proposed to control congestion in WSN at transport layer. Firefly insect produces flashes of short duration through a process called bioluminescence. It is used to attract potential prey or partner or for the issue of warning against predator. Thus intensity of flash becomes an important parameter for the other firefly insects. Firefly Algorithm follows three rules [13]:

1) Fireflies of any gender can attract towards the other firefly;

2) An attractiveness factor is taken into account which depends on the brightness of the flash, as the fireflies' move towards the more attractive fireflies;

3) The brightness of fireflies is calculated through an objective function.

Structure of Firefly Algorithm

In Firefly Algorithm, light intensity and attractiveness are the two important variables. Firefly is attracted toward the other firefly that has brighter flash than itself. The attractiveness is dependent with the light intensity.

Firefly Algorithm

```
Begin:
Initialize algorithm parameters:
MaxGen: the maximal number of generations
γ: the light absorption coefficient
r: the particular distance from the light source
d: the domain space
Define the objective function of f(x), where  x=(x1, ...,xd)
Generate the initial population of fireflies or xi (i=1,2 ,..., n)
Determine the light intensity of Ii at xi via f(xi)
While (t<MaxGen)
        For i = 1 to n (all n fireflies);
                For j=1 to n (n fireflies)
                        If (Ij> Ii),
 move firefly i towards j by
      end if
      Attractiveness varies with distance r via Exp [-γr^2];
      Evaluate new solutions and update light intensity;
 End for j;
        End for i;
       Rank the fireflies and find the current best;
End while;
Post process results and visualization;
End procedure
```

The intensity of the flash light which is also defined as attractiveness is inversely proportional with r distance from the light source. This shows that the attractiveness increases as the distance decreases and *vice-versa.*

$$I(r) = I_0 e^{-\tilde{a} r^2} \tag{1}$$

I = Flash light intensity; I_0 = Initial flash light intensity; $\tilde{a}$ = The light absorption coefficient; r = Distance between firefly i and j.

The flash light intensity is proportional to the attractiveness and is given by β.

$$\beta = \beta_0 e^{-\tilde{a} r^2} \tag{2}$$

β_0 = Attractiveness at r is 0.

Cartesian Distance Formula is used to calculate the distance between two fireflies.

$$r_{ij} = \left| x_i - x_j \right| = \sqrt{\sum_{k=1}^{d} \left(x_{i,k} - x_{j,k} \right)^2} \tag{3}$$

The movement of the less attractive firefly i towards the more attractive one j is given by:

$$\Delta x_i = \beta_0 e^{-\tilde{a} r_{ij}^2} \left(x_j^t - x_i^t \right) + \beta\beta_i, x_i^{t+1} + \Delta x_i \tag{4}$$

In Equation (4), if light absorption coefficient approaches zero then the attractiveness is almost constant otherwise if the coefficient approaches infinity, the attractiveness decreases. The second term in the equation is used for randomization.

In the algorithm described above, light intensity is replaced by Residual Energy of a node, and hence attractiveness ($\hat{a}$) is proportional to energy. $r_{i,j}$ is the distance between two nodes x_i and x_j.

Now, at first the nodes in the network are clustered and each node in the cluster shares its information related to residual energy, its distance from other nodes in the cluster and number of retransmissions of data. This information is used to select the cluster head. After each round this information is updated on each node and re-clustering and cluster head selection is performed.

The bandwidth of the network is limited and each node trying to send its data to the cluster head which further sends its data to the base station. The congestion in the network is occurred because of the data travelling from source to sink. The congestion is detected by monitoring the window size and the queue size of each node. This information is also shared by each node in the network. In case of congestion if the data travelling from source to sink then the nodes lies in the path, which may be either belong to one cluster or to different clusters, have information of other nodes of its cluster and directs the data through other nodes based on the value of queue size and window size. Efforts are being made detect and reduce the congestion of data from each node. The objective is to maximize the throughput of the network by optimizing the objective function.

In firefly based approach (**Figure 1**), the residual energy value plays a major role as this value is shared among other nodes in the network. Number of retransmissions of every node and the distance value between any two nodes in the cluster is calculated. Based on these values and the values of residual energy, a new route is found in the network from source to sink. The nodes which have low energy value attracted towards the high energy node and an attractiveness factor is calculated. The congestion on any node is determined by the same method as described earlier. Based on the remaining queue size value an optimum route is computed by putting the above values in the objective function.

We are comparing our proposed approach with the following two approaches.

CODA: Congestion Detection and Avoidance (CODA) is an energy efficient congestion control mechanism which comprises three schemes [14]:

- Closed-loop multisource regulation;
- Open loop hop-by-hop backpressure;
- Receiver based congestion detection.

MAC layer plays a vital role in the management of data and performance of a CODA based systems. CSMA plays a role in detection and avoidance of congestion.

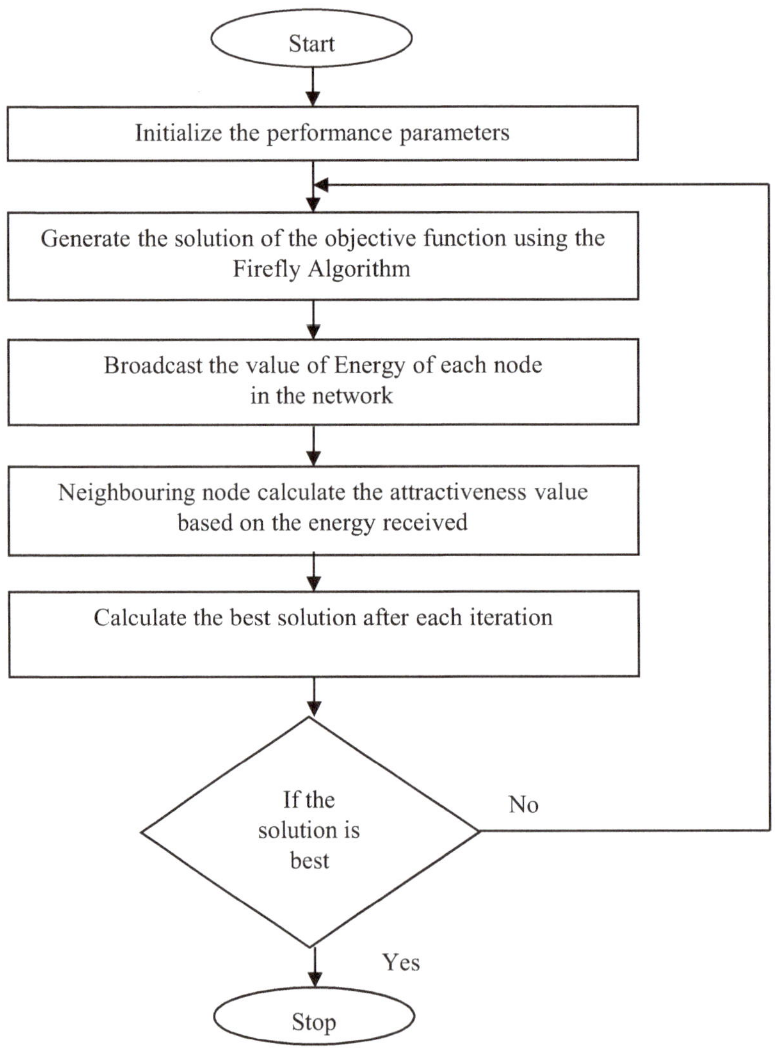

Figure 1. Flow diagram of the proposed methodology.

PSO: Particle Swarm Optimization is a method of finding the mathematical solution of the problem in terms of position and velocity [15]. The PSO iteratively calculates the candidate solution by moving particles in the search space. The particles move from the best known position to the other best position calculated using the formula and updates its value for position. Swarm will follow the path calculated by the particles in the search space.

5. Simulation Environment

For creating the simulation environment, we have taken a 1000 × 1000 in size grid in Network Simulator-ns-2.35 and 50 nodes are placed in the area (grid). As the network is wireless, so we have to assign few wireless network parameters to the node. Each node in the network is configured to wireless channel and Omni-directional antenna type. MAC layer is used in the simulation with standard following IEEE 802.11 and the radio model has been adopted.

6. Results and Discussions

Figure 2 shows that the queue length (number of packets) of each node with respect to the simulation time. The Queue length is inversely proportional to the number of hops and by increasing the number of hops congestion in the network is decreasing. As the number of hops in between source and destination increases, the queue length of each node decreases and packets can travel through another selected nodes.

Figure 3 shows the graph between fitness factor and simulation time for different hop numbers. Fitness factor is the decreasing function. This means that that the optimal route is selected based on the lesser value of fitness function. The fitness function contains various parameters like throughput, packets lost, energy etc. which decreases with the increase in the simulation time. As the number of hops in the network increases the value of the fitness function decreases.

Comparative Analysis

Figure 4 displays the graph between simulation time and throughput of the network. The graph compares Congestion Detection and Avoidance (CODA), Particle Swarm Optimization (PSO) and proposed Firefly Algorithm based approach. In case of Firefly Algorithm the throughput of the network is almost constant and greater than the throughput calculated using the CODA and PSO algorithm. There is a decrease in value for CODA at around 30 seconds of simulation which shows that at the peak of traffic at the middle the performance of the algorithm goes down and packet loss rate in the network increases. **Figure 5** displays the graph between the simulation time and the network lifetime of the nodes in the network. It is clear from the graph that the network lifetime of nodes in case of Firefly Algorithm is greater than CODA and PSO. This is due to the fact that Firefly Algorithm consumes low energy as compared to other algorithms with the increase in simulation time. The following graphs are designed by considering the hop count to be 40.

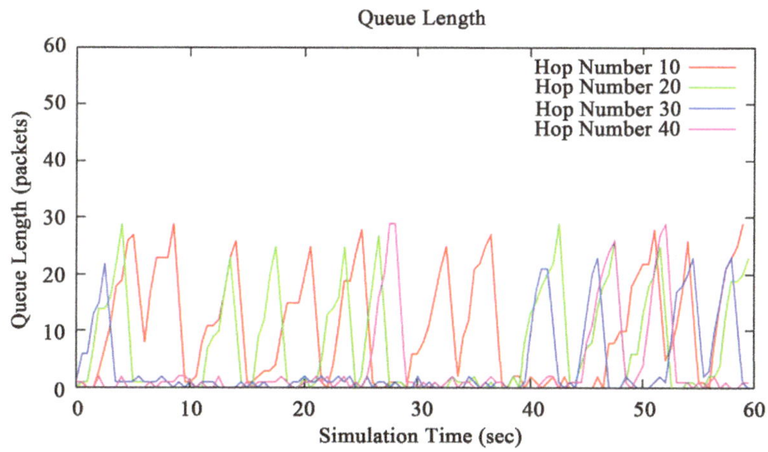

Figure 2. Graph between simulation time and queue length for hop count 10, 20, 30 and 40.

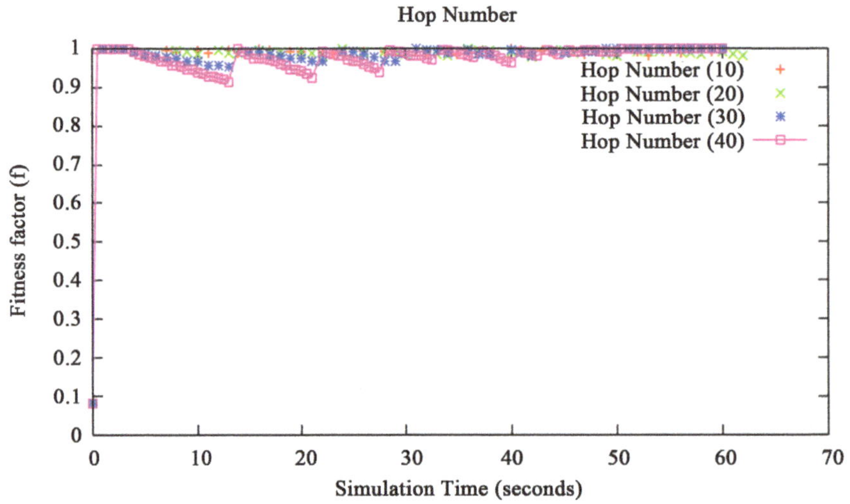

Figure 3. Graph between simulation time and fitness factor.

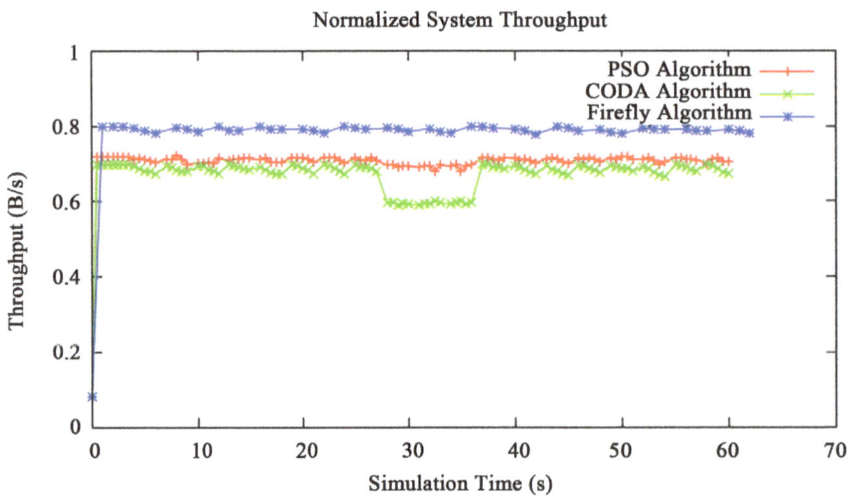

Figure 4. Throughput of the network.

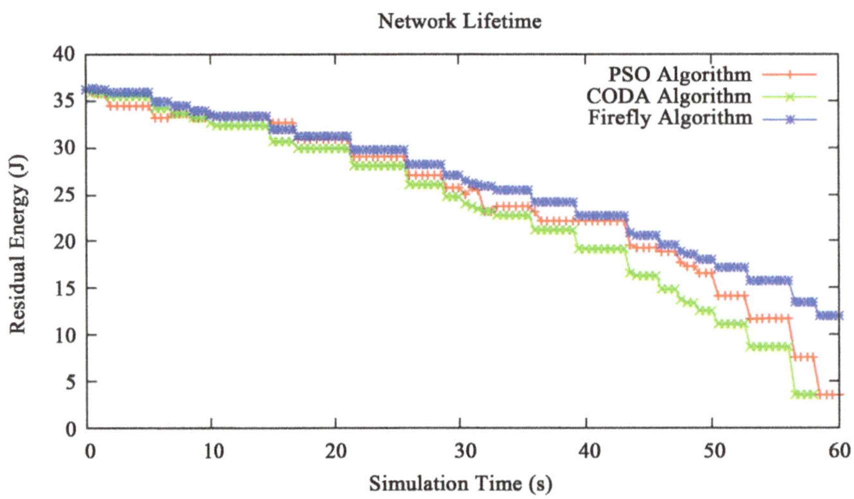

Figure 5. Network lifetime of nodes in network.

7. Conclusion and Future Scope

It is observed from the above results that after the application of Firefly Algorithm, the length of the queue decreases, and the fitness function also tries to decrease by increasing the hop count. The Firefly Algorithm results display the improvement in network lifetime and the throughput of the network. The graphs verify implementation and display much better result as compared to traditional CODA algorithm and PSO algorithm. Network lifetime has measured using the residual energy of node. In future, we can apply other bio inspired algorithms and compare the results with Firefly, PSO and CODA algorithms.

References

[1] Nvs, S. and Gopi, M. (2014) Energy Efficient Clustering Using Jumper Firefly Algorithm in Wireless Sensor Networks. arXiv preprint arXiv:1405.1818.

[2] Ming, X. and Liu, G.Z. (2013) A Multipopulation Firefly Algorithm for Correlated Data Routing in Underwater Wireless Sensor Networks. *International Journal of Distributed Sensor Networks*, **2013**, Article ID: 865154. http://dx.doi.org/10.1155/2013/865154

[3] Luca, S. and Fiorentin, F. (2011) Average TimeSynch: A Consensus-Based Protocol for Clock Synchronization in Wireless Sensor Networks. *Automatica*, **47**, 1878-1886. http://dx.doi.org/10.1016/j.automatica.2011.06.012

[4] Yedavalli, R.K. and Belapurkar, R.K. (2011) Application of Wireless Sensor Networks to Aircraft Control and Health Management Systems. *Journal of Control Theory and Applications*, **9**, 28-33. http://dx.doi.org/10.1007/s11768-011-0242-9

[5] Yick, J., Mukherjee, B. and Ghosal, D. (2008) Wireless Sensor Network Survey. *Computer Networks*, **52**, 2292-2330. http://dx.doi.org/10.1016/j.comnet.2008.04.002

[6] Xu, G.B., Shen, W.M. and Wang, X.B. (2014) Applications of Wireless Sensor Networks in Marine Environment Monitoring: A Survey. *Sensors*, **14**, 16932-16954. http://dx.doi.org/10.3390/s140916932

[7] Akyildiz, I.F., Su, W., Sankarasubramaniam, Y. and Cayirci, E. (2002) Wireless Sensor Networks: A Survey. *Computer Networks*, **38**, 393-422. http://dx.doi.org/10.1016/S1389-1286(01)00302-4

[8] Sandeep, K.E. (2014) Fire-LEACH: A Novel Clustering Protocol for Wireless Sensor Networks Based on Firefly Algorithm, *International Journal of Computer Science Theory and Application*, **1**, 12-17.

[9] Iztok, F., Yang, X.-S, and Brest, J. (2013) A Comprehensive Review of Firefly Algorithms. *Swarm and Evolutionary Computation*, **13**, 34-46. http://dx.doi.org/10.1016/j.swevo.2013.06.001

[10] Prakash, S., Rami Reddy, S.K.L.V. and Kondapalli, S. (2014) Firefly Inspired Energy Aware Cluster Based Tree Formation in WSN. 2014 2*nd International Conference on Information and Communication Technology* (*ICoICT*), Bandung, 28-30 May 2014, 356-360.

[11] Yi, S., Jiang, Q. and Zhang, K. (2012) A Clustering Scheme for Reachback Firefly Synchronicity in Wireless Sensor Networks., 2012 3*rd IEEE International Conference on Network Infrastructure and Digital Content* (*IC-NIDC*), Beijing, Beijing, 21-23 September 2012, 27-31.

[12] Ding, R. and Yang, L. (2010) A Reactive Geographic Routing Protocol for Wireless Sensor Networks. 2010 6*th International Conference on Intelligent Sensors, Sensor Networks and Information Processing (ISSNIP*), Brisbane, 7-10 December 2010, 31-36.

[13] Yang, X.-S. and He, X.S. (2013) Firefly Algorithm: Recent Advances and Applications. *International Journal of Swarm Intelligence*, **1**, 36-50. http://dx.doi.org/10.1504/IJSI.2013.055801

[14] Wan, C.Y., Eisenman, S.B. and Campbell, A.T. (2003) CODA: Congestion Detection and Avoidance in Sensor Networks. *Proceedings of the* 1*st International Conference on Embedded Networked Sensor Systems*, Los Angeles, 5-7 November 2003, 266-279.

[15] Antoniou, P., Pitsillides, A., Blackwell, T., Engelbrecht, A. and Michael, L. (2013) Congestion Control in Wireless Sensor Networks Based on Bird Flocking Behavior Congestion. *Computer Networks*, **57**, 1167-1191. http://dx.doi.org/10.1016/j.comnet.2012.12.008

14

A Secure and Energy-Balanced Routing Scheme for Mobile Wireless Sensor Network

Bei Liu, Yuanming Wu

School of Optoelectronic Information, University of Electronic Science and Technology, Chengdu, China
Email: liubeis@163.com, ymwu@uestc.edu.cn

Abstract

Mobile wireless sensor network (MWSN) has the features of self-organization, multiple-hop and limited energy resources. It is vulnerable to a wide set of security attacks, including those targeting the routing protocol functionality. In this paper, the existing security problems and solutions in MWSN are summarized, and then a trust management system based on neighbor monitoring is proposed. In the trust management system, the trust value is calculated by the neighbor monitoring mechanism, and the direct trust value and the indirect trust value are combined to establish the distributed trust model to detect the malicious nodes. The consistency check algorithm is capable of defending against the attacks on the trust model. In addition, because of the limited energy of the sensor nodes, the energy-balanced algorithm is introduced to prolong the lifespan of MWSN. The residual energy and energy density are considered in the routing decision. Finally, the simulation experiments show that the proposed algorithm can detect the malicious nodes effectively and achieve the energy-balanced goal to prolong the lifespan of MWSN.

Keywords

Mobile Wireless Sensor Network, Network Security, Trust Model, Energy-Balanced

1. Introduction

Mobile wireless sensor network consists of a large number of micro sensor nodes deployed in the monitoring area, and forms a self-organized network system by the wireless multi-hop communication. Mobile wireless sensor network does not require a fixed infrastructure support, and is widely used in military reconnaissance, medical surveillance, environmental monitoring, agricultural breeding and other commercial areas [1]. Wireless sensor network has the following characteristics: 1) limited hardware resources; 2) limited power capacity; 3) no center; 4) self-organization; 5) multi-hop routing; 6) dynamic topology; 7) a large number of nodes and dense distribution. Based on the above characteristics, the design of mobile wireless sensor network which is a special network is challenging.

In data-centric MWSN, because of interference, collision and bad wireless channel, data in the transmission process may make mistakes or miss, resulting in being not correctly sent to the SINK node. The classic encryption and authentication mechanism can resist external attacks such as tapping data and breaking network communication, but can't solve the problem of internal attack. When the normal sensor node is captured and cracked by the enemy, it becomes a malicious node, attacks the network, and damages the normal function of the network. **Table 1** lists common types of malicious node attacks.

Mobile wireless sensor network faces various threats. In [2], a lightweight secure routing protocol is proposed, which can resist the known attacks that break the basic structure of the route. A protocol based on secure routing feedback is proposed [3]. Feedbacks from the neighbor nodes and base station provide the dynamic information of the current network. The secure multicast routing protocol is proposed in [4]. The trust model based on neighbor node monitoring is widely studied in recent years. A node evaluates its neighbor nodes and calculates the trust values as the decision parameters of the routing selection stage [5]. The method of obtaining trust information and calculating the trust value of each node is called trust model that can be classified by the different monitoring behaviors [6]. According to the distribution of trust decision-making mechanism, trust model can be divided into centralized [7], hierarchical [8] or fully distributed [5]. There are a series of behaviors needing to be monitored. Each behavior corresponds to a trust metric. The trust metrics should be comprehensive, including data forwarding [5], location verification [9], data packet integrity and authenticity [8], etc. The various communication behaviors need to be turned into trust values or varying levels of trust. Trust exchange mechanism is proposed [10] [11] to improve the reliability of trust information effectively. However, these mechanisms not only increase the resource consumption, but also make the trust model itself easily attacked. For example, a malicious node deceives the trust exchange mechanism through exchanging false trust information [12]. In the trust model based on direct observation [6] [7] [13], the data that the node itself monitors are not entirely accurate. So the combination of the direct trust values and the indirect trust values obtained by the third party nodes is proposed [14]. This paper proposes a trust management approach from the human social relations to defend against malicious attacks. A node monitors the behaviors of its neighbor nodes to evaluate their trust values and then considers these trust values when the node selects the next hop forwarding node. The trust values can be directly obtained by the node, and can also be provided by other neighbor nodes.

In addition to the security threat caused by malicious node attacks, mobile wireless sensor network needs to consider the problem of limited node energy. The energy factor is considered in the routing algorithm to prolong the lifespan of the network. MDD (Multipath Directed Diffusion) routing algorithm expands the DD (Directed Diffusion) routing algorithm, and establishes multiple paths from the source node to the destination node to balance the energy consumption of each node [15]. The routing algorithm with multiple SINK nodes can reduce the paths of data transmission and save node energy [16]. In [17], LEACH (Low Energy Adaptive Clustering Hierarchy) is proposed which adopts a hierarchical structure and selects cluster heads periodically. Since each node

Table 1. Malicious attacks.

Attack type	Attacker behavior
Selective forwarding, black-hole, grey-hole	A malicious node discards part or all packets
Sinkhole attack	A malicious node tries to attract traffic advertising and fake routing information, and then it does not forward it
Replay attack	A malicious node continues to send the same data repeatedly
Link spoofing attack	A malicious node can deceive the link layer verification mechanism, so that the sender of the packet thinks that the packet has been forwarded successfully
Modification attack	A malicious node modifies the data packets that it forwards
Sybil attack	An attacker presents multiple identities
Collusion attack	Many powerful attackers work in collusion to implement attacks
Flooding attack	A malicious node sends a large number of packets to overwhelm the normal nodes
Bad mouthing attack	A malicious node provides dishonest recommendations

has the same opportunity to be chosen as the cluster head, the algorithm can balance the energy consumption of all nodes. In [18], the EBRP (Energy-Balanced Routing Protocol) routing algorithm is proposed, which can achieve the energy-balanced effect based on the depth, residual energy and energy density. The routing algorithm in this paper considers the energy factor that includes the remaining energy and energy density to prolong the lifespan of the network.

In the rest of the paper, the related work is summarized in Section 2, while in Section 3 a secure and energy-balanced routing scheme is proposed in detail. In Section 4, the simulation experiments for the algorithm are carried out. Finally, conclusions are drawn in Section 5.

2. Problem Formulation

2.1. Security Problem

In the mobile wireless sensor network, all nodes randomly move, the neighbor node set of each node is not fixed. A node monitors its neighbor nodes and evaluates their communication behaviors. When the node detects a malicious node from the neighbor nodes of the node, the node removes the malicious node from the neighbor nodes and no longer communicates with the malicious node. But the malicious node is able to move to a new location to continue to implement the attacks. So, after the node detects the malicious node, the node needs to notify the entire network of the ID of the malicious node. Considering a more complex situation, after the malicious node is detected, the malicious node changes its ID, and then moves to a new location to implement attacks. Because the ID of the malicious node is new, other nodes think that the malicious node is a normal node. As shown in **Figure 1**, this brings serious problems to the monitoring of malicious nodes. In this paper, the tracing technique through neighbor nodes is introduced to defend against this attack.

2.2. Limited Energy Problem

The weighted average of the depth, residual energy and energy density is introduced as the path decision criteria in EBRP. The algorithm can achieve energy-balanced effect in immobile wireless sensor network, but there are some problems in mobile wireless sensor network. In EBRP, the formula for calculating energy density is:

$$ED(i)=\sum_{j}\frac{E(j)}{S(i)} \tag{1}$$

The j means the neighbor node of the node i, $E(j)$ is the residual energy of the node j, $S(i)$ is the communication coverage area of the node i.

Due to node mobility, the neighbor nodes of the node i change constantly, if every time the neighbor nodes change, all values in Formula (1) are calculated again for a sum operation, this calculation is complex. Because

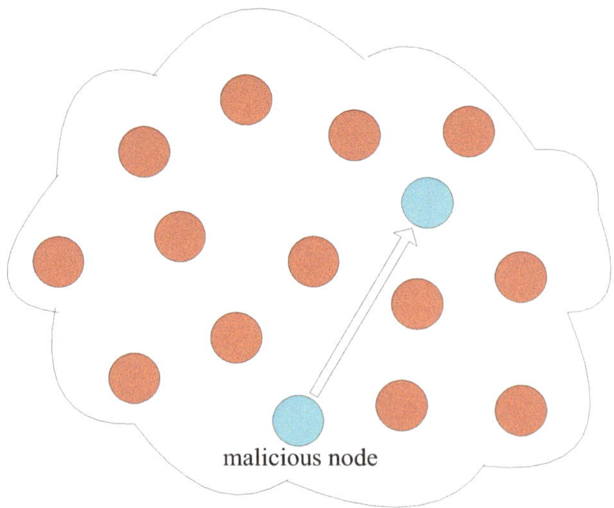

Figure 1. The malicious node moves and changes its ID.

every time the neighbor nodes change, the neighbor node set either adds a new node or reduces an old node. So the energy density only needs to add the value of the new node or subtract the value of the old node. The detailed formula is given in Chapter 3.4.

In addition, Formula (1) for calculating energy density is problematic. When the node j selects the next hop node for forwarding, it selects the forwarding node from the nodes that their depths are less than or equal to the depth of the node j. These nodes are the parent nodes or sibling nodes of the node j. So the node j should not consider all the neighbor nodes in Formula (1), but only the parent nodes and sibling nodes.

3. Secure and Energy-Balanced Routing Algorithm

In this chapter, combining the distributed trust model and the energy-balanced algorithm, a novel, easy to deploy, safe and effective mobile wireless sensor network routing algorithm is proposed. Based on the mechanism of neighbor monitoring and trust exchange, the trust model combines direct trust value with indirect trust value to detect malicious nodes. In addition, the residual energy and energy density of the nodes are considered in the routing decision to make the energy consumption balanced and prolong the lifespan of the network.

3.1. Network Topology

A large number of sensor nodes are distributed in a region randomly and uniformly to collect and forward data, which constitute a dynamic self-organizing network. The SINK node is located in the network to collect data and coordinate network.

3.2. Initialization and Update of the Depth

The SINK node broadcasts routing initialization message that includes message type and depth. The initial depth of each node is infinite. When a node receives the initialization message, if the depth of the message is greater than the depth of the node itself, the node updates the depth of the node itself and increases the depth of the message by 1, and then forwards the message. Eventually, each node has the correct depth, such as **Figure 2**.

Since the nodes are moving, the initial depths change in real time. There are 2 ways to update the depths. One is to update the depths by the SINK node in the global range like the process of route initialization. Because the depths of all nodes are updated, the energy consumption is great. The other one is that the nodes automatically update the depths. In the practical application, the moving distances of the nodes are limited in a short time, and the depths of most nodes are unchanged. When a node moves to a new location, the depth information can be obtained from neighbor nodes of the node. The node takes minimum depth from all depths of its neighbor nodes,

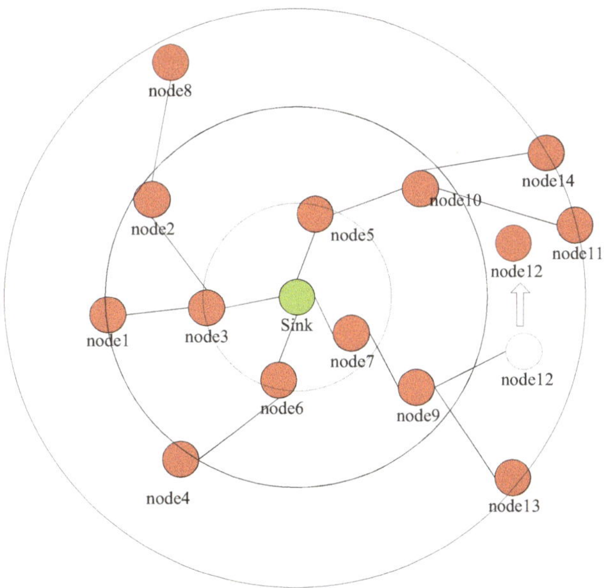

Figure 2. The depths of the nodes.

and then increases the minimum depth by 1 as the depth of the node itself. As shown in **Figure 2**, the node 12 moves to a new place, and updates its depth.

3.3. The Distributed Trust Model

The trust model consists of two modules that are the abnormal behavior detection module and the trust evaluation module. The abnormal behavior detection module establishes a series of abnormal behavior detection rules according to a variety of attack behavior features, and updates the statistics of normal and abnormal behaviors of the node according to the detection results. The trust evaluation module calculates the trust values according to the statistics, and judges whether the node is a malicious node or not. **Table 2** is some symbol descriptions of this paper.

In the trust model, a set of behaviors for node detection are defined. **Table 3** lists the trust metrics, each trust metric corresponding to one or a certain class of malicious attacks.

Specific detection methods are as follows:

1) Data forward: When the node i sends packets to the next hop node j, the node i is in promiscuous mode to overhear the wireless medium to ensure that whether the node j forwards the packets. In addition to the node i, the common neighbor nodes of the node i and the node j also monitor the node j.

2) ACK confirm: After the node i sends packets to the node j, the node i waits for the ACK from the node j. Considering the instability of the wireless communication channel, data retransmission mechanism is necessary, if the node i does not receive the ACK from the node j in a short time, the node i retransmits the packets. After trying the finite number of times, the node i gives up.

3) Data integrity: After the node i sends packets to the node j, the node i does not immediately delete the storage data of the packets, but monitors the node j and compares the forwarding packets of the node j with the storage packets of the node i.

4) Data flooding: If the number of packets received by the node i from the node j exceeds a certain threshold in a period of time, the behavior of the node j is considered abnormal.

Table 2. Symbol description.

Symbol	Description
i, j, k	Sensor node
A_m	Attack type, corresponding trust metric
R_{ij}	The distribution of the communication behavior of the node i about the node j, is a random variable
C_{ij}	The trust value of node i for node j, mathematical expectation for R_{ij}
R_{ij}^{Am}	For trust metric Am, the distribution of the communication behavior of the node i about the node j
C_{ij}^{Am}	For trust metric Am, the trust value of node i for node j, mathematical expectation for R_{ij}^{Am}
α_{ij}^{Am}	For trust metric Am, the number of nodes j normal communication behavior by node j monitor
β_{ij}^{Am}	For trust metric Am, the number of nodes j abnormal communication behavior by node j monitor

Table 3. Trust metrics for different malicious behaviors.

	Trust metric	Malicious behavior
1	Data forward	All types of packet loss attacks (black hole, gray hole, selective forwarding)
2	ACK confirm	The node does not confirm the ACK
3	Data integrity	Data tampering attack
4	Data flooding	Forward a large amount of data in a period of time
5	Data duplication	Delivery the same data repeatedly to the same node
6	Trust value response	In the trust exchange, the trust value is not returned to the requesting node
7	Trust value verification	"Badmouth" attack

5) Data duplication: If the number of the times that the node *i* receives same packet from the node *j* in a period of time exceeds a certain threshold, the behavior of the node *j* is considered abnormal.

6) Trust value response: The node *i* request trust information from the neighbor node *j*, if the neighbor node j responds the trust information to the node *i*, the number of successful times is increased by 1; otherwise the number of failed times is increased by 1.

7) Trust value verification: The trust values provided by the neighbor nodes need to be judged. This part will be given in the trust model.

The abnormal behavior detection module stores the detected results classified by the trust metrics. For a certain trust metric, if the behavior is normal, the α_{ij}^{Am} is increased by 1, otherwise the β_{ij}^{Am} is increased by 1.

There is a special case to consider: a malicious node performs well at the beginning to obtain the high trust value, and then implements attacks. The forgetting factor is introduced to detect the malicious node quickly. The trust value is calculated once every time period Δt, the formulas are as follows:

$$\alpha_{ij}^{Am} = \mu_1 \alpha_{ij}^{Am} + \alpha_{ij}^{Am}(\Delta t) \tag{2}$$

$$\beta_{ij}^{Am} = \mu_2 \beta_{ij}^{Am} + \beta_{ij}^{Am}(\Delta t) \tag{3}$$

$$\mu_1 < \mu_2 \tag{4}$$

μ_1 and μ_2 are the forgetting factor.

For the trust metric *Am*, the direct trust value is:

$$C_{ij}^{Am}\_direct = E\left(R_{ij}^{Am}\right) = E\left(Beta\left(\alpha_{ij}^{Am}+1, \beta_{ij}^{Am}+1\right)\right) = \frac{\alpha_{ij}^{Am}+1}{\alpha_{ij}^{Am}+\beta_{ij}^{Am}+1} \tag{5}$$

For all trust metric, the direct trust value is:

$$C_{ij}\_direct = \frac{1}{n}\left(C_{ij}^{A1} + \gamma C_{ij}^{A2} + \gamma^2 C_{ij}^{A3} + \cdots + \gamma^n C_{ij}^{An}\right) \tag{6}$$

The γ in the Formula (6) indicates the penalty coefficient for multiple attacks. Due to the penalty coefficient, when a malicious node launches a variety of attacks simultaneously, even though the n kinds of trust values are high, the comprehensive trust value is very low.

When $C_{ij}^{Am}\_total$ is below the trust threshold value θ^{Am} or C_{ij}_total is below the trust threshold value θ, the node *i* considers the node *j* as malicious node. The node *i* sends the information about the malicious node *j* to the SINK node and then the SINK node broadcasts the information to all nodes of the network.

After the malicious node is detected, the malicious node may change its ID, and then moves to a new location to implement attacks. The tracing technique through neighbor nodes is required to defend against this attack. In mobile wireless sensor network, the moving path of a node is continuous. A node will not suddenly appear in a new place. Based on the neighbor exchange mechanism, each node broadcasts the neighbor table to its neighbor nodes. In **Figure 3**, the node 3 moves to a new location, before and after the node 3 moves, the node 4 and 5 have always been the neighbors of the node 3. The node 4 and 5 are able to trace the mobile path of node 3 and then provide tracking information to the node 1 and 2. In this way, the node 3 is identified as a normal mobile node according to the neighbor node tracking information. If the node 3 is a malicious node, after it is detected, it changes the ID, and then moves to the new location. All neighbor nodes of the node 3 have no trace information about the node 3. So the neighbor nodes consider the node 3 as a malicious node.

3.4. Energy Factor

The node needs to consider the energy factor when making the routing decision. The energy factor includes the residual energy and energy density.

The detail energy model is shown in **Figure 4**.

Forwarding a k-bit message consumes energy as follows:

$$E_{Tx} = E_{Tx_{elec}}(k) + E_{Tx_{amp}}(k,d) = \begin{cases} kE_{elec} + k\varepsilon_{fs}d^2, & d < d_0 \\ kE_{elec} + k\varepsilon_{mp}d^4, & d \geq d_0 \end{cases} \tag{12}$$

To receive a k-bit message, the energy consumption is computed by:

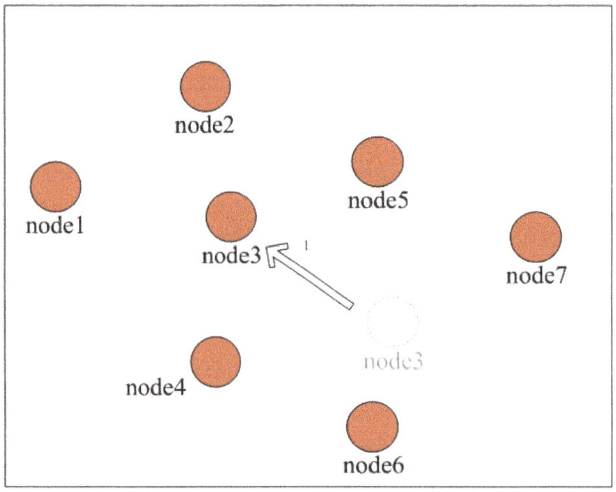

Figure 3. Tracing technique through neighbor nodes.

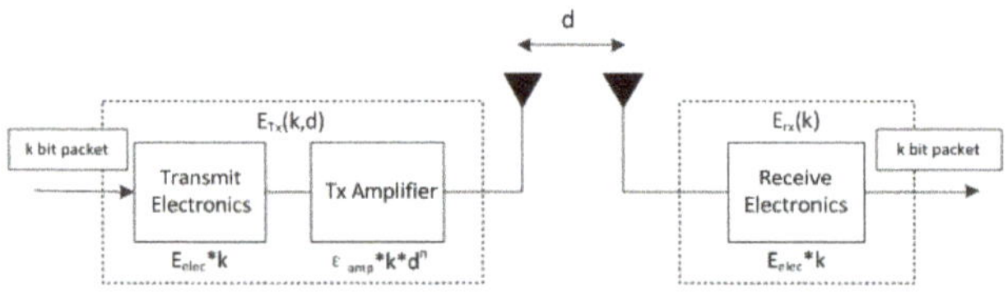

Figure 4. Energy model.

$$E_{Rx} = E_{Rx_{elec}}(k) = kE_{elec} \tag{13}$$

In WSN, energy consumption is caused by communication and calculation, but communication energy consumption is the main. Therefore, Formula (12) and (13) will be used in simulation experiments to estimate the energy consumption of sensor nodes.

Formula (14) is to calculate the energy density of the node:

$$ED(i) = \sum_j \frac{E(j)}{S(i)} \tag{14}$$

j means the parent nodes and sibling nodes that their depths are less than or equal to the depth of the node *j*, *E*(*j*) is the residual energy of the node *j*, *S*(*i*) is the communication coverage area of the node *i*.

When the neighbor nodes of the node *i* change, the neighbor node set either adds a new node or reduces an old node. The energy density only needs to add the value of the new node *k* or subtract the value of the old node *k* in Formula (15).

$$ED(i)\_new = ED(i)\_old \pm \frac{E(k)}{S(i)} \tag{15}$$

The energy factor *EE* is calculated:

$$EE(j) = \alpha * E(j) + \beta * ED(j) \tag{16}$$

$EE(j)$ represents the energy factor of the node *j*. The coefficient α and β mean the proportion of the residual energy and energy density. The Chapter 3.5 will introduce the energy factor in detail in secure and energy-balanced routing algorithm.

3.5. Secure and Energy-Balanced Routing Algorithm

The proposed routing algorithm includes the depth update stage and the data transmission stage. The depth update stage is introduced in detailed in Chapter 3.2. In the data transmission stage the node makes the routing decision to choose the next hop node for forwarding data. When the node *i* forwards data, the routing decision is made according to the following steps:

Step 1: according to the trust model, the neighbor nodes of the node *i* are judged and the malicious nodes are removed.

Step 2: select the nodes that their depths are less than the depth of the node *i*.

Step 3: select the node that has the largest energy factor EE from the nodes that have been selected in the step 2. The selected node becomes the next hop forwarding node of the node *i*.

4. Simulation and Analysis

The experiments of this paper are carried on the simulation platform MATLAB. 200 nodes are randomly deployed in 400 m*400 m rectangular space, and the SINK node is in the center.

4.1. Initialization and Update of the Depth

The **Figure 5** is the depth map after the initialization of the network, the circle represents the sensor node, and the square represents the SINK node. Some nodes are marked with a number that represents the depth of the node.

When a node moves randomly showed in **Figure 6**, the depth of the node is updated periodically showed in **Figure 7**. The depth of the node can be updated correctly and the change of the depth within a period is not more than 1.

4.2. The Distributed Trust Model

Combining direct trust value with indirect trust value, the trust model is simulated. 50 nodes are randomly selected from the network to send 6 packets to the SINK node per second. After 10 seconds 3 malicious nodes are randomly selected to implement selective forward attack. The packet loss rates respectively are 30%, 40% and 50%. After the node with 30% packet loss rate is detected, it changes the ID and moves to a new location to implement selective forward attack again. In **Figure 8**, the data arrival rate is lower than 100% from 11 seconds, which shows that there is packet loss attack in the network. **Figure 9** shows the number of times of malicious node attack. Because from 11 seconds the malicious nodes start to implement selective forwarding attack, the number of times begins adding. After 31 seconds, the numbers of attacks do not increase, it shows that the mali-

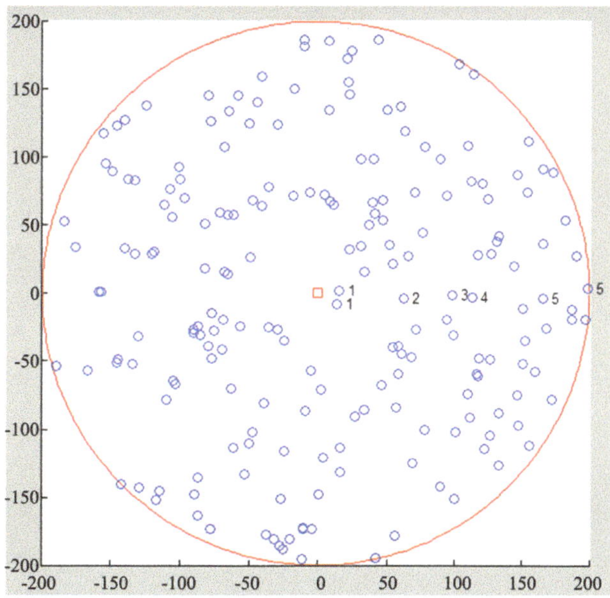

Figure 5. Initialization of the network.

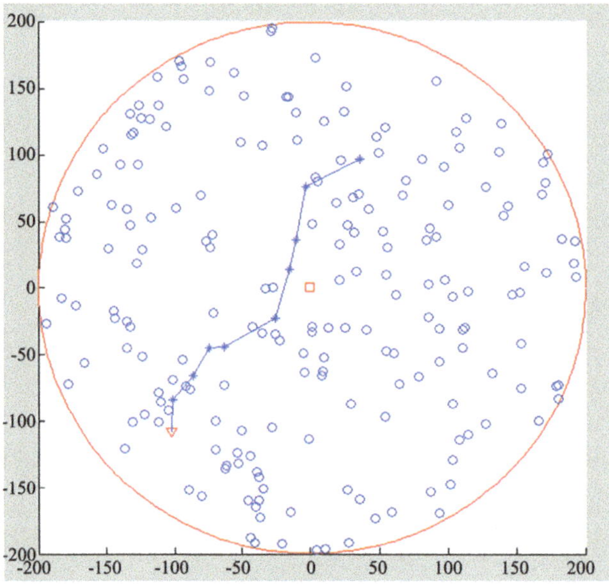

Figure 6. A node moves randomly.

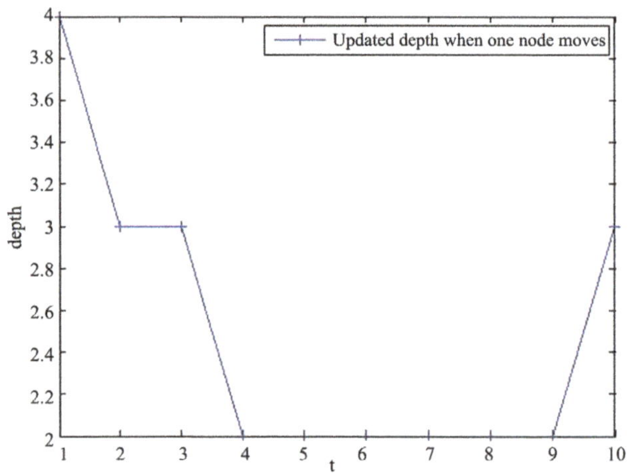

Figure 7. The depth of the node is updated.

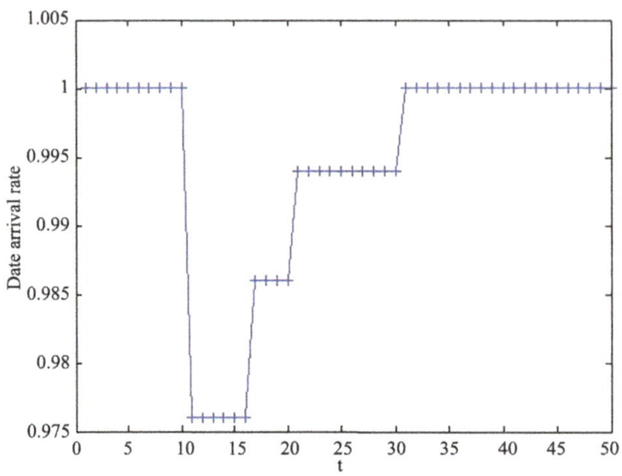

Figure 8. Packet arrival rate.

cious node changes the ID and moves to a new location, the malicious node is still identified as malicious node. **Figure 10** shows the failure rate for malicious node detection. The first malicious node is detected in 17 seconds and second in 21 seconds and third in 31 seconds. When all malicious nodes are detected, the probability of miss is 0. The simulation results show that the trust model can detect the malicious nodes effectively.

4.3. Algorithm Simulation

Malicious nodes implement attacks like Chapter 4.2. The simulation compares the lifespan of 2 algorithms. The algorithm a does not consider the energy factor in the routing decision, only choosing the secure node with minimum depth. The algorithm b is the proposed algorithm in this paper. The result in **Figure 11** shows that the algorithm b with the energy factor can significantly improve the lifespan of the network. Combined with Chapter 4.2, the proposed algorithm is secure and energy-balanced routing algorithm.

5. Secure and Energy-Balanced Routing Algorithm

In this paper, a secure and energy-balanced routing algorithm which supports mobile wireless sensor network is proposed. The proposed algorithm adopts the trust model that realizes a distributed joint detection model combining the direct trust value with the indirect trust value. Meanwhile, the energy-balanced algorithm based on nodes' residual energy and energy density is introduced to prolong the lifespan of the network. Simulation results show that the routing algorithm is capable of detecting malicious nodes successfully. The consistency check

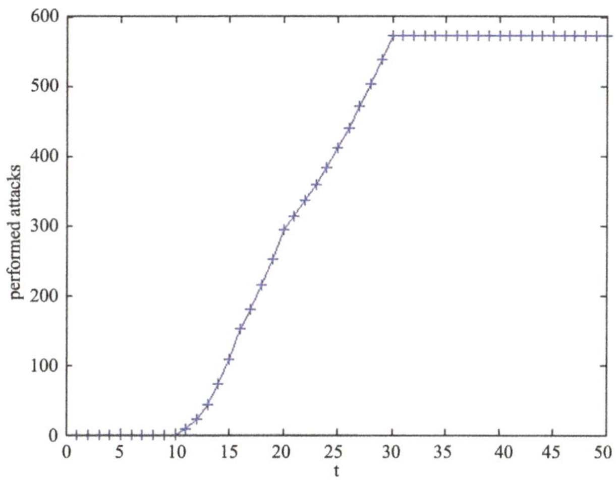

Figure 9. Total numbers of attacks.

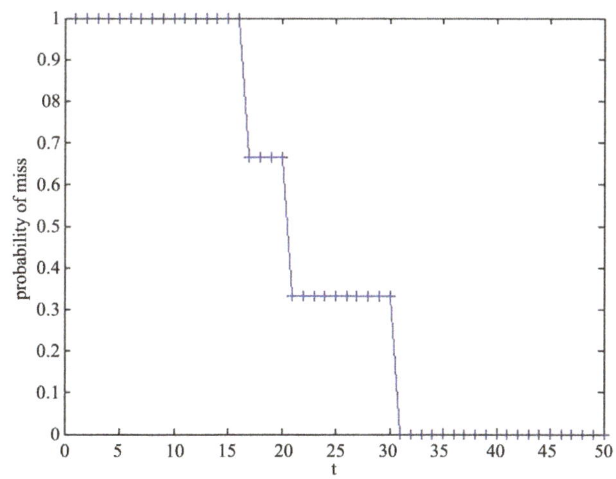

Figure 10. Probability of miss.

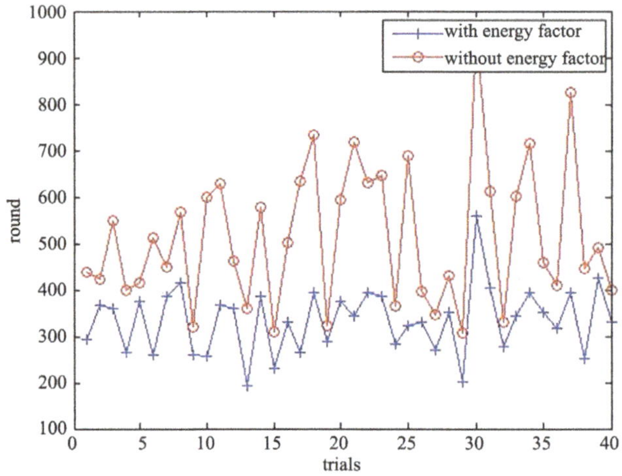

Figure 11. Lifespan of the network.

algorithm of the indirect trust values helps the network to defend against the attacks on the trust model. The algorithm also prolongs the lifespan of the network.

Acknowledgements

The authors wish to acknowledge the anonymous reviewers for their valuable comments and suggestions. We also thank MATLAB Technologies, Inc., for providing us with Wireless Modeler to validate our approaches.

References

[1] Akyildiz, I.F., Melodia, T. and Chowdury, K.R. (2007) Wireless Multimedia Sensor Networks: A Survey. *IEEE Wireless Communications*, **14**, 32-39. http://dx.doi.org/10.1109/MWC.2007.4407225

[2] Deng, J., Han, R. and Mishra, S. (2006) INSENS: Intrusion-Tolerant Routing for Wireless Sensor Networks. *Computer Communications*, **29**, 216-230. http://dx.doi.org/10.1016/j.comcom.2005.05.018

[3] Wood, A.D., Fang, L., Stankovic, J.A., *et al.* (2006) SIGF: A Family of Configurable, Secure Routing Protocols for Wireless Sensor Networks. *Proceedings of the Fourth ACM Workshop on Security of Ad Hoc and Sensor Networks*, Alexandria, 30 October 2006. http://dx.doi.org/10.1145/1180345.1180351

[4] Dong, J., Curtmola, R. and Nita-Rotaru, C. (2011) Secure High-Throughput Multicast Routing in Wireless Mesh Networks. *IEEE Transactions on Mobile Computing*, **10**, 653-668. http://dx.doi.org/10.1109/TMC.2010.194

[5] Hur, J., Lee, Y., Yoon, H., Choi, D. and Jin, S. (2005) Trust Evaluation Model for Wireless Sensor Networks. *Proceedings of Advanced Communication Technology Conference*, *ICACT* 2005, **1**, 491-496.

[6] Ganeriwal, S., Balzano, L.K. and Srivastava, M. (2008) Reputation-Based Framework for High Integrity Sensor Networks. *ACM Transactions on Sensor Networks*, **4**, 3.

[7] Crosby, G., Pissinou, N. and Makki, K. (2006) Location-Aware, Trust-Based Detection and Isolation of Compromised Nodes in Wireless Sensor Networks. *International Journal of Network Security*, **12**, 107-117.

[8] Theodorakopoulos, G. and Baras, J.S. (2006) On Trust Models and Trust Evaluation Metrics for Ad-Hoc Networks. *IEEE Journal on Selected Areas in Communications* (*JSAC*), **24**, 318-328. http://dx.doi.org/10.1109/JSAC.2005.861390

[9] Sun, Y., Yu, W., Han, Z. and Liu, K.J.R. (2006) Information Theoretic Framework of Trust Modeling and Evaluation for Ad Hoc Networks. *IEEE JSAC Special Issue on Security in Wireless Ad Hoc Networks*, **24**, 305-317.

[10] Li, H. and Singhal, M. (2006) A Secure Routing Protocol for Wireless Ad Hoc Networks. *Proceedings of the* 39*th Hawaii International Conference on System Sciences*. http://www.computer.org/csdl/proceedings/hicss/2006/2507/09/250790225a.pdf

[11] Kang, K.D., Liu, K. and Abu-Ghazaleh, N. (2006) Securing Geographic Routing in Wireless Sensor Networks. 9*th Annual NYS Cyber Security Conference*: *Symposium on Information Assurance*, Albany, 14-15 June 2006, 38-49. http://www.albany.edu/iasymposium/proceedings/2006/kang.pdf

[12] Akyildiz, I.F., *et al.* (2002) A Survey on Sensor Networks. *IEEE Communications Magazine*, **40**, 102-114. http://dx.doi.org/10.1109/MCOM.2002.1024422

[13] Lindsey, S. and Raghavendra, C.S. (2002) PEGASIS: Power-Efficient Gathering in Sensor Information Systems. *IEEE Aerospace Conference Proceedings*, **3**, 1125-1130. http://dx.doi.org/10.1109/aero.2002.1035242

[14] Hung, K.S., Lui, K.S. and Kwok, Y.K. (2007) A Trust-Based Geographical Routing Scheme in Sensor Networks. *Proceedings of IEEE Wireless Communications and Networking Conference*, Hongkong, 11-15 March 2007, 3123-3127. http://dx.doi.org/10.1109/wcnc.2007.577

[15] Eghbali, A.N., Javan, N.T. and Dehghan, M. (2007) Load-Balancing Using Multi-Path Directed Diffusion in Wireless Sensor Networks. *Lecture Note on Computer Science*, **4864**, 44-55. http://dx.doi.org/10.1007/978-3-540-77024-4_6

[16] Cicirello, P., Mottola, L. and Picco, G.P. (2007) Efficient Routing from Multiple Sources to Multiple Sinks in Wireless Sensor Networks. *Proceedings of the 4th European Conference on Wireless Sensor Networks* (*EWSN* 2007), Delft, 29-31 January 2007, 34-50. http://dx.doi.org/10.1007/978-3-540-69830-2_3

[17] Heinzelman, W., Chandrakasan, A. and Balakrishnan, H. (2000) Energy-Efficient Communication Protocol for Wireless Microsensor Networks. *Proceedings of the 33rd Hawaii International Conference on System Sciences* (*HICSS*'00), 4-7 January 2000. http://dx.doi.org/10.1109/hicss.2000.926982

[18] Ren, F.Y., *et al.* (2011) EBRP: Energy-Balanced Routing Protocol for Data Gathering in Wireless Sensor Networks. *IEEE Transactions on Parallel and Distributed Systems*, **22**, 2108-2125. http://dx.doi.org/10.1109/TPDS.2011.40

15

Enhancing Performance of Multi-Rate WLANs: Ant Colony Approach

Qiang Ma, Abdullah Al-Dhelaan, Mznah Al-Rodhaan

Department of Computer Science, College of Computer & Information Sciences, King Saud University, Riyadh, Saudi Arabia

Email: qiang.ma@student.ksu.edu.sa, dhelaan@ksu.edu.sa, rodhaan@ksu.edu.sa

Abstract

The trade-off between users' fairness and network throughput may be unacceptable in a multi-rate 802.11 WLAN environment. In this paper, we will design a new intuitive simplified mathematical model called simplified coefficient of variation (SCV) to closely reflect our topic. Through controlling the power of Access Points, SCV can optimize and improve the performance. Since our topic is a NP-hard problem, we use Ant Colony Algorithm to solve our model in a practical scenario. The simulation shows excellent results indicating that our model is efficient and superior to an existing method. Also we use software SAS to further reveal the relationships among the three indicators to illustrate the essence of our approach and an existing algorithm.

Keywords

Power Control, Proportional Fairness, Ant Colony, Optimization, Coefficient of Variation, Heuristic Function

1. Introduction

The rapid development of the Internet and the progress of wireless technology are making wireless networks play an increasingly important role in many areas. This is particularly true for the IEEE 802.11 wireless local area network (WLAN) technology. With its development, the increasing demands of service quality and a sharp rise in the number of user groups, the fairness and congestion etc. problems have become heavily concentrated in some places like offices, meeting rooms and other crowded places. In this case, many access points may be allocated, but without an overall channel or power planning, this will result in a large amount of co-channel interference, load imbalance, and network throughput decline, which will degrade the user experience.

Currently, many valuable solutions have been developed to solve above problems of WLAN, and are mainly focused on the following two aspects:

(a) Wireless channel planning. Through different methods, the limited channel resources will be reasonably assigned to all access points (APs) to make it possible to reduce co-channel interference and network overhead in order to improve overall network throughput.

(b) Power control to achieve load balancing. Through increasing or decreasing the AP's power to adjust the signal strength of the AP, thus it changes the access topology of the user-AP in the network in order to reduce the scheduling overhead, and improve load balance, fairness etc.

This paper involves both aspects above. The rest of the paper is organized as follows: related work is discussed in Section 2, and Section 3 shows our motivation. After that, a brief introduction of ant colony algorithm as background will be provided in Section 4. Then we start to explain the new model SCV and implement it using ant colony algorithm, the simulation modeling using Matlab is explained in Section 5. After this, we give comparison and SAS analysis in Section 6 and then draw the conclusions in Section 7.

2. Related Work

According to IEEE 802.11, a high-density WLAN deployment environment offers a short distance between APs and users since each user will be surrounded by many APs. In this case, each user will connect with the AP by the strongest received signal strength indicator (RSSI) by default. We know that the users are not uniformly distributed in an area, which makes some APs connect more users than the other APs. This will produce the load imbalance problem, as some APs are hungry while some APs are overloaded. This situation results in unfair use of resources.

As a part of our research, the basic solution has been introduced in [1].

To solve interference problem, the author in [2] proposed an adaptive neuro fuzzy distributed power control algorithm to maximize the minimum Carrier to Interference Ratio (CIR) among all of co-channel users in each channel of system. The simulation results showed better performance comparing with a classical method in terms of convergence.

To solve congestion problem, the authors in [3] proposed an adaptive load balance allocation strategy for small antenna based wireless networks that can enhance the traffic-carrying capacity for variations in traffic. Their simulation results were confirmed to be efficiency.

In order to improve the throughput, the authors in [4] introduced three strategies which were Dirty Paper Coding (DPC) strategy, Noise strategy, and Opportunistic Interference Cancellation (OIC) strategy. Also they proposed corresponding optimal power control algorithms for each strategy. The simulation results showed their aims were achieved.

The popular 802.11 MAC protocol provides equal transmission chances to all users, which may achieve throughput-based fairness if all users have the same frame size during a cycle [5]-[8]. Recent studies have shown that time-based fairness is much better than throughput-based fairness in multi-rate WLANs [9].

So far, we have two fairness criteria factors that are widely used in network management: proportional fairness [7] which allocates bandwidth to users in proportion to their bit rates to maximize the sum of the bandwidth utilities of the users, and max-min fairness [10] which allocates throughput as equally as possible through maximizing the minimum throughput. Proportional fairness and time-based fairness are equivalent in multi-rate WLANs when all users have the same weight [11]. The equivalence of max-min fairness and throughput-based fairness under the same condition (integral association) was proved in [12].

The authors in [13] proposed a new algorithm called Power Control for AP (PCAP) to optimize the network utility by maximizing the average of the AP utility and minimizing the variance of the AP utility, the result directly maximized the "throughput" as its target, and then they started to calculate the users' fairness indicator—"J" (Jain's fairness index [14]) as their byproduct. Though the result showed significant improving, the "J" is unreasonable to be a dependent variable. We will analyze the relationship between these two variables.

According to IEEE802.11, AP's transmission power can be changed in an allowable range, this technique is called power control. Some previous studies, such as [15] [16], have assumed that the user-AP associated topology will not change when adjusting the power of APs, this assumption is not the reality. On the contrary, some papers have noticed this phenomenon and developed techniques called cell breathing [17].

To enhance load-balancing and network throughput, a variable polyhedron genetic algorithm (GA) is proposed in [18]. Which not only provides an AP service availability guarantee but also yields a near-optimal beacon range for each AP when the number of evolutions is large enough. Their simulation study indicates that the algorithm is superior over the default 802.11 AP association model.

Also the problem researches like oblem. Similarly, in order to solve such trade off problem, there are many researches, such as [19], the authors proposed an algorithm that transformed the trade off problem into a monotonic optimization problem. Also the problem was solved with geometric programming in [20], but it is not suitable for the low Signal to Interference Ratio (SIR) case. While in [21], the authors proposed a centralized algorithm called Non-Linear Approximation Optimization for Proportional Fairness to derive the user-AP association via relaxation, and also gave a distributed heuristic algorithm called Best Performance First which provides an AP selection criterion for new comers.Regarding the tradeoff problem, also some other valuable solutions are given. Like [22], the authors derived the stability region of a multiuser multichannel WLAN system and determined the throughput optimal channel switching scheme within a certain class of schemes. In [23], the authors proposed an extension of the AP aggregation algorithm to ensure the minimum average throughput for each host in the field.

In [24], the authors proposed a novel AP association approach LBAA, taking AP's load-balancing, Wireless Mesh Networks' (WMNs') multi-hop characteristic, and user's RSSI into consideration to solve network congestion and performance degradation problem in Wireless Mesh Networks (WMNs).

In [25], to solve the fairness problem in Wireless Mesh Networks (WMNs), the authors proposed a probabilistic approach to provide proportional fairness without solving global non-linear and non-concave optimization. Their simulation result showed that the proposed scheme works better than the standard IEEE 802.11 based EDCA MAC in terms of fairness and throughput.

In summary, there are many valuable research papers related to the above trade off problem. Up to date, none of those researches in term of the tradeoff between fairness and throughput could offer a rational, clearly designed mathematical model which can be easily and widely implemented using the well known AI algorithms. Thus our research aims to fill such gap; we will design a reasonable mathematical model, which is SCV. It should have a simple form, easy to be understood. The model itself should be highly effective and efficient to match our topic, and should be a bridge to link our topic with AI algorithms. So we choose one of representative AI algorithms—Ant Colony, and design its detail to implement the SCV model. At the end, we will compare its performance with an existing algorithm-PCAP in [13], also through SAS analysis, we will reveal the essence of our approach which has not been discussed at such micro level in the previous research.

In this paper, the contributions are listed as follows: 1) we describe the "trade-off" using "J of user" and "J of AP", which refer to the fairness of users and fairness of APs respectively. Then we design our target function using many skills to deal with those complex formulas, after that we get our simplified coefficient of variation (SCV) model, which is a clear mathematical function to solve such trade-off problem. This is the core contribution of our paper; **2)** we define the problem as an informed search NP-hard problem and apply Ant Colony algorithm to solve the SCV model; 3) we use multi-channel allocation to improve the transmission rate; 4) we use Statistical Analysis System (SAS) for analysis to reveal the relationships of three indicators and the essence of algorithms; 5) SCV opens a door for many AI algorithms; it is a bridge between Network & AI.

3. Motivation

3.1. The Essence of PCAP: throughput

From our SAS analysis in **Figure 3**, three indicators (Juser: J of user; Jap: J of AP; Tpt: relative throughput) show that J of AP can represent throughput (value > 0.8, so it is highly linear related).

Through our statistics calculation, PCAP focus on J of AP only, which means it only focus on throughput. This is a deficiency of Target Function design, which is not well reflecting our topic.

3.2. The Essence of SCV

The problem is defined as a NP-hard problem since we apply a practical scenario that includes 20 APs, each AP has 10 levels of power, so the state space of the problem will be 10^{20}, making it neither solvable nor verifiable in polynomial time, so it is NP-hard.

From the computation theory, we know that we cannot get an accurate solution. Comparing with other NP-hard problems such as TSP (Traveling Salesman Problem), we get some heuristic methods. Since existing models either using complex definition such as utility or involving many parameters such as channel gain, those models are not clear enough to apply informed search techniques, so first we need to build a clear, simplified model SCV, and then apply the Ant Colony algorithm to solve the model.

Since our topic is: "J of user (fairness of users) & throughput", which means to make balance between these two parameters. Obviously the two parameters have different units, then we have to convert the "throughput" to "J of AP" (already explained, it can represent throughput, with high linear relation).

Then our SCV gives a new designed target function:

$$F = \left(1/J_{users} - 1\right) + \omega\left(1/J_{aps} - 1\right)$$

which reflects the balance of two parameters (J of user & throughput), and we will rewrite to get its final form *f*.

4. Ant Colony Algorithm

This part is clearly explained in [26]. The Ant Colony algorithm is a method from the field of Swarm Intelligence. Ant Colony Algorithm, Elite Ant System, and Rank-based Ant System compose the Ant Colony Optimization methods.

The Ant Colony algorithm is inspired by the foraging behavior of ants, specifically the pheromone communication among ants regarding a short path between the living place and a food source.

Ants initially wander randomly around their environment. Once the food is located, the ants that pass by the food will begin laying down pheromone in the environment. Numerous trips between the food and the living place are performed, and if the same path is followed that leads to food then additional pheromone is laid down. And also at the same time, the pheromone decays in the environment, so that older paths are less likely to be followed. Other ants may discover the same path that has strongest pheromone and also lay down pheromone when they are passing by. The positive feedback process makes more ants produce paths that are further refined through use.

The goal of this strategy is to exploit heuristic information to construct candidate solutions and hold the information as the history. Paths are constructed in a probabilistic step-wise manner. The probability of selecting a path is determined by both heuristic contribution and historical contribution. History is updated proportional to the quality of candidate paths and is uniformly decreased ensuring the most recent and useful information is retained. We will use 2 important formulas:

(a) After each ant conducts its tour of the trail, the pheromone is updated using the following formula:

$$T_{ij}\left(t+n\right) = p * T_{ij}\left(t\right) + \Delta t_{ij}$$

$$\Delta T_{ij}^{k} = \begin{cases} \dfrac{Q}{L_k} & \text{if the k}^{\text{th}} \text{ ant uses edge}\left(i,j\right) \text{ in its tour } \left(\text{between time } t \text{ and } t+1\right) \\ 0 & \text{otherwise} \end{cases}$$

Q is a constant and L_k is the tour's length of the kth ant.

(b) Transition probability:

$$p_{ij}^{k}\left(t\right) = \begin{cases} \dfrac{\left[T_{ij}\left(t\right)\right]^{\alpha} \cdot \left[n_{ij}\right]^{\beta}}{\sum_{k\in \text{allowed}_k} \left[T_{ik}\left(t\right)\right]^{\alpha} \cdot \left[n_{ik}\right]^{\beta}} & \text{if } j \in \text{allowed}_k \\ 0 & \text{otherwise} \end{cases}$$

where α and β are controlling parameters that control the relative importance of trail versus visibility.

5. Model Design and Simulation

Now we are going to explain our SCV model and implement it in Ant Colony Algorithm.

5.1. The Way APs Attract Users

The user will select the strongest Received Signal Strength Indicator (RSSI) as default. In the model [27], $\text{RSSI} = aP/X^{\alpha}$ where "*a*" is a constant factor, "*P*" is received power by user, "*X*" is distance between user and selected AP, while "α" has different value in different scenarios, generally between 1.6 and 6.5 [28]. The formula only determines the association matrix of user-AP. In practice, the general power range of the AP is 10 dBm - 30 dBm, *i.e.* 1 mw ~1 w, here we adopt $\alpha = 3$ for indoor case. From the formula, the value of "*a*" does not affect the association results, to simplify the mathematical form, we take $a = 1$, so our model adopts a simplified form:

$$\text{RSSI} = P/X^3 \quad (1)$$

5.2. Study the SINR [r_{ij}] of the User [i]

We use r_{ij} to denote SINR (Signal to Interference plus Noise Ratio). Assuming the user [i] connects to AP[j], the power of AP[j] is P_j. Wherein "g" are channel gains, A_i is a set of all APs within the same channel of AP[j]. N_{0j} is an additive white Gaussian noise generated by AP[j].

$$r_{ij} = \frac{g_{ij} p_j}{\sum_{k \in A_i \cap k \neq j} g_{ik} p_k + N_{0j}} \quad (2)$$

It is worth noting that N_{0j} can be adjusted to an exact value [29] [30]. So we can set a constant $\mu > 0$,

$$\mu = \frac{g_{ij}}{\sum_{k \in A_i \cap k \neq j} g_{ik} p_k + N_{0j}} \quad (3)$$

5.3. Study the Relationship between User [i]'s Transmission Rate v_i and Its SINR [r_{ij}]

From **Table 1** in [13], v_i denotes user's transmission rate and r_{ij} is SINR. We see the monotonically increasing relationship between the two variables. Here we might assume that two variables meet the linear relationship as an approximation, $v_i = \beta r_{ij}$, $\beta > 0$ is a constant of proportionality. Then connect this to (2) and (3) we have:

$$v_i = \beta r_{ij} = \beta \mu p_j = \lambda p_j \quad (4)$$

So λ is a constant:

$$\lambda = \beta \mu \quad (5)$$

5.4. Study the Effective Speed V_i

Let $N[j]$ denote the total number of users which connect with AP[j]. P_j is AP[j]'s power. Because the users are time based sharing the chance of AP[j], so we use $\overline{v_i}$ to denote the effective speed of v_i:

$$\overline{v_i} = \frac{\lambda p_j}{N[j]} \quad (6)$$

5.5. Study the AP's Power

According to the simulation result in [13], we know that usually 10 Levels of AP power will be enough to achieve a good result. Therefore, in our model, p_{max} and p_{min} have relationship as following: $p_{max}/p_{min}=10$, p_{max} will be the basis of calculation, since we need to increase the p_j, so the 10 APs' power value is in **Table 2**.

Note here the unit of power is "mw", not "dBm". Since $P_j = P_{min} l_j$ ($l_j = 1, 2, \cdots, 10$), here l_j denotes the level of AP's power, so the Formula (6) can be rewritten as follows:

Table 1. v_i - r_{ij} relationship.

r_{ij} (dB)	6 - 7.8	7.8 - 9	9 - 10.8	10.8 - 17	17 - 18.8	18.8 - 24	24 - 24.6	24.6-
V_i (Mbps)	6	9	12	18	24	36	48	54

Table 2. Level-value relationship.

Level	1	2	…	10
Value	p_{min}	$2p_{min}$	…	$10p_{min}$

$$\overline{v_i} = \frac{\lambda p_{\min} l_j}{N[j]} = \lambda p_{\min} \frac{l_j}{N[j]} \tag{7}$$

Let M be the total number of users and N be the total number of APs. From statistics we know that the Expectation of $\overline{V}_i$ for all users is denoted as $E(\overline{V}_i)$, and Variance of $\overline{V}_i$ for all users is denoted as $S^2(\overline{V}_i)$. We have the following ($i = 1,2,\cdots,M$; $j = 1,2,\cdots,N$):

$$E\left(\overline{v_i}\right) = \lambda p_{\min} E\left(\frac{l_j}{N[j]}\right) \tag{8}$$

$$S^2\left(\overline{v_i}\right) = \left(\lambda p_{\min}\right)^2 S^2\left(\frac{l_j}{N[j]}\right) \tag{9}$$

Let $b[i]$ denote the effective transmission speed from user[i] to AP[j], we have $b[i] = \overline{V}_i$. Moreover, let $U[j]$ denote the transmission speed from the AP[j] to backbone. The Expectation of $b[i]$ is denoted as: $E\left(b[i]\right)$, and Variance of $b[i]$ is denoted as: $S^2\left(b[i]\right)$, the Expectation of $U[j]$ is denoted as: $E\left(U[j]\right)$, and Variance of $U[j]$ is denoted as: $S^2\left(U[j]\right)$, so continue we have formulas as following:

$$E\left(b[i]\right) = E\left(\overline{v_i}\right) = \lambda p_{\min} E\left(\frac{l_j}{N[j]}\right) \tag{10}$$

$$S^2\left(b[i]\right) = S^2\left(\overline{v_i}\right) = \left(\lambda p_{\min}\right)^2 S^2\left(\frac{l_j}{N[j]}\right) \tag{11}$$

$$E\left(U[j]\right) = E\left(\sum_{user[i]\to AP[j]} b[i]\right) = E\left(\lambda p_{\min} N[j] \frac{l_j}{N[j]}\right) = \lambda p_{\min} E\left(l_j\right) \tag{12}$$

$$S^2\left(U[j]\right) = S^2\left(\sum_{user[i]\to AP[j]} b[i]\right) = S^2\left(\lambda p_{\min} N[j] \frac{l_j}{N[j]}\right) = \left(\lambda p_{\min}\right)^2 S^2\left(l_j\right) \tag{13}$$

Here cvusers denotes the coefficient of variation of transmission speed of all users, and cvAPs denotes the coefficient of variation of transmission speed of all APs, we have:

$$\begin{aligned} \text{cvusers}^2 &= \frac{S^2\left(b[i]\right)}{E^2\left(b[i]\right)} = \frac{\left(\lambda p_{\min}\right)^2 S^2\left(\frac{l_j}{N[j]}\right)}{\left(\lambda p_{\min}\right)^2 E^2\left(\frac{l_j}{N[j]}\right)} = \frac{S^2\left(\frac{l_j}{N[j]}\right)}{E^2\left(\frac{l_j}{N[j]}\right)} \\ &= \frac{M \sum\limits_{i=1,b[i]\to AP[j]}^{M} \frac{l_j^2}{N^2[j]}}{\left(\sum\limits_{j=1}^{N} l_j\right)^2} - 1 = \frac{M \sum\limits_{j=1}^{N} \frac{l_j^2}{N[j]}}{\left(\sum\limits_{j=1}^{N} l_j\right)^2} - 1 = \frac{1}{J_{users}} - 1 \end{aligned} \tag{14}$$

$$\text{cvAPs}^2 = \frac{S^2\left(U[j]\right)}{E^2\left(U[j]\right)} = \frac{\left(\lambda p_{\min}\right)^2 S^2\left(l_j\right)}{\left(\lambda p_{\min}\right)^2 E^2\left(l_j\right)} = \frac{S^2\left(l_j\right)}{E^2\left(l_j\right)} = \frac{N \sum\limits_{j=1}^{N} l_j^2}{\left(\sum\limits_{j=1}^{N} l_j\right)^2} - 1 = \frac{1}{J_{APs}} - 1 \tag{15}$$

Note, here we adopt the definition of J in [13], where we have $J_x = \left(\sum\limits_{i=1}^{n} x_i\right)^2 \Big/ n \sum\limits_{i=1}^{n} x_i^2$, above is the relationship between J and the square of coefficient of variation.

5.6. Cost Function Construction

According to our topic, we need a function that can describe the tradeoff between fairness of users and throughput of network. In [13], the algorithm is divided into two steps: increase average value and decrease variance value of AP utility to increase throughput of network. They are equal to decreasing cvAPs or cvAPs^2. So increasing J of users is equal to decreasing cvusers^2.

Let F denote a target function as follows: $F = \text{cvusers}^2 + \omega\left(\text{cvAPs}^2\right)$, ω is a weight proportion factor, it is very important reflecting our requirement that how to make the balance between fairness and throughput, it is a quantifiable indicator

Here we do some mathematical derivation to illustrate how we get a reasonable value of ω. Considering the static grouping problem: *m* numbers are average divided by *n* groups, therefore each group has *m/n* numbers. Given that the expectation of total numbers is σ, and their variance is s^2. Cvnumbers denotes the coefficient of variation of m numbers. Cvgroups denotes the coefficient of variation of n groups. Defined as before, we have:

$$E\left(\text{group}[i]\right) = E\left(\sum_{\text{number}[j]\in\text{group}[i]} \text{number}[j]\right) = \sum_{\text{number}[j]\in\text{group}[i]} E\left(\text{number}[j]\right) = \frac{m}{n}\sigma$$

$$S^2\left(\text{group}[i]\right) = S^2\left(\sum_{\text{number}[j]\in\text{group}[i]} \text{number}[j]\right) = \sum_{\text{number}[j]\in\text{group}[i]} S^2\left(\text{number}[j]\right) = \frac{m}{n}s^2$$

$$\text{cvnumbers}^2 = \frac{s^2}{\sigma^2}$$

$$\text{cvgroups}^2 = \frac{S^2\left(\text{group}[i]\right)}{E^2\left(\text{group}[i]\right)} = \frac{\frac{m}{n}s^2}{\left(\frac{m}{n}\right)^2\sigma^2} = \frac{1}{\frac{m}{n}}\frac{s^2}{\sigma^2} = \frac{1}{\frac{m}{n}}\text{cvnumbers}^2 \tag{16}$$

From the above, it means cvgroups^2 is much smaller than cvnumbers^2, comparing this example to our function "*F*", in function "*F*" we should amplify the small part since two parts have relationship. So we decide to give value to ω, let $\omega = M/N$.

$$\begin{aligned} F &= \text{cvusers}^2 + \omega\left(\text{cvAPs}^2\right) = \frac{M\sum_{j=1}^{N}\frac{l_j^2}{N[j]}}{\left(\sum_{j=1}^{N} l_j\right)^2} - 1 + \frac{M}{N}\left(\frac{N\sum_{j=1}^{N} l_j^2}{\left(\sum_{j=1}^{N} l_j\right)^2} - 1\right) \\ &= M\frac{\left(\sum_{j=1}^{N}\frac{l_j^2}{N[j]} + \sum_{j=1}^{N} l_j^2\right)}{\left(\sum_{j=1}^{N} l_j\right)^2} - \left(1 + \frac{M}{N}\right) = Mf - \left(1 + \frac{M}{N}\right) \end{aligned} \tag{17}$$

where in

$$f = \frac{\left(\sum_{j=1}^{N}\frac{l_j^2}{N[j]} + \sum_{j=1}^{N} l_j^2\right)}{\left(\sum_{j=1}^{N} l_j\right)^2} = \frac{\sum_{j=1}^{N}\left(l_j^2\left(1 + \frac{1}{N[j]}\right)\right)}{\left(\sum_{j=1}^{N} l_j\right)^2} \tag{18}$$

Note that M and N are constants as defined before. M is total number of users, N is total number of APs. When "F" goes to minimum, it is equal to "f" goes to minimum. So (18) will be our simplified target function to achieve the purpose of the tradeoff between Fairness (users) and throughput (network).

5.7. Throughput

From Formula (12) we know that:

$$\text{Throughput}_{\text{real}} = \sum_{j=1}^{N} U[j] = \lambda p_{\min} \sum_{j=1}^{N} l_j = \lambda p_{\min} \text{Throughput}_{\text{relative}} \tag{19}$$

$$\text{Throughput}_{\text{relative}} = \sum_{j=1}^{N} l_j \tag{20}$$

$\text{throughput}_{\text{real}}$ is real throughput, and $\text{throughput}_{\text{relative}}$ is relative throughput. Since the $\lambda p_{\min}$ is constant, we use $\text{throughput}_{\text{relative}}$ to represent $\text{throughput}_{\text{real}}$.

5.8. Ant Colony Design & Simulation

In this part we are going to place a total number of $N = 20$ APs on a 4 by 5 grid, with each AP on a grid point. The coverage area of each AP can stretch across the whole area. The distance between two adjacent APs is set to 100 meters. The maximum transmission power of each AP is set to 20 dBm (100 mw), so the minimum transmission power of each AP is set to 100/10 = 10 mw = 10 dBm.

We arrange $M = 200$ users randomly distributed in the whole area. According to [31], a separation of four channels can be used without reducing the performance, so the possibilities could be opened to channels 1, 5, 9 and 13. In this paper we decide to use these channels in order to get a bigger $\text{Throughput}_{\text{real}}$.

Let $AP_j \rightarrow Ci$ denote AP_j using channel i, we use 1, 5, 9, 13 these channels to configure the network as in **Table 3**.

5.9. Ant Colony Algorithm Design

(a) To initialize 200 users' and 20 APs' coordinates. NC_max is the maximum number of iterations, m is the number of ants, Alpha is a parameter to characterize the importance of pheromone, Beta is a parameter to characterize the importance of heuristic information, and Rho is a pheromone evaporation coefficient. Also we need to initialize pheromone matrix Tau: Tau = ones (10, 10, 19). We calculate the average distance for each AP to all users then to construct a row vector called xaverage (1 × 20): xaverage = sum (Distance)/200.

(b) Heuristic Matrix Eta Design: $\bar{x}$ is the average distance from AP to all the users, then in order to obtain an attractive balance, based on our attractive model in Formula (1), the ratio of two power levels is equal to the ratio of the two $(\bar{x})^3$, so that both APs may have the similar chance to attract users. Heuristic matrix is Eta (10 × 10 × 19), from the 1st AP to 20th AP, which has total of 19 pages, and each page is a matrix (10 level × 10 level) describing the heuristic information from all power levels of current AP to all power levels of the next AP, and we set the value of these heuristic transition points as "1". We introduce a new mechanism: we give such heuristic matrix a quantitative parameter defined as "similar", namely the degree of similarity between non-heuristic points and the heuristic points, and we set the value of non-heuristic points in the heuristic matrix as "similar×1". Then we finish the design of heuristic matrix Eta.

Table 3. AP-channel relationship.

$AP_1 \rightarrow C1$	$AP_2 \rightarrow C9$	$AP_3 \rightarrow C1$	$AP_4 \rightarrow C9$	$AP_5 \rightarrow C1$
$AP_6 \rightarrow C5$	$AP_7 \rightarrow C13$	$AP_8 \rightarrow C5$	$AP_9 \rightarrow C13$	$AP_{10} \rightarrow C5$
$AP_{11} \rightarrow C9$	$AP_{12} \rightarrow C1$	$AP_{13} \rightarrow C9$	$AP_{14} \rightarrow C1$	$AP_{15} \rightarrow C9$
$AP_{16} \rightarrow C13$	$AP_{17} \rightarrow C5$	$AP_{18} \rightarrow C13$	$AP_{19} \rightarrow C5$	$AP_{20} \rightarrow C13$

(c) We set totally 10 ants at AP_1, and each ant is in a grid. Here the grid denotes one power level, so there are 10 grids (10 power levels). Initially, all the ants are in AP_1 (city1), then 10 ants will move to AP_2 (city2) together but choose the different grid (power level) according to the probability function: P = (Tau.^Alpha). * (Eta.^Beta), then repeat the procedure until moving to AP_{20} (city20) to complete their travel. Then we start to calculate the cost function of 10 paths (a path is defined as the route from AP_1 to AP_{20}, and the value of each AP is a power level) and select the minimum cost, then record such path. After this, finally we will update the pheromone which includes decreasing all of the paths' pheromone according to parameter Rho and increasing all of the paths' pheromone traveled by ants.

(d) Repeat step(c) until reaching the NC_max (a constant), and output all cycles' results.

6. Result Analysis

In this part, some definitions are as following. m is the number of ants, Alpha is a parameter to characterize the importance of pheromone, Beta is a parameter to characterize the importance of heuristic information, and Rho is a pheromone evaporation coefficient. NC_max is a constant that describes the number of iterations. Q and "similar" are constants according to the algorithm.

Let $m = 10$; Alpha = 1; Beta = 2; Rho = 0.1; NC_max = 200; $Q = 100$; similar = 0.9.

6.1. Simulation Analysis

(a) The solution of ant colony algorithm has many parameters, so we need to run multiple times on the same data set and same parameters and select the average result, also to search for the more suitable parameters' values. Here we use $\text{Throughput}_{\text{relative}}$ to represent $\text{Throughput}_{\text{real}}$. We know that the maximum of $\text{Throughput}_{\text{relative}}$ is: $10 \times 20 = 200$, but it will never be achieved, at least because of users' distribution.

(b) All parameters can affect the calculation result. Generally, it is suitable to set Alpha = 1; Beta = 2; Rho = 0.1; and similar = 0.9 to get better results.

(c) Strictly speaking the convergence of ant colony algorithm is defined as that the pheromone in the iteration process without change, or at least the obtained path without change. This is usually in the theoretical analysis. In practical applications due to a number of random factors, it is very difficult to achieve this status. So as long as the actual optimal solutions have vibration around the average straight line after each iteration, this is considered as convergence, and such average line is called convergence line. It can be seen from **Figure 1** and **Figure 2**, after traveling 200 times, all of the parameters have been around the horizontal line with slight vibration, and these mean convergence.

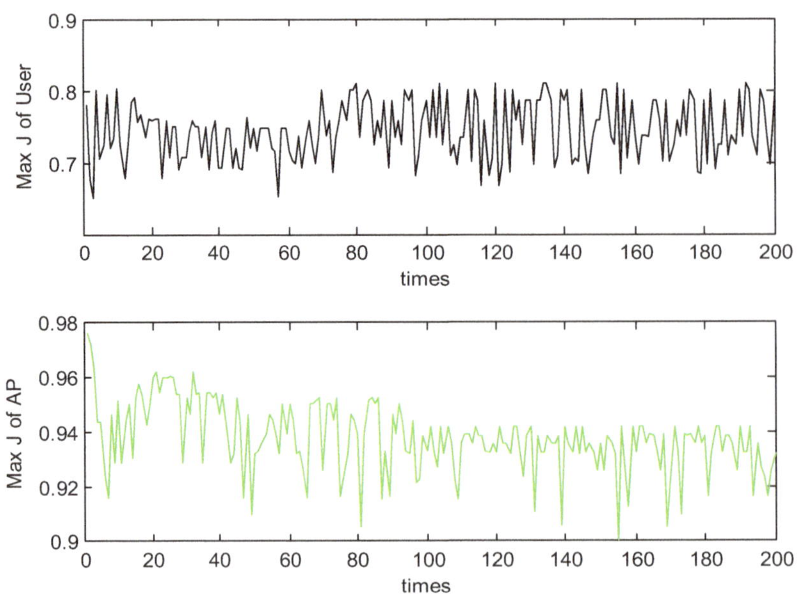

Figure 1. {Max J of user, Max J of AP}-times plot.

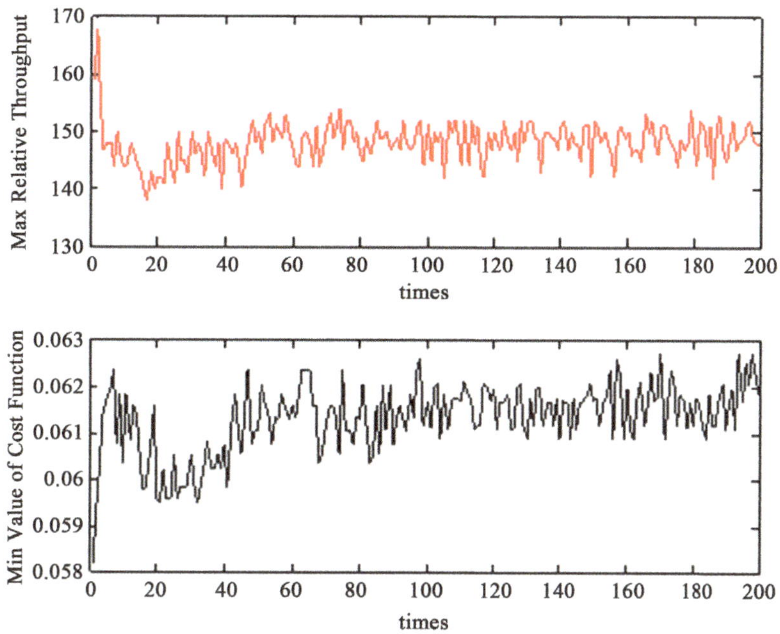

Figure 2. {Relative throughput, Min value of cost function}-times plot.

6.2. SAS Analysis

We use the samples from experimental data to study the correlation coefficients among these indicators. Wherein Juser denotes J of user, Jap denotes J of AP, Tpt denotes $\text{Throughput}_{\text{relative}}$, cost denotes the f in (18).

Figure 3 shows that at Alpha = 0.05 significance level, all the p-values are less than 0.05, we reject the H_0 and accept H_1 that these variables are linearly related, wherein the Tpt-(Jap, Juser) have highly significant linear correlations, while correlations of Jap-Juser is weak. We compared the degree of concentration of those data points in **Figure 4** and **Figure 5**. It is clear that data points are more concentrated in **Figure 5**. This means the linear correlation of Tpt-Jap is much higher than the linear correlation of Tpt-Juser, which also proves the effectiveness of SCV model (coefficient of Tpt-Jap > 0.8, Tpt-Juser = 0.17, so it is more effective to use J of AP whereas not J of user to represent the throughput).

6.3. Comparison Analysis

We select average case in **Figure 1**, at the 100th time, the J of User is almost equal to 0.75, and corresponding J of AP is almost equal to 0.94, the $\text{Throughput}_{\text{relative}}$ is almost equal to 145, since its maximum value is 200 as mentioned before, then the throughput of the network is equal to 145/200 = 72.5% of the network bandwidth. And the corresponding cost of f is almost equal to 0.061.

Here we want to compare our solution with PCAP in **Table 4** from [13], since we use different definitions to denote throughput of AP and throughput of network, we have to use the indirect method to illustrate some issues. According to [13], we can transfer and calculate their J of AP and their throughput percentage of network bandwidth:

$$cvAPs^2 = \frac{S^2(U[j])}{E^2(U[j])} = \frac{\log(2.98\times10^9)}{\left(\log(3.82\times10^9)\right)^2} = 0.1 = \frac{1}{J_{APs}} - 1 \quad (21)$$

So their

$$J_{APs} = 0.9 = J_{users} \quad (22)$$

And we have:

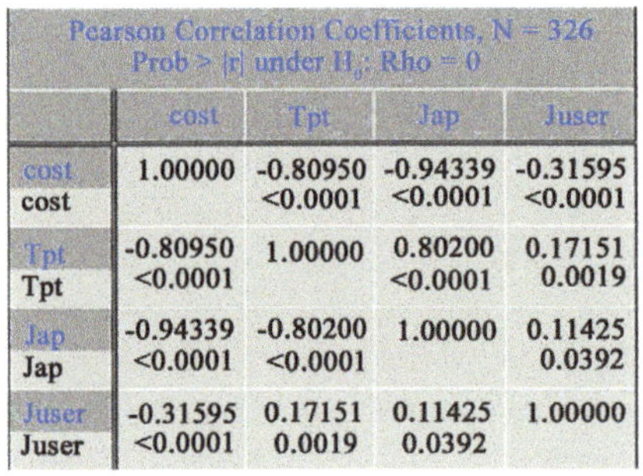

Pearson Correlation Coefficients, N = 326
Prob > |r| under H_0: Rho = 0

	cost	Tpt	Jap	Juser
cost	1.00000	-0.80950 <0.0001	-0.94339 <0.0001	-0.31595 <0.0001
Tpt	-0.80950 <0.0001	1.00000	0.80200 <0.0001	0.17151 0.0019
Jap	-0.94339 <0.0001	-0.80200 <0.0001	1.00000	0.11425 0.0392
Juser	-0.31595 <0.0001	0.17151 0.0019	0.11425 0.0392	1.00000

Figure 3. Correlation coefficients.

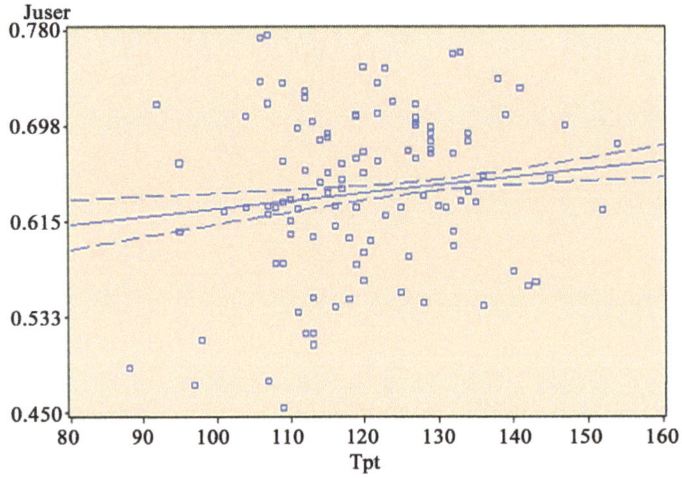

Figure 4. Tpt-Juser linear regression.

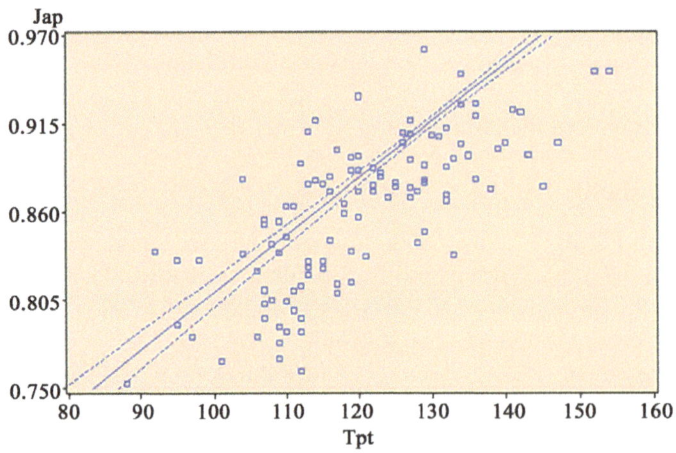

Figure 5. Tpt-Jap linear regression.

Table 4. The statistics of the results.

Algorithm	Average AP Utility $(\bar{U}^a)$	AP Utility Variance (σ_u)	Total Network Utility	Jain's Fairness Index	Average Power (dBm)
PCAP	3.82×10^9	2.98×10^9	117.42	0.90	17.15
MARL	2.67×10^9	5.00×10^9	102.30	0.85	17.28
SR	1.46×10^9	12.80×10^9	99.66	0.75	18.06
SSF	1.11×10^9	27.31×10^9	79.59	0.37	20.00

$$U \le n\log\left(\overline{U^a}\right) = 20\log\left(3.82\times10^9\right) = 191.64 = U_{\max} \tag{23}$$

Then their throughput percentage of network bandwidth is:

$$U_{\text{network-utility}} / U_{\max} = 117.42/191.64 \approx 61.3\% \tag{24}$$

In **Figure 1**, our J of AP is superior to theirs in (22). From the throughput point of view, our throughput percentage of network bandwidth is 72.5% > 61.3% in (24), so our method is better than PCAP. But from the fairness of users (J of user) point of view, PCAP is better than ours since 0.75 < 0.9 in (22).

According to (17), we convert (22) into our function F, we have:

$$\begin{aligned} F_{PCAP} &= \left[\left(1/J_{users}\right)-1\right]+\left(M/N\right)\left[\left(1/J_{APs}\right)-1\right] \\ &= \left[\left(1/0.9\right)-1\right]+\left(300/16\right)\left[\left(1/0.9\right)-1\right] = 2.17 \end{aligned} \tag{25}$$

$$F_{SCV} = Mf - \left(1+M/N\right) = 200\times0.061-\left(1+200/20\right) = 1.2 \tag{26}$$

So the overall performance depends on the requirement of administrators, which indicator they most concern. We define the value of "F" as the overall performance criteria of algorithm, note smaller "F" is better, and then from (25) and (26) we know that our SCV model is much better than PCAP. The above comparison analysis result is in **Table 5**.

Theoretically, our design of target function "F" in (17) is more simple and rational than PCAP algorithm, since we joint consider the J of user and throughput (represented by J of AP), we regard them as two variables to reflect our topic which is a balance problem. While the target of PCAP is the throughput, the authors used two sub-algorithms to achieve J of AP only, and then got their by-product: J of user.

Technically, our SCV math model is a door that leads this problem to AI algorithms. The clear target function "F" is easy to be applied to other AI algorithms, while PCAP cannot.

7. Conclusions

The objective of this paper is to improve the trade-off between user fairness (J of user) and network throughput (represented by J of AP) via power control in multi-rate WLANs.

In this article, we first construct a new simplified model called SCV. The goal of the model is to derive a target function "F" in (17) and its simplified form "f" (18) as our key foundation. Then we use Ant Colony Algorithm to solve our model, and we conduct a simulation in Matlab. After that we give analysis of our SCV model and simulation results confirm that our model is efficient and superior to PCAP in some aspects and overall performance under new criteria designed for such specific problem. In addition, based on the data samples from the state space, we use SAS to conduct correlationship analysis mainly among three indicators, and reveal their relationships.

SCV (Target function F) opens a door for many AI algorithms to apply in this problem; it is a bridge between Network & AI.

Our future work is to derive a more accurate target function, and adjusts the values of parameters to find more suitable combination so that to improve the performance. Also we are working on other AI solutions based on SCV model.

Table 5. Comparison result.

	PCAP	SCV-ant
J of user (↑win)	0.9	0.75
J of AP (↑win)	0.9	0.94
throughput % (↑win)	61.3%	72.5%
Function "F" value (↓win)	2.17	1.2

Acknowledgements

The authors extend their appreciation to the Deanship of Scientific Research at King Saud University for funding this work through research group No. RGP-264.

References

[1] Ma, Q., Al-Dhelaan, A. and Al-Rodhaan, M. (2015) Using Hopfield Neural Network to Improve the Performance of Multi-rate WLANs. *Proceedings of 4th WSEAS International Conference on Circuits, Systems, Communications, Computers and Applications*, Kuala Lumpur, May 2015, 170-178.

[2] Asoodeh, S. (2008) New Algorithm for Power Control in Cellular Communication with ANFIS. *WSEAS Transactions on Communications*, **7**, 8-14.

[3] Chen, J.-S., Wang, N.-C., Hong, Z.-W. and Chang, Y.-W. (2009) An Adaptive Load Balance Allocation Strategy for Small Antenna Based Wireless Networks. *WSEAS Transactions on Communications*, **8**, 588-597.

[4] Hu, Q. and Tang, Z.Z. (2010) Study on Power and Rate Control Algorithm for Cognitive Wireless Networks. *WSEAS Transactions on Communications*, **9**, 281-289.

[5] Tan, G. and Guttag, J. (2004) Time-Based Fairness Improves Performance in Multi-Rate WLANs. *Proceedings of Usenix Annual Technical Conference*, USENIX Association Berkeley, 23.

[6] Heusse, M., Rousseau, F., Berger Sabbatel, G. and Duda, A. (2003) Performance Anomaly of 802.11b. *Proceedings of IEEE INFOCOM*, San Francisco, 30 March-3 April 2003, 836-843. http://dx.doi.org/10.1109/infcom.2003.1208921

[7] Kelly, F.P. (1997) Charging and Rate Control for Elastic Traffic. *European Transactions on Telecommunications*, **8**, 33-37. http://dx.doi.org/10.1002/ett.4460080106

[8] Banchs, A., Serrano, P. and Oliver, H. (2007) Proportional Fair Throughput Allocation in Multi-Rate IEEE 802.11e Wireless LANs. *Wireless Networks*, **13**, 649-662. http://dx.doi.org/10.1007/s11276-006-6972-9

[9] Babu, A.V. and Jacob, L. (2005) Performance Analysis of IEEE 802.11 Multi-Rate WLANs: Time Based Fairness vs. Throughput Based Fairness. *Proceeding of IEEE International Conference on Wireless Networks, Communications, and Mobile Computing*, Sheraton Maui Resort Maui, June 2005, 203-208.

[10] Bertsekas, D. and Gallager, R. (1987) Data Networks. Prentice-Hall, Upper Saddle River.

[11] Li, W., Cui, Y., Wang, S.L. and Cheng, X.Z. (2010) Approximate Optimization for Proportional Fair AP Association in Multi-Rate WLANs. *Proceedings of 5th International Conference on Wireless Algorithms, Systems, and Applications*, Beijing, 15-17 August 2010, 36-46.

[12] Bejerano, Y., Han, S.J. and Li, L. (2007) Fairness and Load Balancing in Wireless LANs Using Association Control. *IEEE/ACM Transactions on Networking*, **15**, 560-573. http://dx.doi.org/10.1109/TNET.2007.893680

[13] Li, W., Cui, Y., Cheng, X.Z., Al Rodhaan, M.A. and Al Dhelaan, A. (2011) Achieving Proportional Fairness via AP Power Control in Multi-Rate WLANs. *IEEE Transactions on Wireless Communications*, **10**, 3784-3792. http://dx.doi.org/10.1109/twc.2011.091411.101899

[14] Jain, R., Chiu, D.M. and Hawe, W.R. (1984) A Quantitative Measure of Fairness and Discrimination for Resource Allocation in Shared Computer System. *Digital Equipment*, Tech. Dec-Tr-301.

[15] Mhatre, V.P., Papagiannaki, K. and Baccelli, F. (2007) Interference Mitigation through Power Control in High Density 802.11 WLANs. 26*th IEEE International Conference on Computer Communications, INFOCOM*, Anchorage, 6-12 May 2007, 535-543. http://dx.doi.org/10.1109/infcom.2007.69

[16] Hasu, V. and Koivo, V. (2006) Fair Transmission Rate Allocation: A Power Control Feasibility Approach. 10*th IEEE Singapore International Conference on Communication Systems, ICCS*, Singapore, 30 October-1 November 2006, 1-5. http://dx.doi.org/10.1109/iccs.2006.301457

[17] Bejerano, Y. and Han, S.J. (2009) Cell Breathing Techniques for Load Balancing in Wireless LANs. *IEEE Transactions on Mobile Computing*, **8**, 735-749. http://dx.doi.org/10.1109/TMC.2009.50

[18] Wang, S.L., Huang, J.H., Cheng, X.Z. and Chen, B. (2014) Coverage Adjustment for Load Balancing with an AP Service Availability Guarantee in WLANs. *Wireless Networks*, **20**, 475-491. http://dx.doi.org/10.1007/s11276-013-0615-8

[19] Qian, L.P. and Jun, Y. (2009) Monotonic Optimization for Non-Concave Power Control in Multiuser Multicarrier Network Systems. *Proceedings of IEEE INFOCOM* 2009, Rio de Janeiro, 19-25 April 2009, 172-180. http://dx.doi.org/10.1109/INFCOM.2009.5061919

[20] Chiang, M., Tan, C.W., Palomar, D.P., O'Neill, D. and Julian, D. (2007) Power Control by Geometric Programming. *IEEE Transactions on Wireless Communications*, **6**, 2640-2651. http://dx.doi.org/10.1109/TWC.2007.05960

[21] Li, W., Wang, S.L., Cui, Y., Cheng, X.Z., Xin, R., Al-Rodhaan, M.A. and Al-Dhelaan, A. (2014) AP Association for

Proportional Fairness in Multirate WLANs. *IEEE/ACM Transactions on Networking*, **22**, 191-202. http://dx.doi.org/10.1109/tnet.2013.2245145

[22] Wang, Q.S. and Liu, M.Y. (2013) Throughput Optimal Switching in Multichannel WLANs. *IEEE Transactions on Mobile Computing*, **12**, 2470-2482. http://dx.doi.org/10.1109/tmc.2012.228

[23] Islam, M.E., Funabiki, N., Nakanishi, T. and Watanabe, K. (2013) An Extension of Access-Point Aggregation Algorithm to Ensure Minimum Host Throughput for Wireless Local Area Networks. 2013 1*st International Symposium on Computing and Networking*, Matsuyama, 4-6 December 2013, 141-147. http://dx.doi.org/10.1109/CANDAR.2013.27

[24] Cui, Y., Ma, T.Z., Liu, J.C. and Das, S. (2013) Load-Balanced AP Association in Multi-Hop Wireless Mesh Networks. *The Journal of Supercomputing*, **65**, 383-409. http://dx.doi.org/10.1007/s11227-010-0519-7

[25] Chakraborty, S., Swain, P. and Nandi, S. (2013) Proportional Fairness in MAC Layer Channel Access of IEEE 802.11s EDCA Based Wireless Mesh Networks. *Ad Hoc Networks*, **11**, 570-584. http://dx.doi.org/10.1016/j.adhoc.2012.08.003

[26] Brownlee, J. (2011) Clever Algorithms: Nature Inspired Programming Recipes. LuLu, Raleigh.

[27] Ding, X.L., Li, F.H., Li, H.W., Jiang, Y. and Wu, J.P. (2007) Dynamic Load Balancing Mechanism in WLAN Based on Power Control and Location Information. *Journal of Xiamen University* (*Natural Science*), **46**, 150-155.

[28] Goldsmith, A. (2004) Wireless Communications. Stanford University, Stanford.

[29] Yin, Z.Q., Shi, C.H., Chen, M.S. and Liu, S.Z. (2008) A White and Gaussian White Noise Generator with Adjustable Parameters. *Fire Control and Command Control*, **33**, 109-111.

[30] Wang, P.Y., Zhai, L.L. and Shi, J.F. (2013) Design of Gaussian White Noise Generator with Adjustable Parameters Based on FPGA. *Shipboard Electronic Countermeasure*, **36**, 113-115.

[31] Villegas, E.G., López-Aguilera, E., Vidal, R. and Paradells, J. (2007) Effect of Adjacent-Channel Interference in IEEE 802.11 WLANs. 2*nd International Conference on Cognitive Radio Oriented Wireless Networks and Communications*, Orlando, 1-3 August 2007, 118-125. http://dx.doi.org/10.1109/CROWNCOM.2007.4549783

16

Cars as a Diffuse Network of Road-Environment Monitoring Nodes

Marco Allegretti[1], Silvano Bertoldo[2]

[1]Envisens Technologies s.r.l., Torino, Italy
[2]CINFAI (Consorzio Interuniversitario Nazionale per la Fisica delle Atmosfere e delle Idrosfere), c/o Dipartimento di Elettronica e Telecomunicazioni, Politecnico di Torino, Torino, Italy
Email: marco.allegretti@polito.it, silvano.bertoldo@polito.it

Abstract

The present paper aims to describe the conceptual idea to use cars as sensors to measure and acquire data related road environment. The parameters are collected using only standard equipment commonly installed and operative on commercial cars. Real sensors and car sub-systems (e.g. thermometers, accelerometers, ABS, ESP, and GPS) together with other "implicit" sensors (e.g. fog lights, windscreen wipers) acquire and contain information. They are shared inside an in-vehicle communication network using mainly the standard CAN bus and can be collected by a simple central node. This node can also be available on the market without too expensive costs thanks to some companies which business is devoted to car fleet monitoring. All the collected data are then geolocalized using a standard GPS receiver and sent to a remote elaboration unit, exploiting mobile network technologies such as GPRS or UMTS. A large number of cars, connected together in a diffuse Wireless Sensor Network, allow the elaboration unit to realize some info-layers put at the disposal of a car driver. Traffic, state of the road and other information about the weather can be received by car drivers using an ad hoc developed mobile application for smartphone which can give punctual information related to a specific route, previously set on the mobile phone navigator. The description of some experimental activities is presented, some technical points will be addressed and some examples of applications of the network of cars "as sensors" will be given.

Keywords

Cars, Meteorological Sensor, Traffic Sensor, Weather, Wireless Sensor Network, WSN, Cars as a Sensor, Commercial Sensors, Car Sensors, Road Environment Sensors, Environment Monitoring

1. Introduction

Detailed information about traffic, state of the road and weather along a specific route a car is going to take, are essential to avoid dangerous situation and improve safety.

Following a scheme we have already implemented for specific information concerning rain distribution (described below), we are developing a more general approach using only standard equipment commonly installed and operative in commercial cars.

Let us assume that a driver is choosing on the navigator of his smartphone the route he intends to follow. A specific app installed on his own smartphone is checking its position and, at the same time, is consulting a database containing a set of specific georeferenced information about the surrounding area and, in particular, along the road indicated by the navigator: the app gets the data from the database and reports them on the map which is presented to the driver.

A similar experimental system has been already realized although limited to rain data. Since 2004 the Remote Sensing Group (RSG), the local unit of Consorzio Interuniversitario per la Fisicadelle Atmosfere e delle Idrosfere (CINFAI) at Politecnico di Torino, started to develop a first fully operational (although experimental) network of short range X-band mini weather radars. Their low cost, the lack of any special requirements for installation, practically maintenance-free and the possibility to be remotely controlled with software adjournments, allow their installations almost everywhere. Arain image map, adjourned each minute with a high resolution within a range of 30 km, is produced for real time monitoring of weather [1].

Two mobile applications for Android© systems have been already developed by Envisens Technologies s.r.l. (EST). The first, already available for free on Google store©, is called Meteoradar-IT and allow users to have a look at the precipitation in real time over the areas monitored by the different radars already installed (**Figure 1**). The second is Meteoradar-IT-OnRoute, it is experimental only, and it checks on real time for the presence of rain over the route set on the navigator. It is an example of car driver service to communicate a possible alarm if a severe rainfall event occurs on a predefined route. Because the system is also able to provide projection son the movement of the rain fields in the next 30 minutes, the car driver can be also informed in advance of the rain situation along his chosen route for the next 30 minutes.

The idea to use the moving cars to acquire meteorological and traffic data is not presented here for the first time. Especially in the USA, some experiments and discussions were already addressed since 2006 [2]. They are based on dedicated meteorological sensors installed onboard to the car [3], on the interconnection between the car and other infrastructure to improve the safety, and on the acquisition of cars parameters to improve drivers' safety [4]. The innovation presented here is that data and information are acquired using the "car as a sensor" as it is and, at least to start, using only commercial fleet of cars already connected in a network for commercial and insurance purposes.

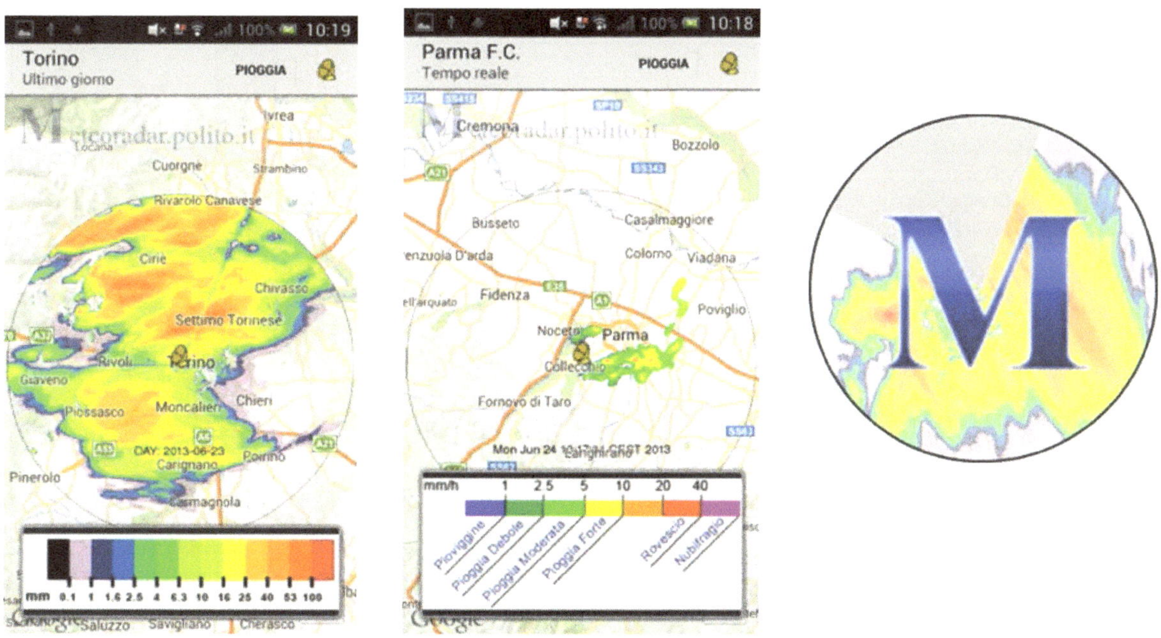

Figure 1. The meteoradar-IT application.

The new conceptual idea of a Wireless Sensor Network (WSN) for meteorological purposes, traffic and roads monitoring can be technically feasible because, presently, cars are already provided with a set of sensors, systems, and technologies which are commonly used to improve car and passengers' safety and comfort: ABS (Antilock Breaking System), ESP (Electronic Stability Comfort), tire pressure control and accelerometers, thermometers to give the passengers the indication of the temperature outside the vehicle and possibly turn on or off the automatic air-conditioner and rain sensors, on newer cars, to automatically turn on the windscreen wipers.

The set of information from the sensor are collected inside an in-vehicle car network. Furthermore more recent cars are already equipped with an integrated GPS navigator, thus knowing the position of the car itself.

Of course, all these data need to be centralized in order to enter in the construction of the database, accessed by any user. However in most European countries numerous fleets of cars already presents devices for transmitting data for fleet management purposes, namely car position plus parameters like sudden strong deceleration, to a central node.

In other words, while in the future most cars may transmit their on-board information to some centralized collector, for the time being, it is possible to take advantage of the telecommunication structures already operative for many cars fleet (**Figure 2**).

It is to note that the central collector on the WSN car can be also already installed in common commercial cars.

2. Cars as a Sensor Node

As reported in a paper of 2008 [5], a modern car can be equipped with more than 100 different sensors and the entire automotive systems application for sensors can be divided in three main areas: power train, chassis, and body.

Most of the sensors of the three areas are not useful, directly or not directly, for road environment monitoring. However:

- Accelerometers (mainly used for motion detection and impact measurements) can be used to monitor the state of the streets.

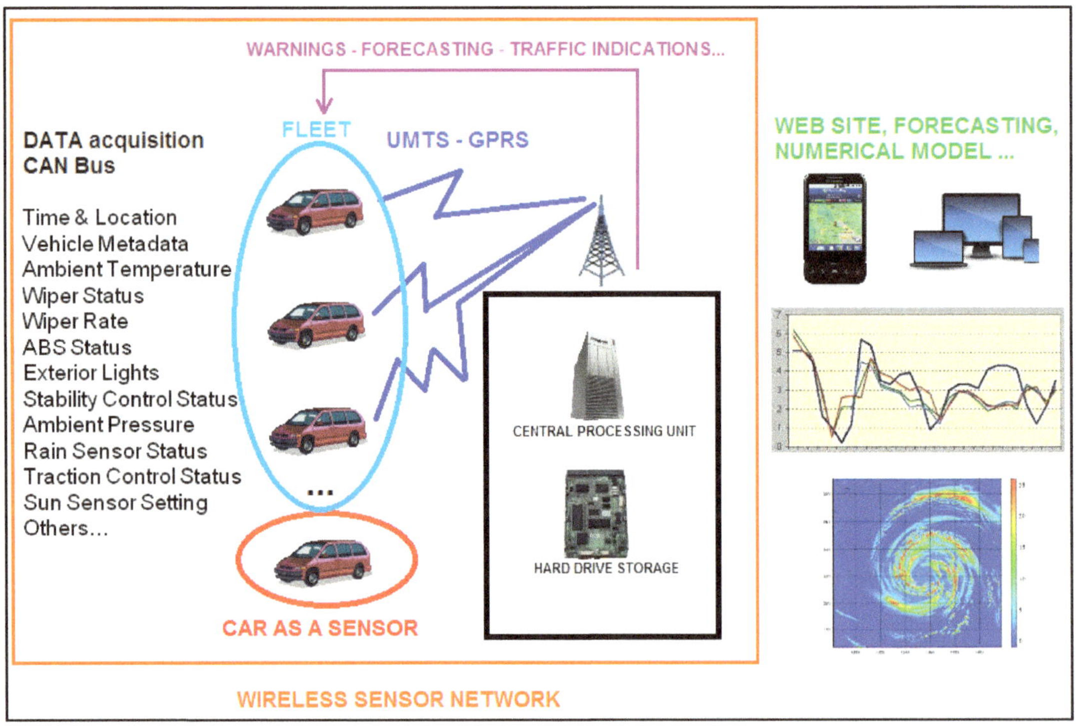

Figure 2. Operational scheme of the proposed diffused Wireless Sensor Network, from the sensors (*i.e.* the cars) to the final elaborated products (improved numerical model, traffic indication, weather indication, web sites…).

- Speed monitor can be used, after re-analysis, for traffic estimate.
- Angular and linear position sensors or gyroscope (to measure, for instance, steering angle, mirror adjustment, seat position and light adjustment) can be used together with the accelerometers to evaluate the state of the roads.
- Sensors restricted to the luxury cars, such as radar and video sensors to assist the driver during the journey and during parking operations, can give additional information.
- External temperature sensors are especially important when near zero degree Celsius are measured.
- Pressure sensors are essential to characterize the atmosphere.
- Rain sensors, present on some modern vehicles, are capable to detect when it is raining and automatically activate the windscreen wipers

Other potentially useful complex systems, made by a set of sensors, are ABS and ESP. They work when some criticality is detected on the road in order to improve the safety of the car, at first glance. Their information seems useless because criticalities on the road may be due not only to ice, water and other meteorological events but also to the way the car is used and its stability problems. However, the occurrence of the same output from different cars, in the same road section, at the same time, is certainly signaling specific road problems.

It is to be underlined that data acquired and measured by the sensor and the systems above are available on the CAN (Controlled Area Network) bus (see paragraph 2.3) of the car: it is consequently technically simple to collect them for future use.

2.1. "Implicit" Sensors and Systems

Together with the specific sensors and subsystems of immediate use for road environment monitoring, other devices and car apparatus can be used as "implicit" sensors.

The fog lights are the first example of this kind of "implicit" sensors: if a set of cars has the fog lights on in a road section, it gives evidence of fog presence in that specific road section.

The windscreen wipers are another example: if they are turned on for a certain interval of time and on a certain number of cars, they indicate that it is raining; their speed gives also an indication of the intensity of the rain (see Section 5).

In order to have reliable information coming from this kind of sensors, it is of course necessary to consider a set of car and not a single one. Indeed the fog lights or windscreen wipers may have been turned on for different reasons by a single driver. Therefore considering the information coming from a single car may lead to a misunderstanding of the meteorological situation of a specific area.

The tire pressure sensors are the last example of "implicit" sensor. You can note that the pressure is measured with a real pressure sensor. But the measured pressure can be put in relation with the temperature of the air and the temperature of the road. The tire pressure sensors implicitly can act as road temperature sensors giving information related to the state of the road and the surrounding environment.

2.2. Communication on the In-Vehicle Network

The most recent research trend concerning the communications on the in-vehicle network, also called Smart Car Sensor Network (SCSN), is mainly twofold: to improve the performance of the communication with new standards and protocols and to implement innovative applications devoted to improve safety and reliability of the current ones [6].

In 1994 the Society for Automotive Engineers (SAE) classified the communication protocols into three categories (A, B, C) on the basis of the data transmission speed and their capabilities to manage different functions. A fourth category (D) was added later [7] [8]. Class A networks are characterized by a data rate lower than 10 kb/s, class B network between 10 and 125 kb/s, class C up to 1 Mb/s and class D includes all the networks requiring more than 1 Mb/s.

CAN was one of the first automotive control networks and nowadays it is the de facto standard for in-vehicle communication network. It was developed by Robert Bosch in the middle of 1980s and it has become ISO standard in 1994 [9]. CAN is a serial, event-triggered, broadcast bus, able to offer different transmission rates (up to 1 Mb/s), flexibility, robustness, bounded delay, cheapness, and simplicity. It also implements several error detection techniques and also an energy saving mechanism (e.g. nodes sleep until a CAN message wakes them). It embraces all the categories of communication protocols on board the cars.

Readers can refer to [6] for a complete list and a detailed analysis of the traditional protocol for the SCSN communication.

For what concern the usage of a car as a road environment sensor the CAN bus is very suitable. In fact all the car sensors are organized in a SCSN able to collect the acquired measurements and making them available on the traditional CAN bus, which allows also a sufficient data rate. It is simply necessary to connect to the CAN bus also the access point capable to acquire all the information of interest.

3. A WSN of "Cars as Sensor"

3.1. The Cars: The Network Nodes

According to what we have been seeing above, the car may be considered as a sensor itself.

The data can be acquired by a central node, called Access Point (AP), already existing in the cars fleet, sending, for the time being, position and, in many cases, excess deceleration to some centralized unit via UMTS or GPRS. Such existing APs are already connected to the CAN bus and need to be reprogrammed (in some cases the reprogramming operations may be performed remotely).

Of course, in developing the acquisition program, one has to carefully analyze the raw data, programming also the final amount of data to be transmitted as a function of car speed and data nature. If, for instance, a car is running at 100 km/h, transmitting, let us say, 10 data each 30 seconds, it implies giving road environment information along and approximately 1000 m track. Another example is related to when the on-board system realizes the car is entering in a fog bank: the corresponding information has to be transmitted immediately.

A structure of the message sent by the car to the central unit can be very simple. It must contain sensor information (temperature, accelerometers data, indication of speed of windscreen wipers…) together with coordinates and, possibly speed, of the car.

3.2. The Central Elaboration Unit

The central elaboration unit is devoted to the storage of the data and their elaboration. It should be connected to internet (and possibly to the mobile phone network) in order to receive all the data coming from the cars. The central unit is the central node of the network. It can be made by different hardware units, PCs and servers and it should be located in a safe place. It should perform different tasks:

- It acts as database server, because a well-designed database can be implemented to collect all the data coming from a huge set of cars in the referring area.
- It acts as a web server, in order to provide web services that can be browsed by citizens and drivers.
- It acts as an application server to deploy data to mobile apps developed for Android© and Mac IOS© operative systems (e.g. for indication of serious meteorological events on a predefined route, for traffic jam…).
- It allows specific and ad-hoc developed services to run (e.g. forecast services).
- It executes specific ad-hoc routine developed to signal a possible warning due to severe meteorological situation over a segment of a specific area (e.g. fog alert, rainfall alert, traffic jam alert…).
- It gives authorized users the possibility to access to the database and download data.
- In case of meteorological usage of the cars as sensors, it integrates data coming not only from cars, but also from other weather monitoring tools (weather radars, rain gauges, weather stations…) in order to run improved forecast algorithm working on different types of data.

All the parameters measured by the cars and made available on internet are part of the so called "Internet of Things" world. The cars become intelligent embedded systems capable to share some of their characteristics and parameters and communicate the information to other car drivers who have the mobile applications, in order to improve their safety.

4. Examples of Application

4.1. Rain Monitoring

Different examples of applications of the network of cars as sensors can be reported (traffic monitoring, monitoring the state of the roads, fog communication) but one of the most significant is related to rain monitoring, thanks also to the join expertise of EST and CINFAI.

Accurate observation of precipitations in terms of intensity and location is important for many applications: hydro-geological risk management, hydroelectric power, road maintenance, emergency planning, and tourism are only some examples of a wider number of fields. For example in [10] it is shown the importance of high-resolution rainfall data using X-band radar network for what concern monitoring operation of extreme intense events and in complex orography environment.

There are severaloperative techniques for rainfall intensity measurements such standard C-band or S-band weather radar, satellites, microwave links [11], acoustic rain gauges [12] and even radio links from cellular communication [13].

The usage of rain sensor installed on the car and the indication of the functioning and speed of the windscreen wipers may be another innovative method to monitor rain fields. The cars are then used as a huge number of rain gauges.

Some models and experiments about using the moving cars as rain gauges are presented in [14] [15] with very good results. They showed also that is possible to extrapolate a W-R (windscreen wipers speed-rain) relationship which gives an indication of the amount of rain over a specific area.

Radars suffer time variable bias in rainfall estimation [16] and should be calibrated often in order to have a better rain estimation with a point precipitation network. Results in [17] have shown that a high number of possibly inaccurate measurement devices (moving cars used as rain gauges) provide more reliable areal rainfall estimations than a lower number of precise measurement devices (stationary gauges). Therefore the cars may also be used to improve standard radar estimation and to calibrate X-band weather radars.

4.2. Fog Alert

The presence of fog along a road is often a cause of car accidents. The fog lights of the cars, as previously said, can be used as implicit sensors. Let us consider a set of cars where the information related to the fog lights on or off is acquired by the central node of the in-vehicle WSN and sent to the central elaboration unit. Let us consider also that two cars are moving in the same direction and the driver of the second car is equipped also with a smartphone with the application capable to signal the presence of fog (similar to the one described in Section 2). The first car turn on the fog lights, the information is sent to the central elaboration unit and the driver of the second car, who has previously set his route on the application, receive the alert for the fog in the next 10 kilometers on his smartphone, so he can reduce its speed and turn on the fog lights on his car.

4.3. Traffic Jam Alert

The communication of a traffic jam or a car queue on a highway can be very useful for drivers. Let us consider to have a set of car connected together in a WSN as previously described. All the cars are equipped with a GPS and some accelerometers capable to detect position, speed and accelerations. Combining such data it is possible to detect the traffic on a specific segment of road. If a car driver has previously installed on his smartphone the ad-hoc developed applications can have the indication of the traffic considering information coming from a huge set of vehicle moving in the same direction.

The following **Figure 3** shows an example of indication of heavy traffic considering GPS information. It is the results of an experimental project between EST, CINFAI and another company devoted to the production of car equipment for car fleet control for assurance purposes in 2013. The same technology can be adopted in the proposed WSN.

4.4. Data Input to Meteorological Models

Detailed data are the most important input information to improve the local meteorological forecast and the need of large amount of meteorological and climatological information is always growing due to the increasing complexity of the numerical model. Real time forecasting over local areas, potentially provides continuous, high-resolution and large-coverage information, are of high practical use: think of hydrogeological risk management, civil protection activities, road maintenance and safety.

A meteorological model runs starting from some initial conditions. It gives as output a forecast of the evolution of the weather within a predefined interval of time by setting specific boundary conditions. Both initial and boundary conditions are often set using the measurements made by weather stations installed on the ground.

Figure 3. App with real time traffic indication along a user predefined route with data coming from a fleet of cars. Purple color indicates heavy traffic. Icon below refer to other information related to the user defined route (e.g. weather alert, fog alert).

Moreover transfer schemes commonly used in meteorological models, often do not account for small scale variability of the earth surface, because measurements of the atmospheric conditions do not exist in such a high spatial resolution to force the models. A high resolution grid of data coming from the ground improve a lot the performances of the models or validate the models themselves, and ad hoc WSNs can be used for these purposes [17].

The cars, used as road environment sensors are capable to acquire meteorological parameters (rain intensity, temperature, humidity…). The acquired information can be used as further input data for meteorological models.

5. Conclusions and Outlooks

The feasibility of WSN made by cars was described. Each car is a sensor node of the WSN and exploits only the sensors and the sub-systems commonly carried on board without adding any other component, possibly except from a simple access point capable to acquire the desired information already available the car internal CAN buses. All the information is sent to a central elaboration unit which made them available to car drivers with some mobile applications.

Future projects and researches should be focused on the hardware realization of an ad hoc access point to be installed on the car to acquire and integrate data measured by different sensors. Moreover, the central elaboration unit of the WSN will be designed and some specific services will be implemented. The WSN should be realized, put in test and eventually should be made fully operational.

Acknowledgements

The present paper represents the result of a research activity promoted by Envisens Technologies s.r.l., spinoff of Politecnico di Torino (www.envisens.com) together with the local research unit of CINFAI (Consorzio Interuniversitario Nazionale per la Fisicadelle Atmosfere e delle Idrosfere, www.cinfai.it) at DET (Dipartimento di

Elettronica e Telecomunicazioni, www.det.polito.it) of Politecnico di Torino.

References

[1] Gabella, M., Notarpietro, R., Bertoldo, S., Prato, A., Lucianaz, C., Rorato, O., Allegretti, M. and Perona, G. (2012) A Network of Portable, Low-Cost, X-Band Radars. In: Bech, J., Ed., *Doppler Radar Observations—Weather Radar, Wind Profiler, Ionospheric Radar, and Other Advanced Applications*, InTech, 175-202.

[2] Stern, A.D., Shah, V.P., Biesecker, K.J. and Yeung, C. (2007) Vehicle as Mobile Sensing Platforms for Meteorological Observations: A First Look. *Proceedings of the* 87*th* *AMS Annual Meeting*, 23*rd* *Conference on IIPS*, San Antonio, 15-18 January 2007. https://ams.confex.com/ams/87ANNUAL/techprogram/paper_118986.htm

[3] Crawford, D. (2012) Pioneering Sensors Collect Weather Data from Moving Vehicles. *ITS International*, Swanley. http://www.itsinternational.com/sections/nafta/features/pioneering-sensors-collect-weather-data-from-moving-vehicles

[4] Drobot, S., Chapman, M., Shuler, E., Wiener, G., Mahoney, W., Pisano, P. and McKeever B. (2009) Improving Road Weather Hazard Products with Vehicle Probe—The Vehicle Probe Data Translator Quality Check Procedures. *Transportation Research Record—Journal of Transportation Research Board*, **2169/2010**, 128-140.

[5] Fleming, W.J. (2008) New Automotive Sensors—A Review. *IEEE Sensors Journal*, **8**, 1900-1921. http://dx.doi.org/10.1109/JSEN.2008.2006452

[6] D'Orazio, L., Visintainer, F. and Darin, M. (2011) Sensor Networks on the Car: State of the Art and Future Challenges. *Proceedings of Design, Automation & Test in Europe Conference & Exhibition* (*DATE*), Grenoble, 14-18 March 2011, 1-6.

[7] (1994) J2056/1 Class C Application Requirements Classifications. SAE Handbook. Soc. Automotive Eng. (SAE), Warrendale.

[8] (1994) J2056/2 Survey of Known Protocols. SAE Handbook, Vol. 2, Soc. Automotive Eng. (SAE), Warrendale.

[9] (1994) Road Vehicle, Low Speed Serial Data Communication—Part 2: Low Speed Controller Area Network. International Standard ISO 11519-2:1994.

[10] Allegretti, M., Bertoldo, S., Prato, A., Lucianaz, C., Rorato, O., Notarpietro, R. and Gabella, M. (2012) X-Band Mini Radar for Observing and Monitoring Rainfall Events. *Atmospheric and Climate Sciences*, **2**, 290-297. http://dx.doi.org/10.4236/acs.2012.23026

[11] Upton, G.J.G., Holt, A.R., Cummings, R.J., Rahimi, A.R. and Goddard, J.W.F. (2005) Microwave Links: The Future for Urban Rainfall Measurement? *Atmospheric Research*, **77**, 300-312. http://dx.doi.org/10.1016/j.atmosres.2004.10.009

[12] De Jong, S. (2010) Low Cost Disdrometer. Master Thesis Report, TU Delft, Delft.

[13] Leijnse, H., Uijlenhoet, R. and Stricker, J.N.M. (2007) Rainfall Measurement Using Radio Links from Cellular Communication Networks, *Water Resources Research*, **43**, Article ID: W03201. http://dx.doi.org/10.1029/2006WR005631

[14] Haberland, U. and Sester, M. (2010) Areal Rainfall Estimation Using Moving Cars as Rain Gauges—A Modelling Study. *Hydrology and Earth System Sciences*, **14**, 1139-1151. http://dx.doi.org/10.5194/hess-14-1139-2010

[15] Rabie, E., Haberland, U., Sester, M. and Fitzner, D. (2013) Rainfall Estimation Using Moving Cars as Rain Gauges—Laboratory Experiments. *Hydrology and Earth System Sciences*, **17**, 4701-4712. http://dx.doi.org/10.5194/hess-17-4701-2013

[16] Krajewski, W.F. and Smith, J.A. (2002) Radar Hydrology: Rainfall Estimation. *Advances in Water Resources*, **25**, 1387-1394. http://dx.doi.org/10.1016/S0309-1708(02)00062-3

[17] Lengfeld, K., Zacharias, S. and Ament, F. (2013) Using a Small Scale Wireless Sensor Network for Model Validation: Two Case Studies. *Meteorologische Zeitschrift*, **22**, 283-295. http://dx.doi.org/10.1127/0941-2948/2013/0422

17

A Novel Data Preparation Approach for Target Based Association Rule Mining

Prachi Singh, Rajendra Kumar Gupta

Madhav Institute of Technology & Science Gwalior, Gwalior, India
Email: Singh.prachi5@gmail.com, rkgiiitm@gmail.com

Abstract

Target based association rules extract the correlation between targets in and around strategically specified regions. These rules can be used to predict and estimate the future results and find the faulty targets, etc. To extract these rules efficiently and effectively data preparation is required. The existing mechanisms of data preparation have the problem of redundancy in the data. As a consequence, extra energy is required for the sensors which used to monitor the targets. In this paper, authors propose a novel data preparation approach for target based association rule mining from the point of coverage wireless sensor networks, which reduces the redundancy in the data and thus enhances the performance of the network.

Keywords

Knowledge Discovery, Point of Coverage Wireless Sensor Networks

1. Introduction

Knowledge discovery process has been proved a very effective technique to find out hidden, previously unknown and useful information (knowledge) from the wireless sensor networks [1]-[4]. It is a complex process which involves various steps such as domain understanding, knowledge definition, data preparation (data cleaning, data reduction, etc.) and finally data mining. Association rule mining has become the one of the popular technique of knowledge discovery which is introduced by Agrawal *et al.* [5] in 1993 to find association among objects from the transactional databases. This technique can also be used to find patterns from wireless sensor networks, which can be used to predict future results and missing values and find faulty nodes. Using this, Samarah *et al.* [6] [7] have proposed a technique that is known as the sensor association rules to find the relation between the nodes in the wireless sensor networks. In this technique the sensor itself acts as the main object, rather than the sensors' readings.

In order to prepare the data required for obtaining sensor association rules, each sensor node should profile its activity over time and notify to the sink about the timestamps of detected events. This technique uses the meta-data, which describes the sensors' behavior, not the data reported by the sensor about the surrounding environment [6]. The extensions of this technique have been used in target based association rules (TARs) that can be used to enhance the Quality of Service (QoS) of a point of coverage wireless sensor network (POCW) [8].

POCW is a network that consists of a set of targets, a set of sensor nodes and a sink node, as shown in **Figure 1**. In this the sensor nodes deployed in the region, around the targets, which are responsible for monitoring the activities around the targets. These types of networks are generally used in military applications for monitoring the border areas over time. The main objective of TARs is to discover the relationship among a set of targets that are monitored by wireless sensor networks and are used to enhance the Quality of Service of the network, to predict the source of the future events, etc. [8] [9].

To generate TARs efficiently, preparation of Target Database is required. Target Database consists of the collection of activity sets of targets that are sent by the sensor nodes to sink node. This activity data can be gathered using existing routing protocols, or using data collection algorithms. Existing data collection algorithms are not designed for mining purpose and they do not consider the redundancy existing between targets' activity data. That is why special data collection algorithms are required for mining.

Figure 2 shows the main steps of target based association rule mining (TARM). It has six steps. The first three are used to perform data preparation; the fourth and the fifth are used to perform mining, and the sixth one is used to perform analysis on generated patterns.

In this paper authors propose a novel data preparation approach for target based association rule mining from the point of coverage wireless sensor networks, which reduces the redundancy in the data and thus enhances the performance of the network.

This paper is organized as follows: Section 2 explains the preliminaries of target based association rules; Section 3 reviews some of the related works; Section 4 introduces the mechanism for preparing the data to generate the TARs; Section 5 includes the experimental results to evaluate the performance; Section 6 concludes this paper.

2. Preliminaries of Target Based Association Rule Mining

As mentioned in the previous section, target based association rules are based on the common intervals of events occurrences around the targets. To find these associations, time is assumed to be divided into number of time slots, of size λ. Each sensor node is responsible for profile activities of the targets for a given profile time (T_{prof}) in its sensing range. Profiling time is the time period that is determined by the application to profile the activities of the targets. Sensor nodes also have the set of buffers to store the activities of each target. Each buffer will include an entry for each time slot within the given profile time and have size equal to ($T_{prof/\lambda}$). Each sensor profiles targets into its sensing range over time, if any event is detected around any target in the current time slot, an entry is set into the buffer in that time slot. A set of slot numbers for which there is a set bit in the buffer, called as the activity set (AS), is then send to the sink node, at the end of the profile time. The sink node, then prepares the target database, T by accumulating the activity sets of each target.

Definition: Target based association rules are defined as P'→ P" where P', P" ⊂ T and P' ∩ P" = ∅. Here P'and P" are the frequent patterns of the targets and T is the set of targets in the network. A pattern P is said to be frequent if Support (P) ≥ Min_Sup , where support is given by $\text{Support}(\text{P}) = \forall l_i \in \text{P} \left| \bigcap \text{AS}(T_j) \right|$.

Here $\text{AS}(T_j)$ is the activity set of target *j*. An example of target association rule could be ($T_1 \rightarrow T_2$, λ, 60%). Here T_1 and T_2 are the sets of the targets. This rule says that if an event is detected around targets from T_1 within λ, then there is 60% chance that the event will be detected around the target from T_2 within the same time interval [8].

3. Related Work

In the traditional operational scenario of POCW, all the sensor nodes have to be active all the time. When an event is detected at any particular target then a notification message is sent by the sensors, covering that target, to the sink. Sink maintains a target database, accumulates all the notifications of each target and then store them in the target database for further processing. This mechanism is called as the "All nodes based mechanism" which sounds simple but requires lot of energy to send the notification messages to the sink. Here one more issue is the redundancy, because a target may covered by multiple sensor nodes then, sensors will send multiple redundant notification messages to the sink about the same target. It consumes both extra time and energy [8].

The above issues were addressed by Samarah *et al*, they proposed a mechanism in which a global schedule

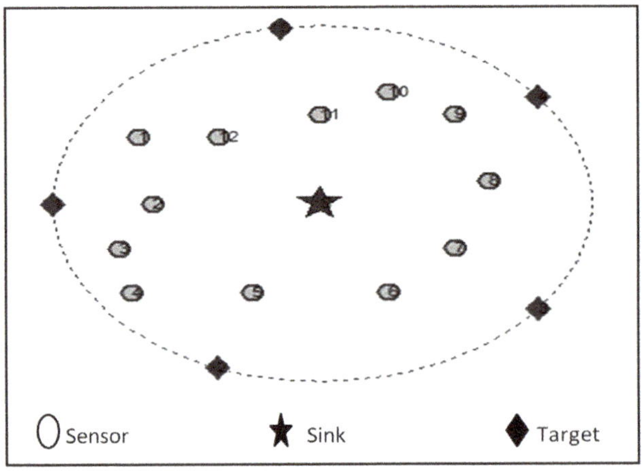

Figure 1. Point of coverage wireless sensor network.

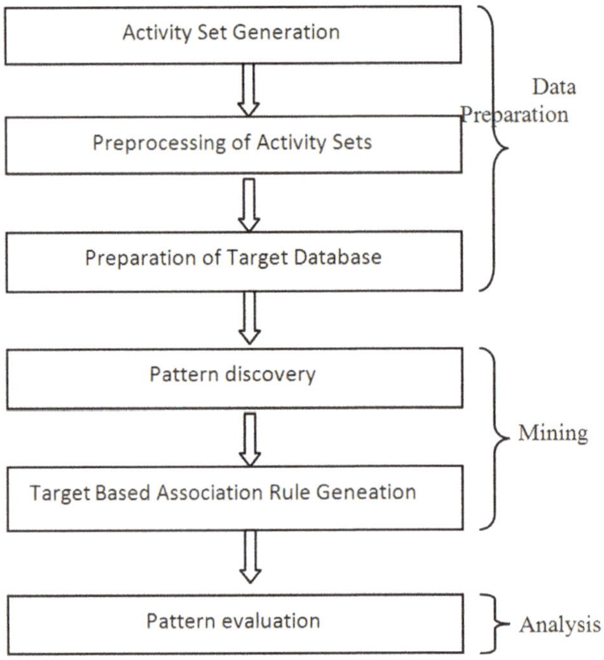

Figure 2. Flow diagram of TARM.

is broadcasted by the sink node which consists of activity time of the sensor node (for which a sensor node should be active for profile the activities) and the set of targets that should be covered by the sensor node. Then the sensor node, during its activity time profiles the targets for activities and prepares a activity set which is a binary bit pattern, "1" for detected event in the particular time slot and "0" for others. This activity set is then sent to the sink. It reduces the required energy to prepare the data for the mining process than the previous one (All-nodes-based mechanism) but there may be the situation where multiple messages have the similar activity sets for different targets in the sensor node. Hence it is the waste of the energy to send the similar or redundant information to the sink for each target by the sensor nodes. In the proposed work this redundancy can be reduces by applying a new parameter refer to as similarity support (sim_sup), on the prepared activity sets which reduces the required energy for data preparation mechanism.

4. Proposed Work

In this section an efficient data preparation mechanism is proposed for target based association rule mining, called as Reduce Redundancy (RR) schedule buffer based data preparation mechanism.

To define this mechanism it is assumed that a global schedule has been defined by the sink. The global schedule specifies the set of targets that should be monitored by each sensor and the activity time of each sensor. Each sensor contains number of buffers to store targets activities, which is divided into number of time slots λ for capture the activity at targets T in given profiling period T_{prof}. During profiling period when an event is detected at a particular target T_i, an entry is set into current time slot of that target's buffer B(T_i). At the end of this profiling period, sensor scans its buffers for each target and prepare the activity set AS_{T_i}. The Activity Set is a set of time slot identifiers for which there is a set bit in the targets' buffer. If the cardinality of activity set is greater or equal to the minimum support (min_sup) then it finds the similarity between buffer entries of targets by performing the bit and operation between the buffers of each target. If the result is greater or equal to the similarity support (sim_sup) then it sends a combine message of targets to the sink which satisfies the similarity support. Otherwise it sends individual message of each target to the sink.

Algorithm 1 gives a formal description for this mechanism.

Algorithm 1. RR-Schedule-Buffer-based-Data-Preparation.

Sink Node:

1. Broadcast global schedule;
2. Broadcast mining parameters (T_{prof} , λ, min_sup, sim_sup);
3. For each received activity set AS_{Ti};
4. Insert (AS_{Ti}, TD);
5. End for.

Sensor Node:

1. Upon receiving mining parameters and node schedule;
2. Slot number = 1;
3. Time = current time;
4. While (current time < time + T_{prof});
5. If (current time < time + (slot number * λ));
6. If there is a detected event
    a. Let;
    b. Set;
    c. Else;
    d. Slot number = slot number + 1;
    e. End if;
7. End while;
8. For each target T_i within the sensor's coverage;
9. AS(T_i);
10. If cardinality;
11. If
    a. Send (sensor id, T_i, T_j, AS(T_i)) to the sink;
    b. Else;
    c. Send (sensor id, T_i, AS(T_i));
    d. End if;
12. End if;
13. End for.

5. Experimental Results

In this section, we discuss a comparison analysis of the data preparation process for point of coverage wireless sensor network, using schedule buffer based and RR-schedule buffer based mechanisms. Two metrics are used to evaluate performance of the network. First is the number of messages, required to report the activity sets of targets to the sink node, and second is the average energy consumption for each sensor node. These metrics are generated based on a simulator using Matlab version 9. In simulation we have assumed that targets are located at fixed points in a grid 100 × 100 cm. We have used 10 targets and 25 sensors deployed randomly around the targets. For calculating the energy consumption we have used the first order radio model introduced in [10]. Equations (1) and (2) show the energy consumption for sending k bit message across distance d [11] [12].

$$E_{TX}(\mathrm{k,d}) = E_{elec} * \mathrm{k} + E_{amp} * \mathrm{k} * \mathrm{d}^2 \quad (1)$$

$$E_{RX}(\mathrm{k}) = E_{elec} * \mathrm{k} \quad (2)$$

Here E_{TX} (k,d), represents the energy consumption for sending k bit message across d distance. E_{elec} represents the energy consumption to run the transmitter circuit, which is 50 nJ/bit. E_{amp} represents the energy consumption for the transmitter's amplifier. E_{RX} (k) represents energy consumption to run receiver circuit. We have assumed k = 10 bits and d = 30 cm. For storage device we assumed a Toshiba 16MB NAND flash memory which costs 0.136 μJ to read, write and erase a bit [11] [12]. Equation (3) shows the average energy consumption of data preparation for target based association rule mining, in μJ.

$$E_{avg} = E_{TX}(\mathrm{k,d}) + E_{RX}(\mathrm{k}) + 136 \quad (3)$$

Figure 3 shows the number of messages to report the activity sets to the sink node at 50% minimum support and 50% similarity support over different profiling times. As we can see the number of messages generated by RR-schedule buffer based mechanism is 40% to 60% less than the schedule buffer based mechanism. **Figure 4** shows the number of messages required to report the activity sets to the sink node at 50% minimum support and 70% similarity support over different profiling times. Here we can see clearly that the number of messages generated by RR schedule buffer based mechanism is 60% to 85% less than the schedule buffer based mechanism. **Figure 5** shows the average energy consumption at 50% minimum support and 50% similarity support, over different profiling time. We can see clearly the average energy consumption of RR schedule buffer based is 50% to 60% less than the schedule buffer based mechanism. **Figure 6** shows the average energy consumption at 50% minimum support and 70% similarity support, over different profiling times. We can see the average energy consumption by RR schedule buffer based mechanism is 60% to 90% less than the schedule buffer based

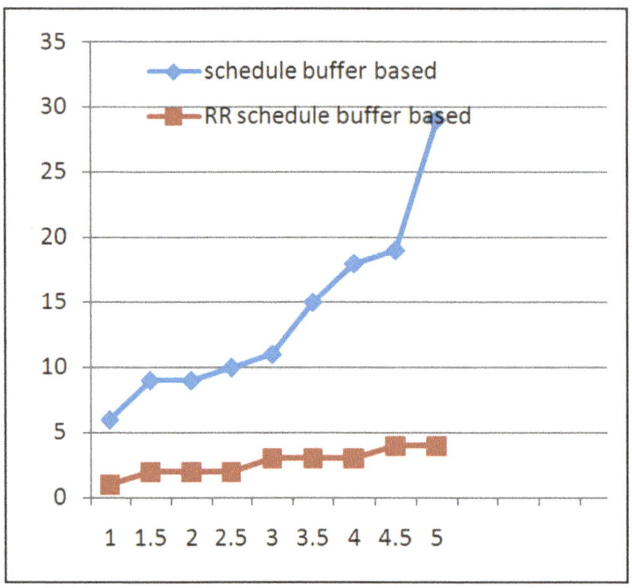

Figure 3. Number of messages (50% min_sup, 50% sim_sup).

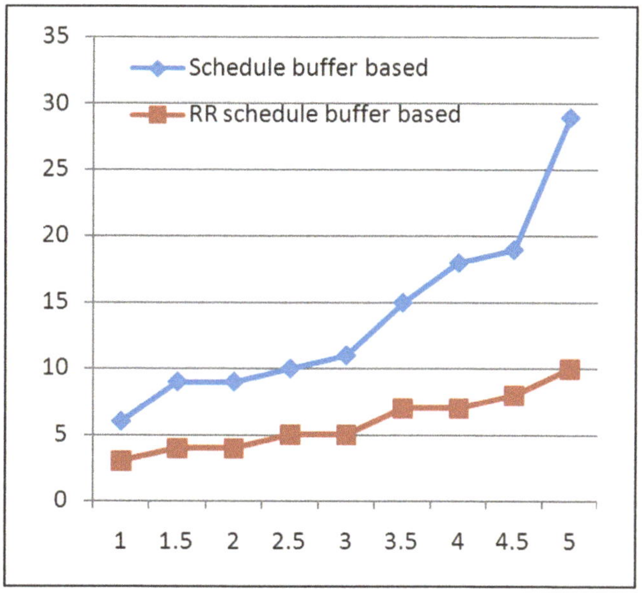

Figure 4. Number of messages (50% min_sup, 70% sim_sup).

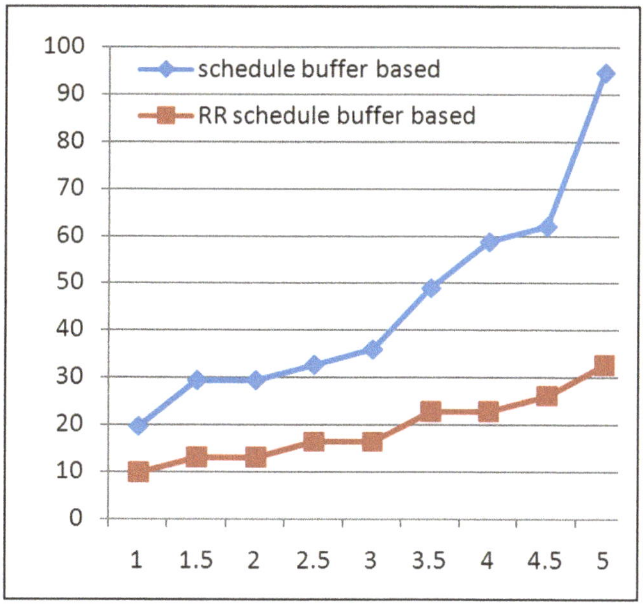

Figure 5. Average energy consumption (50% min_sup, 50% sim_sup).

mechanism.

6. Conclusion & Future Work

Target based association rule mining is a knowledge discovery process which has been proven to be very important and useful in point of coverage wireless sensor networks (POCW). It finds the correlation between the set of targets around a network. In this paper an efficient mechanism is proposed to collect data for mining the target based association rules. Results indicate clearly that the proposed mechanism outperforms existing mechanisms. The target based association rules are very useful for estimation and prediction for future situations and results in a sensor network. Although the proposed mechanism is more effective in terms of reduction in redundancy than the existing mechanisms, it puts an extra load on sensor nodes. Effective distribution of loads on

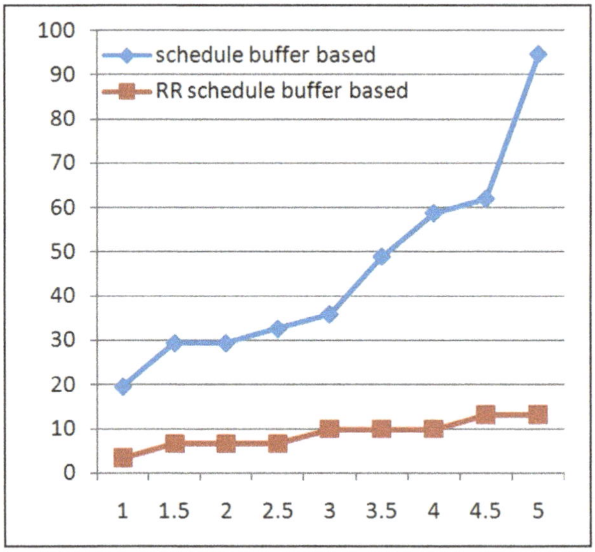

Figure 6. Average energy consumption (50% min_sup, 70% sim_sup).

sensor nodes can be the scope for future work.

References

[1] Stankovic, S.V., Rakocevic, G., Kojic, N. and Mllicev, D. (2012) A Classification and Comparison of Data Mining Algorithms for Wireless Sensor Networks. 2012 *IEEE International Conference on Industrial Technology*, Athens, 19-21 March 2012, 265-270.

[2] Halatchev, M. and Gruenwald, L. (2005) Estimating Missing Values in Related Sensor Data Streams. *Advances in Data Management*.

[3] Romer, K. (2006) Distributed Mining of Spatio-Temporal Event Patterns in Sensor Networks, NCCR-MICS.

[4] Loo, K.K., Tong, I., Kao, B. and Cheung, D. (2005) Online Algorithms for Mining Inter-Stream Association from Large Sensor Networks. HKU CS Tech Report.

[5] Agrawal, R., Imielinski, T. and Swami, A.N. (1993) Mining Association Rules between Sets of Items in Large databases. *Proceedings of the* 1993 *ACM SIGMOD International Conference on Management of Data*, **22**, 207-216.

[6] Boukerchem, A. and Samarah, S. (2007) A New Representation Structure for Mining Association Rules from Wireless Sensor Networks. University of Ottawa, Ottawa.

[7] Boukerche, A. and Samara, S. (2008) A Novel Algorithm for Mining Association Rules in Wireless ad Hoc Sensor Networks. *IEEE Transactions on Parallel and Distributed Systems*, **19**, 865-877. http://dx.doi.org/10.1109/TPDS.2007.70789

[8] Samarah, S., Habyalimana, A.S. and Boukerche, A. (2009) Target-Based Association Rules for Point-of-Coverage Wireless Sensor Networks. *IEEE Symposium on Computers and Communications*, Sousse, 5-8 July 2009, 938-943.

[9] Samarah, S. and Boukerche, A. (2011) Target Association Rules: A New Behavioral Patterns for Point of Coverage Wireless Sensor Networks. *IEEE Transactions on Computers*, **60**, 879-889. http://dx.doi.org/10.1109/TC.2010.227

[10] Heinzelman, W.R., Chandrakasan, A. and Balakrishnan, H. (2000) Energy-Efficient Communication Protocol for Wireless Microsensor Networks. *Proceedings of the* 33*rd Hawaii International Conference on System Sciences*, 4-7 January 2000, 8020-8030.

[11] Mathur, G., Desnoyers, P., Ganesan, D. and Shenoy, P. (2006) Ultra-Low Power Data Storage for Sensor Networks. *Proceeding of the fifth IEEE/ACM Conference on Information Processing in Sensor Networks*, Nashville, 19-21 April 2006, 374-381.

[12] Toshiba 128-MBIT (16M8BITS/8Mx16BITS) CMOS NAND E2PROM.

18

Multipath Grid-Based Enabled Geographic Routing for Wireless Sensor Networks

Bassel Arafeh*, Khaled Day, Abderezak Touzene, Nasser Alzeidi

Department of Computer Science, Sultan Qaboos University, Al Khoudh, Oman
Email: *arafeh@squ.edu.om, kday@squ.edu.om, touzene@squ.edu.om, alzidi@squ.edu.om

Abstract

This work proposes an efficient disjoint multipath geographic routing algorithm for dense wireless sensor networks (WSN), called Multipath Grid-based Enabled Geographic Routing (MGEGR). The proposed algorithm relies on the construction of a 2-D logical grid in the geographical region of deployment. The objective of the proposed scheme is to determine optimal or near-optimal (within a defined constant) multiple disjoint paths (multipath) from a source node to the sink, in order to enhance the reliability of the network. The determined multiple disjoint paths would be used by the source node in a round-robin way to balance the traffic across the disjoint paths, and to avoid discovered paths with cell holes. The proposed scheme limits the use of broadcasting to the process of gateway election within each cell, and the process of maintaining the table of neighbors of each gateway. Our simulation results show the effectiveness and scalability of our routing scheme with increased network size compared to on-demand routing protocols.

Keywords

Wireless Sensor Networks, Mobile Ad Hoc Networks, Clustering Algorithms, Disjoint Multipath Routing, Grid-Based Routing, Geographic Routing

1. Introduction

Wireless Sensor Networks (WSNs) have been the focus of attention of many researchers, application developers and users in the recent years [1]. WSN have the potential for many applications in wide spectrum of areas, such as military target tracking and surveillance, natural disaster management, biomedical health monitoring, hazardous environment exploration, and seismic sensing. The unstructured nature of WSNs, in which sensor nodes may be deployed in an ad hoc manner into a field, imposes many challenges to network designers who are developing routing protocols for such networks [2]. In particular, the design of routing algorithms for WSNs has to

*Corresponding author.

solve many issues due to their special characteristics, that distinguish them from Mobile Ad-hoc NETworks (MANETs), such as the inability to have a global addressing scheme, requirements of most applications to have the sensor nodes send data to a particular sink node, redundancy in the generated data traffic by multiple sensor nodes, and the limited capabilities of sensor nodes in terms of transmission power, on-board energy, processing capacity, and storage space.

In general, a routing technique can be either a single path or multiple paths constructed between a source and the destination. A single path routing approach is usually simpler to construct and less costly. However, single path routing in WSNs is not providing the required reliability for forwarding data. Since a single path routing may select the same forwarding nodes each time a source node transmits a packet to the sink, the limited power resource of those nodes would be depleted earlier than other nodes and cause a network partition. Besides, sensor nodes are very vulnerable devices to failures due to many factors, such as unreliable wireless communication, or unpredictable harsh environment for deployment. On the other hand, multipath routing is characterized by the existence of redundant paths that can solve the problems of reliability, load balancing, as well as security in WSNs.

This work proposes an efficient disjoint multipath geographic routing algorithm for dense wireless sensor networks (WSN), called Multipath Grid-based Enabled Geographic Routing (MGEGR). The proposed algorithm relies on the construction of a 2-D logical grid in the geographical region of deployment. The objective of the proposed scheme is to determine optimal or near-optimal (within a defined constant) multiple disjoint paths (multipath) from a source node to the sink, in order to enhance the reliability of the network. The determined multiple disjoint paths would be used by the source node in a round-robin way to balance the traffic across the disjoint paths, and to avoid discovered paths with cell holes. Packet forwarding is handled in each grid cell by an elected gateway node, using unicasting communication. The determination of the multipath depends on the relative location of the destination cell with respect to the source, and the geographic direction of forwarding. The proposed scheme limits the use of broadcasting in the WSN to the process of gateway election in each grid cell, and the process of constructing and maintaining the table of neighboring gateways in adjacent grid cells by each gateway. Our simulation results show the effectiveness and scalability of our routing scheme with increased network size compared to on-demand routing protocols.

The remainder of the paper is organized as follows. Section 2 presents the related work. Section 3 presents the network model including our assumptions, the construction of the logical grid, the gateway election process, and the method of constructing multiple disjoint paths in the proposed protocol. Section 4 describes the cases of MGEGR routing paths. Section 5 presents the simulation model, the performance metrics and the performance evaluation results. Finally, Section 6 is the conclusion.

2. Related Work

Akkaya and Younis [2] have classified routing protocols for WSNs into three main categories, as data-centric, hierarchical or location-based. Our proposed routing scheme can be categorized as a combination of hierarchical (clustering) and location-based. The main objective of the hierarchical routing protocols is grouping the sensor nodes into clusters, in order to achieve the network scalability objective. In such configurations, each cluster would have a cluster head (called gateway in this work) through which all sensor nodes in the same cluster transmit their packets to the sink. Accordingly, all multi-hop routing paths are formed as paths connecting cluster-heads from the source cluster to the destination cluster. Eventually, clustering enables conserving the communication bandwidth, because it restricts transmission to be among cluster-head nodes, and prevents redundant broadcasting. For more details on clustering algorithms for WSNs, see the survey paper by Abbasi and Younis [3]. Our proposed algorithm achieves clustering the sensor nodes based on the construction of 2-D logical grid in the geographical region of deployment, similar to the definition in [4]. The main advantage of the clustering of sensor nodes based on the construction of a logical grid, in comparison to other methods, is the ability of the network to utilize location-based routing paths using 2-D grid coordinates system. Many researchers have proposed grid-based routing for MANETs or WSNs in order to conserve energy and achieve network scalability. In this section, we will give first an overview of proposed single path grid-based routing in the literature, followed by an overview of proposed multipath grid-based routing.

2.1. Grid-Based Routing

Liao *et al.* [5] presented an on-demand location-aware routing protocol, called GRID, which assumed the deployment region as a virtual 2-D grid of square cells. To reduce the overhead of reactive routing discovery, a leader election process is performed among the nodes located in each cell to select a node as a cluster-head. The protocol has high overhead cost, because it still relies on route discovery using flooding, and does not consider the existence of holes. Yu *et al.* [6] proposed a grid-clustering routing protocol, called GROUP that the authors claim it provides scalability and energy efficiency for packet routing in large-scale WSNs. The GROUP algorithm assumes the WSN has multiple sinks. The algorithm proceeds after nodes deployment into electing one sink as the primary sink (PS). The PS is responsible of initiating the process of constructing grid-clustering periodically by broadcasting a GS-election command in its radio range. The PS elects a Grid Seed (GS) node from its neighbours based on the one having the highest residual energy among the other neighbours. The elected GS node is assumed at one crossing point of the constructed cluster. Accordingly, the new GS node would broadcast a command to its neighbours to elect cluster heads in its four down-stream directions. Each elected cluster head stores the GS's information in its cache as its upstream head. The process of electing cluster heads will be done iteratively until all cluster heads are elected in the region, which makes the election process costly. Akl *et al.* [7] presented a non-uniform grid-based coordinated routing scheme in wireless sensor networks. The non-uniform grid-based routing was derived from the uniform grid-based coordinated protocol. The non-uniform grid-based routing protocol utilizes the same scheme for electing grid coordinators (cluster heads) as in the uniform one. It also follows the same scheme for conserving energy through putting all non-coordinators into sleeping mode. The source node transmits the data to the sink through the active coordinators. The authors argue that uniform grid-based is efficient when the distribution of nodes in the region is uniform. Their simulation results show that by varying the node density the non-uniform grid-based coordinated routing is more efficient than the uniform one. However, the major drawbacks of the protocol are still its reliance on the process of flooding for discovering routes, and no considerations for the existence of holes.

Shrestha *et al.* [8] proposed a multi-grid based routing protocol, called MGRR. The MGRR protocol employs a multi-grid routing scheme, in which it uses adaptively varying cell sizes. The protocol performs on-demand route discovery, starting with the smallest cell size. If a neighbour cannot be found using the current cell size, then the cell of the next larger grid is selected to locate a neighbouring node. The MGRR constructs local graphs for each grid size. The edges of each local graph are weighted with their proposed reliability metric. These local graphs are merged to form a global graph at each node. The route discovery process tries to discover a connected path of reliable cells in the global graph. The source node utilizes Dijkestra's algorithm on the global graph to find a path towards the cell where the destination is located. The drawback of the MGRR protocol is its high cost of computation for building a global graph by a source node in order to find a path to the destination.

2.2. Multipath Grid-Based Routing

There have been many multipath routing protocols proposed in the literature for WSNs. Sha *et al.* [9] paper provides a survey of proposed multipath routing protocols for WSNs. Sha *et al.* classify the existing multipath routing schemes for WSNs into three categories. These are the infrastructure based, non-infrastructure based, and the coding based. The rationale of their classification is based on the method of design and the techniques used to discover the multiple paths for forwarding data. Although the multipath grid-based routing approach is not mentioned in Sha *et al.* survey, our proposed MGEGR protocol is belonging to the hierarchy approach of the infrastructure based category; since the MGEGR protocol relies on constructing a logical 2-D grid as an infrastructure for determining the multiple paths from a source gateway to the sink.

Wu *et al.* [10] have proposed a multipath protocol called Energy-Aware Grid Multipath routing (EAGMR) to solve the problem of power depletion due to the limitations of nodes batteries in MANETs. The construction of the multiple paths in the EAGMR scheme is based on broadcasting route request (RREQ) messages within a rectangular area confined by the source and destination nodes. The EAGMR scheme defines a path cost metric based on the energy consumed in transmitting and receiving by each path link. The determination of the set of discovered multiple paths are done by the destination. The destination waits a threshold period of time after receiving the first RREQ packet, before it starts to make replies for all received RREQs till that moment. Accordingly, the source gateway would probabilistically select one of the discovered multiple paths.

Wu *et al.* [11] have proposed a grid-based routing algorithm called Grid-based Stable Backup Routing Algorithm (GSBRA). The objectives of the proposed algorithm are to improve the link stability and enhance the network reliability. The authors propose achieving the objectives by utilizing two elected grid heads in each cell.

The authors define a parameter called Grid Head Stability in order to determine an elected Primary Grid Head (PGH) in each cell. A secondary Backup Grid Head (BGH) is elected by the PGH based on a mechanism for predicting the maximum link expiration time between the PGH and its 1-hop neighbours. The PGH is responsible of delegating its role to the BGH, when it cannot act as a cluster head due to any reason. The GSBRA utilizes the same method described in EAGMR protocol for discovering multiple paths [10]. However, the GSBRA scheme utilizes a parameter which predicts the path stability for selecting one path out of the discovered multiple paths.

Banimelhem and Khasawneh [12] proposed a grid-based multipath routing scheme called Grid-based Multipath with Congestion Avoidance Routing (GMCAR). The objective of the GMCAR scheme is to provide support for QoS traffic in WSNs. The GMCAR scheme differentiates between two types of grid cells: boundary and non-boundary. The scheme considers the sink location is always at one of the corners of the region. The GMCAR scheme constructs multiple diagonal paths towards the sink for the non-boundary cells, while it constructs a single path to the sink for the boundary cells through the downstream boundary cells. The protocol assigns a weight to every forwarding grid head candidate in the routing table. The weight is defined based on both the hop count and the grid cell density, and used by the GMCAR scheme to rank every grid cell.

Compared with the previous schemes, our proposed scheme considers the location of the sink to be anywhere in the region. Besides, our proposed scheme uses all determined multiple paths for load balancing, extending the lifetime of the network, and avoiding discovered holes. Our proposed MGEGR uses geographic routing protocol, which does not rely on flooding in order to discover the multiple paths from the source to the destination like all the other discussed protocols. Our scheme utilizes the geographic location of the destination to determine the number of multiple paths and their directions. However, our proposed scheme does not address the issue of congestion. The MGEGR protocol assumes that gateways have sufficient buffer space to sustain the required workload.

3. Network Model

3.1. Assumptions

In this work, we consider that all sensor nodes are deployed over a rectangular field, which is determined by its length and width. We assume the sensor nodes are static in the geographic region of deployment. All sensor nodes have identical initial power sources, identical processing and communication capabilities, and equipped with one or more sensors of various types. Each sensor node is equipped with some localization service mechanism for determining its location and the location of other nodes. The base station (*i.e.*, sink node) is static and has unlimited power. Sensor nodes communicate wirelessly with one another using one-to-all broadcasting. We assume the packet size is fixed, and there is no data compression or in-network data aggregations are performed; *i.e.*, all packets have to be forwarded to the base station. Each sensor node is aware of its address, referred to as its Node Identification (NID).

3.2. Construction of the Logical Grid

The network area is constructed as a logical grid by dividing the geographic region covering the sensing field into virtual uniform rectangular cells. Each grid cell is identified by a unique pair of numbers, called Grid Identification (GID), representing the grid cell's coordinates. We assume the geographic location of the grid origin (x_o, y_o) and the grid cell side, d, are determined by the base station at the initialization phase of the network. Two grid cells are called neighbouring cells, if they have a common side or a common corner. Therefore, each grid cell has eight neighbouring cells. The value of a grid cell side, d, is calculated based on the radio transmission range, R, of a sensor node. In order to allow each sensor node to communicate directly with all nodes located in the neighbouring grid cells, the grid cell side, d, must satisfy the condition $R \geq 2d\sqrt{2}$.

Initially, each sensor node should find out its geographic location (X_i, Y_i), and then uses broadcasting to send its location and Node Identification (NID) to the base station. Accordingly, the base station calculates the grid cell side, d, and broadcasts d, its geographic location (X_b, Y_b), and the geographic location of the grid origin (X_o, Y_o) to all sensor nodes. After receiving the broadcasted message from the base station, each sensor node calculates its grid coordinates (GID_i) and the grid coordinates of the base station (GID_b) as follows:

$$GID_i = \left(\left\lfloor \frac{X_i - X_o}{d} \right\rfloor, \left\lfloor \frac{Y_i - Y_o}{d} \right\rfloor \right) \quad (1)$$

$$GID_b = \left(\left\lfloor \frac{X_b - X_o}{d} \right\rfloor, \left\lfloor \frac{Y_b - Y_o}{d} \right\rfloor \right) \quad (2)$$

We assume the communication overhead between the nodes and the sink would be negligible compared to the life time of the network, and therefore this factor will not be addressed in this work furthermore. In this work, we will refer to a logical grid coordinates of a cell by lower-case letters, as (x, y); while the reference to absolute coordinates of a node will be referred to by upper-case letters, as (X, Y). We assume the origin, (x_o, y_o), of the grid coordinates system is located at the bottom left corner of the region. The rectangular region is defined by the origin (x_o, y_o) and the Top right corner point (x_T, y_T).

3.3. Gateway Election Process

In this work, the method for electing a gateway is based on selecting the sensor node having the highest remaining energy level among all other sensor nodes in the same grid cell, and if there is a tie-break, the node having highest NID is elected. When a gateway is elected, all other sensor nodes will be considered alive but Inactive. Initially, each sensor node nominates itself as a gateway by broadcasting a Local-HELLO message containing its node address (NID), grid coordinates (GID), node energy level, a flag set for gateway Election REQuest, EREQ, and a status flag bit set to indicate its status as a gateway. Only sensor nodes located in the same grid cell are considered in the election process in that cell. Each sensor node compares its energy level and its grid coordinates with the information contained in the Local-HELLO message it receives. A sensor node that finds itself having a lower energy level than any other node would withdraw from the election process, by not sending Local-HELLO messages. However, if there is a tie, the node with the highest NID declares itself as the cell gateway. The sensor node satisfying this condition declares itself as a gateway by setting the status flag bit in aLocal-HELLO message and broadcasts it to all nodes in the same cell. An elected gateway should build its table of neighbour gateways in adjacent cells. The table of neighbour gateways is constructed and maintained by HELLO messages broadcasted among the gateways in the adjacent grid cells. The election process is repeated periodically by the current gateway, unless its energy is depleted, or it becomes *Dead* due to failure. In case there is no re-election request is heard from the gateway after several periods, each Inactive node in the cell switches itself independently to Active mode, nominates itself as a gateway, and enters into the process for gateway election again. In normal operation, a gateway initiates a request for a gateway election periodically by re-broadcasting a Local-HELLO message with the EREQ flag set. The result of the process would be either re-electing the same current gateway, or finding another node with an energy level higher than that of the current one instead. When all the sensor nodes in a cell have their energy depleted, then no gateway can be elected in that cell to carry out the responsibilities of a router to the traffic transmitted by other nodes. We will call such a cell as forming a *routing hol*e in the path of any traffic towards the destination.

3.4. Construction Method of the Multipath Grid-Based Enabled Geographic Routes

The construction of the set of disjoint paths from a gateway to the base station is performed in a distributed manner by each gateway node. Let the source and destination grid coordinates be (x_s, y_s) and (x_d, y_d), respectively. Then, let δx and δy be the difference between the source and destination in the x-coordinate and y-coordinate, respectively, and they are defined as:

$$\delta x = x_d - x_s, \text{ and } \delta y = y_d - y_s$$

Our approach for the determination of a grid routing path from a source cell at (x_s, y_s) to a destination cell at (x_d, y_d) is based on the relative location of the destination cell with respect to the source, and the direction of forwarding. The relative location of the destination with respect to the source can be defined by the slope of the virtual line connecting the source to the destination. In general, the direction of forwarding can be found to be one of the possible eight geographic directions from the source cell to the destination. A decision to select a grid routing path can be determined according to the forwarding direction as defined by the conditions in **Table 1**.

It is obvious that the East and West forwarding directions represent the situation of a line of slope 0 connecting the source to the destination. While the case of North and South forwarding directions represent the situation of a line with a slope of infinity connecting the source to the destination. For the other forwarding directions, we consider three possibilities for determining the grid routing paths based on the absolute slope, $|m|$, these are

Table 1. Conditions for forwarding directions.

Forwarding Direction	Condition
East	$\delta x > 0, \delta y = 0$
West	$\delta x < 0, \delta y = 0$
North	$\delta x = 0, \delta y > 0$
South	$\delta x = 0, \delta y < 0$
North-East	$\delta x > 0, \delta y > 0$
South-East	$\delta x > 0, \delta y < 0$
North-West	$\delta x < 0, \delta y > 0$
South-West	$\delta x < 0, \delta y < 0$

defined as follows:

i. $|m| = 1: \ |\delta x| = |\delta y|$

ii. $0 < |m| < 1: \ |\delta x| > |\delta y|$

iii. $1 < |m| < \infty: \ |\delta x| < |\delta y|$

Therefore, there are16 possibilities for determining the grid routing paths based on the relative location of the destination cell with respect to the source. In this work, we emphasize the construction of disjoint multiple paths from a gateway to the sink in order to distribute the load of transmitted traffic, or avoid paths with holes. However, we restrict the set of disjoint paths to be in the range 4 - 5, in order to establish simple and uniform method for constructing disjoint routing paths with optimal or near-optimal path lengths. In case the source or the destination is at a boundary location, the number of disjoint paths drops down to three. The goal of the construction method is to devise a subset of disjoint paths from a source gateway to the destination that can have a maximum path length within a constant, c, from the optimal path length, l_{opt}. In the following, we define the optimal length and characterize the subset of disjoint paths from a source gateway S that can satisfy the objective of a maximum path length of $l_{opt} + 2$.

Given a source gateway located at $S = (x_s, y_s)$ and a destination gateway located at $D = (x_d, y_d)$, the optimal path length, l_{opt}, from a source to a destination is defined as:

$$l_{opt} = \max\{|x_d - x_s|, |y_d - y_s|\} = \max\{|\delta x|, |\delta y|\} \tag{3}$$

Mathematically, we can describe a path direction from S to one of its neighbouring cells V by a vector as,

$$\boldsymbol{SV} = a\boldsymbol{i} + b\boldsymbol{j} \tag{4}$$

where, a and b can take a value from the set $\{-1, 0, 1\}$. Accordingly, the eight directions out of a source gateway S can be specified by vectors representing all geographic directions (e.g., the direction vector to the East is $\boldsymbol{SV} = \boldsymbol{i}$, and the direction vector to North-East is $\boldsymbol{SV} = \boldsymbol{i} + \boldsymbol{j}$). The direction of forwarding from S to D can be represented by a vector $\boldsymbol{SD}$ defined as

$$\boldsymbol{SD} = (x_d - x_s)\boldsymbol{i} + (y_d - y_s)\boldsymbol{j} = \delta x.\boldsymbol{i} + \delta y.\boldsymbol{j} \tag{5}$$

where, $\boldsymbol{i}$ and $\boldsymbol{j}$ are the unit vectors along the x and y Grid coordinates system. Accordingly, the subset of neighbouring grid cells, N_F, that are facing the destination and can satisfy the objective of a maximum path length of $l_{opt} + 2$ to the destination is defined as:

$$N_F = \{V_n = (x_s + a, y_s + b)\}$$

Such that,

$$\boldsymbol{SV}_n \cdot \boldsymbol{SD} \geq 0 \tag{6}$$

where, a and b are two constants from the set $\{-1, 0, 1\}$. We will refer to the subset N_F as the subset of source neighbours facing the destination, or the subset of *starting gateways*. It can be shown that the size of the subset N_F is either 4 or 5 neighbours depending on the direction of the vector $\boldsymbol{SD}$. In order to achieve optimal or near-optimal path lengths, the construction of disjoint paths is constrained in this work to at most three path segments. The subset of disjoint paths, P, that can be constructed from the subset N_F can have a maximum path length $l_{\max} = l_{opt} + 2$.

We call the subset of neighbouring grid cells of the destination at the end of disjoint paths as the *ending gateways*. For disjoint paths with two or three segments, there are one or two gateways at which the path ends a segment and starts another one, respectively. We call those gateways as the *pivot gateways*, in general. There are three types of pivot gateways. The pivot gateway at the end of a one hop from a starting gateway is called the *starting pivot gateway*, the one at the beginning of a one hop from an ending gateway is called the *ending pivot gateway*, and the one in the middle is called the *intermediate or actual pivot gateway*.

4. Cases of MGEGR Routing Paths

In this work, we assume a continuous data delivery model, where each sensor node would send data to the sink periodically. In the MGEGR scheme the disjoint paths to the destination are utilized by a source gateway in a round-robin manner, so that energy consumption due to forwarding packets would be balanced over the nodes of the WSN, and consequently would prolong the lifetime of the WSN. For brevity, we describe the methods for constructing the disjoint paths of the forwarding directions to the East, and North-East in the next sections. The construction of the disjoint paths in the other forwarding directions can be obtained from the described ones by considering the symmetry between the various quadrants and octants around the translated coordinates' axes at the source cell. Each disjoint path is identified by its starting gateway, its ending gateway, and one or more pivot gateways. In all cases, paths are identified by a Path Identifier, called PID, (e.g., P1 - P5) sequentially in a clockwise way based on the positions of their starting gateways with respect to the source cell. Also, each path segment is identified by a Path Segment Identifier, called PSID, (e.g. 0, 1, and 2).

4.1. Neighbourhood Notations

The moves out of a cell (x, y) are determined by the coordinates of the neighbouring cells for a cell (x, y). These are: $(x + 1, y)$, $(x - 1, y)$, $(x, y + 1)$, $(x, y - 1)$, $(x + 1, y + 1)$, $(x - 1, y + 1)$, $(x + 1, y - 1)$, $(x - 1, y - 1)$. We will refer to the neighbouring cells for a cell (x, y) as <x+, y> (or <x+> for short), <x−>, <y+>, <y−>, <x+, y+>, <x−, y+>, <x+, y−>, <x−, y−>, respectively. We will denote the coordinates of a cell that is a neighbour to a neighbouring cell of (x, y) in the same way. For example, the neighbour cell in the east direction to the <x+> neighbour of (x, y) is <x++>, and the neighbour cell in the south-east direction to the <x+, y+> neighbour of (x, y) is <x++, y+−>. We will be using the described notations for facilitating the specifications tables of the disjoint routing paths and their methods of construction in the next sections.

4.2. Disjoint Routing Paths towards the East

The set of the disjoint routing paths towards the East is illustrated by **Figure 1**. The determination of the starting, starting pivot, ending pivot, and ending gateways of the set of disjoint multiple paths are defined by **Table 2**, under the condition that $\delta x > 1$. In this case, there are up to 5 defined disjoint paths P1 - P5. The routing paths are characterized by at most three path segments. In the case the source and the destination are located at one of the boundaries, say $y_s \leq 1$, or $y_d \geq y_T - 1$, the set of disjoint paths would be adjusted accordingly. The disjoint routing paths towards the West are obtained from this case by reversing all $x+$ notations by $x-$ and vice versa in **Table 2**, and all directions are reversed in **Figure 1**. The lengths of the disjoint paths, l_p, are in the range $|\delta x| \leq l_p \leq |\delta x| + 2$.

4.3. Disjoint Routing Paths towards the North-East

The disjoint routing paths towards the North-East (South-East, North-West, or South-West) are determined based on one of the three conditions characterizing the absolute slope, $|m|$, of the virtual line connecting the source to the destination. These conditions are described in this section, in general, using the North-East direction as an example.

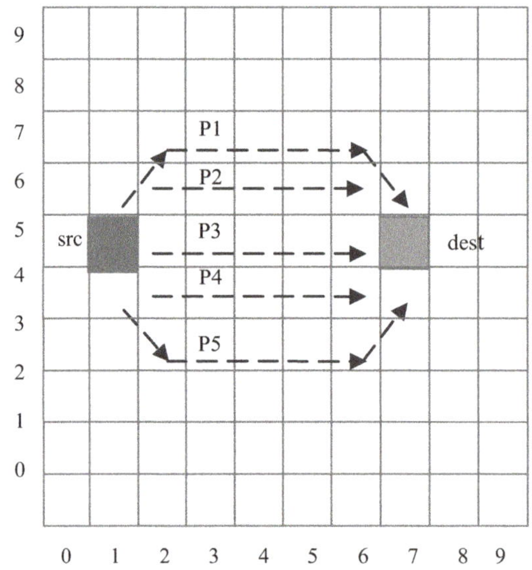

Figure 1. Example of disjoint paths towards the east.

Table 2. Disjoint routing paths towards the east.

Path	Starting Gateway	Starting Pivot Gateway	Ending Pivot Gateway	Ending Gateway
P1	$<y_s+>$	$<x_s+, y_s++>$	$<x_d-, y_d++>$	$<y_d+>$
P2	$<x_s+, y_s+>$			$<x_d-, y_d+>$
P3	$<x_s+>$			$<x_d->$
P4	$<x_s+, y_s->$			$<x_d-, y_d->$
P5	$<y_s->$	$<x_s+, y_s-->$	$<x_d-, y_d-->$	$<y_d->$

4.3.1. Condition 1: $|m| = 1$

The set of the disjoint routing paths towards the North-East for $|m| = 1$ is illustrated by **Figure 2**. The determination of the starting, and ending gateways of the set of disjoint multiple paths are defined by **Table 3**, under the conditions that $\delta x = \delta y$ and $\delta y > 1$. In this case, each path proceeds towards the destination directly to one of the ending neighbours, without any pivot gateways. The lengths of the disjoint paths, l_p, are in the range $|\delta x| \leq l_p \leq |\delta x| + 2$. Under this condition, if the source or destination is located at one of the boundaries (*i.e.*, $x_s = 0$, $y_s = 0$, $x_d = x_T$, or $y_d = y_T$), then the set of disjoint paths would be adjusted accordingly. The disjoint routing paths towards the South-West are obtained from **Table 3** by using reflection operations by changing the $y+$ to $y-$, $x+$ to $x-$, and vice versa, and by reversing all the directions in **Figure 2**. Similarly, the disjoint paths towards the North-West are obtained from **Table 3** using reflection operations by changing the $x+$ to $x-$ and vice versa, in both the starting and ending gateway columns. While the disjoint paths towards the South-East are obtained from **Table 3** using reflection operations by changing the $y+$ to $y-$ and vice versa, in both the starting and ending gateway columns.

4.3.2. Condition 2: $1 < |m| < \infty$

The set of the disjoint routing paths towards the North-East for $1 < |m| < \infty$ (*i.e.*, $|\delta x| < |\delta y|$) is illustrated by **Figure 3**. The determination of the starting, starting pivot and ending gateways of the set of disjoint multiple paths are defined by **Table 4** under the condition that $\delta x > 1$. The disjoint routing paths under this condition are constructed from at most three path segments, as shown in **Figure 3**. Because $|m| > 1$, the first (or second) path segment direction proceeds towards a point at which the remaining distance to the destination in the y-direction is equal to the distance to the destination in the x-direction. The point at which this equality can be achieved is referred to as the Virtual Pivot cell. The virtual pivot cell coordinates, (x_p, y_p), are defined as follows: $x_p = x_s$, $\delta x_p = \delta y_p = x_d - x_s$, $y_p = y_d - \delta y_p$. Based on the determination of the virtual pivot cell, we can define four actual

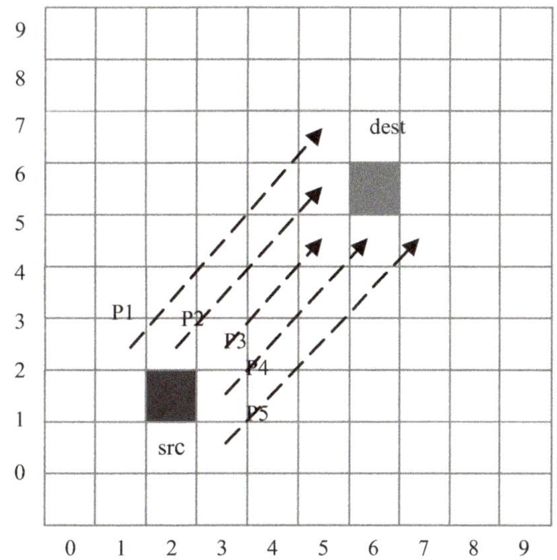

Figure 2. Example of disjoint paths towards the north east with $m = 1$.

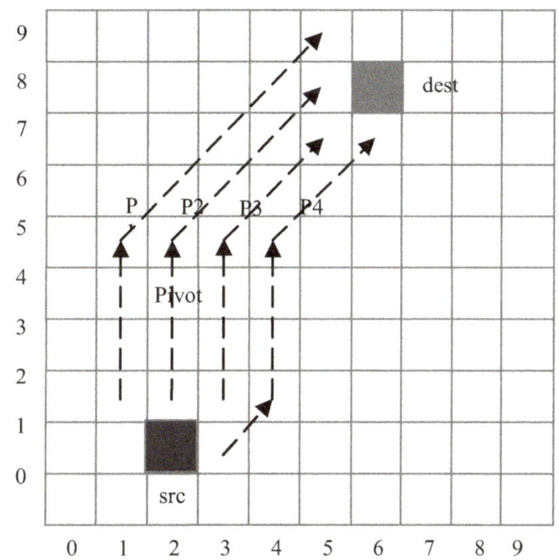

Figure 3. Example of disjoint paths towards the north east with $m > 1$.

Table 3. Disjoint routing paths towards the north-east, when $m = 1$.

Path	Starting Gateway	Ending Gateway
P1	$\langle x_s-, y_s+\rangle$	$\langle x_d-, y_d+\rangle$
P2	$\langle x_s, y_s+\rangle$	$\langle x_d-, y_d\rangle$
P3	$\langle x_s+, y_s+\rangle$	$\langle x_d-, y_d-\rangle$
P4	$\langle x_s+, y_s\rangle$	$\langle x_d, y_d-\rangle$
P5	$\langle x_s+, y_s-\rangle$	$\langle x_d+, y_d-\rangle$

(intermediate) pivot gateways, one for each disjoint path. The coordinates of the actual pivot gateways are resembling the starting gateways for paths P1 - P3, and the starting pivot gateway for P4 in **Table 4**. Therefore, the actual pivot gateway coordinates relative to the virtual pivot are as follows: $\langle x_p-, y_p+\rangle$, $\langle x_p, y_p+\rangle$, $\langle x_p+, y_p+\rangle$, $\langle x_p+, y_p\rangle$. Accordingly, the second (or third) path segment of a disjoint routing path would start at the actual pivot gateway, and proceeds towards one of the ending gateways, based on the absolute slope of $|m| = 1$. In this case, the direction of the second routing path segment (or third in case of P4) is North-East.

Under this condition, if the source x-coordinates is at $x_s = 0$, or destination y-coordinates at $y_d = y_T$, then the set of disjoint paths would be adjusted accordingly. The same approach applies for determining the disjoint routing paths for the other directions. The lengths of the disjoint paths, l_p, are in the range $|\delta y| \leq l_p \leq |\delta y| + 2$. The disjoint routing paths towards the South-West are obtained from **Table 4** by changing $x+$ to $x-$, and vice versa, and $y+$ to $y-$, and vice versa, for both starting and ending columns. In a similar way, we can apply the procedure on the starting pivot and intermediate pivot gateway columns. The disjoint routing paths for the North-West direction can be obtained from **Table 4** by changing all entries with $x+$ to $x-$, and vice versa, in all the columns. Similarly, we can obtain the disjoint routing paths for the South-East from **Table 4** by changing all entries with $y+$ to $y-$, and vice versa, in all columns.

4.3.3. Condition 3: $0 < |m| < 1$

The set of the disjoint routing paths towards the North-East for $0 < |m| < 1$ (*i.e.*, $|\delta x| > |\delta y|$) is illustrated by **Figure 4**. The determination of the starting, initial pivot and ending gateways of the set of disjoint multiple paths are defined by **Table 5**, under the condition that $\delta y > 1$. The construction of the disjoint routing paths under this

Table 4. Disjoint routing paths towards the north east, when $m > 1$.

Path	Starting Gateway	Starting Pivot Gateway	Actual Pivot Gateway	Ending Gateway
P1	$\langle x_s-, y_s+\rangle$		$\langle x_p-, y_p+\rangle$	$\langle x_d-, y_d+\rangle$
P2	$\langle x_s, y_s+\rangle$		$\langle x_p, y_p+\rangle$	$\langle x_d-, y_d\rangle$
P3	$\langle x_s+, y_s+\rangle$		$\langle x_p+, y_p+\rangle$	$\langle x_d-, y_d-\rangle$
P4	$\langle x_s+, y_s\rangle$	$\langle x_s++, y_s+\rangle$	$\langle x_p++, y_p+\rangle$	$\langle x_d, y_d-\rangle$

Table 5. Disjoint routing paths towards the north east, when $0 < |m| < 1$.

Path	Starting Gateway	Starting Pivot Gateway	Actual Pivot Gateway	Ending Gateway
P1	$\langle x_s, y_s+\rangle$	$\langle x_s+, y_s++\rangle$	$\langle x_p+, y_p++\rangle$	$\langle x_d-, y_d\rangle$
P2	$\langle x_s+, y_s+\rangle$		$\langle x_p+, y_p+\rangle$	$\langle x_d-, y_d-\rangle$
P3	$\langle x_s+, y_s\rangle$		$\langle x_p+, y_p\rangle$	$\langle x_d, y_d-\rangle$
P4	$\langle x_s+, y_s-\rangle$		$\langle x_p+, y_p-\rangle$	$\langle x_d+, y_d-\rangle$

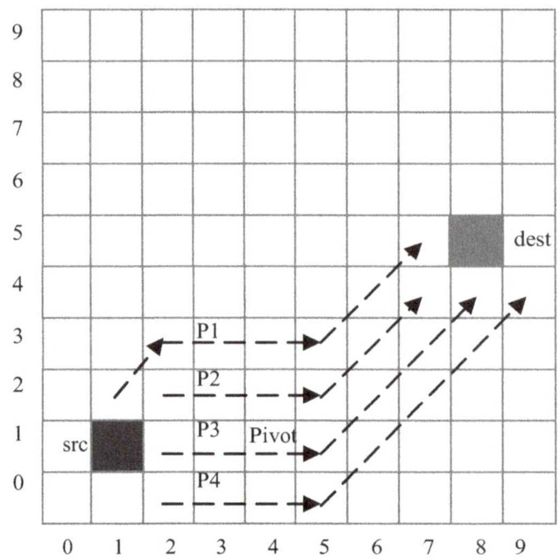

Figure 4. Example of disjoint paths towards the north east with $0 < m < 1$.

condition is similar to the ones described for $1<|m|<\infty$. However, the first or second path segment direction proceeds towards a virtual pivot cell at which the remaining distance to the destination in the x-direction is equal to the distance to the destination in the y-direction. The virtual pivot cell coordinates, (x_p, y_p), are defined as follows: $y_p = y_s$, $\delta x_p = \delta y_p = y_d - y_s$, and $x_p = x_d - \delta x_p$. Similarly, the coordinates of the actual pivot gateways are resembling the starting gateways for paths P2 - P4, and the initial pivot gateway for P1 in **Table 5**. Therefore, the actual pivot gateways relative to the virtual pivot are as follows:$<x_p+, y_p++>$, $<x_p+, y_p+>$, $<x_p+, y_p>$, $<x_p+, y_p->$. Accordingly, the second (third for P1) path segment of a disjoint routing path would start at the actual pivot gateway, and proceeds towards one of the ending gateways, based on the absolute slope of $|m| = 1$. The lengths of the disjoint paths, l_p, are in the range $|\delta x| \leq l_p \leq |\delta x| + 2$. The disjoint routing paths towards the South-West are obtained from this case by changing $x+$ to $x-$, and vice versa, and $y+$ to $y-$, and vice versa, for all the columns. The disjoint routing paths for the North-West direction can be obtained from **Table 5** by changing all entries with $x+$ to $x-$, and vice versa, in all the columns. Similarly, we can obtain the disjoint routing paths for the South-East from **Table 5** by changing all entries with $y+$ to $y-$, and vice versa, in all columns.

5. Performance Evaluation

In this section, we present our evaluation of the MGEGR algorithm by simulation. We have implemented the MGEGR protocol as a module within the QualNet simulator software. We have also implemented a version of the protocol with deterministic unipath routing, referred to as Grid-base Enabled Geographic routing (GEGR) [13]. The GEGR scheme differs from MGEGR in using deterministically one of the shortest paths to the destination based on the relative location of the sink with respect to the source gateway. Simulations were conducted using QualNet to assess our proposed MGEGR protocol with respect to GEGR and two well-known MANET routing algorithms, Dynamic Mobile Ad-hoc Network On-demand (DYMO) [14] and Location-Aided Routing (LAR) [15]. The DYMO protocol simplifies the operations of AODV with no added features. The LAR protocol is an on-demand routing protocol similar to DSR [16]. It exploits location information of the destination by the source node. The QualNet LAR protocol is based on scheme 1 in [15], which limits the search for a route to the so-called request zone. The request zone in scheme 1 is the smallest rectangle that includes the current location of the source and the expected zone of the destination, at the time of route discovery.The MGEGR and GEGR algorithms were evaluated against the DYMO and LAR routing algorithms in series of simulation runs based on random nodes deployment. Random deployment is defined as placing the nodes randomly on the terrain. In addition, the MGEGR and GEGR algorithms were evaluated under different failing ratios. In this section, we define the performance metrics, the simulation results for MGEGR, GEGR, LAR, and DYMO assuming no failures, and the simulation results for GEGR and MGEGR based on varying a failing ratio between 0.1 and 0.3. The parameters of the simulation scenarios are summarized in **Table 6**.

5.1. Performance Metrics

In this work, we have focused on three performance metrics for comparisons among MGEGR, GEGR DYMO and LAR protocols. These are the packet delivery ratio (PDR), the end-to-end delay (sec.), and the average energy consumption (mJoule). The PDR is defined as the ratio of the total number of packets received by the destination (*i.e.* sink) to the total number of packets transmitted by all nodes. The end-to-end delay is the average duration taken by a packet starting from the time it is transmitted by the source to the time it is being received by the sink. The end-to-end delay includes the waiting times in interface queues, processing times, retransmission delays by the MAC protocol, and propagation time. The average energy consumption refers to the total energy consumed in the network during transmission and reception. The energy consumed during idle time is not counted. In this work, we will use the average energy consumption level of the network as an indicator of the cost of the routing protocol. We have counted the number of discovered holes as an indication to network partitions. In this work, the lifetime of the network is defined as the first time at which a source or intermediate gateway discovers that it does not have a path to the destination. In case of the GEGR scheme, a network partition is recorded when the first time a neighbouring grid hole or dead gateway is discovered by a source or intermediate gateway. In the case of MGEGR, a network partition is recorded when a source gateway loses all defined paths to the sink by the protocol.

Table 6. Simulation parameters.

Parameter	Value
Frequency	2.4 GHz
Path-loss model	Two-ray
Physical model	IEEE 802.11(a)
Bandwidth	9 Mbps
Battery-model	Linear
Battery initial charge	200 mAhr
Antenna model	Omni-directional
Transmission range	150 m
Cell side	50 m
Simulation time	900 sec
Grid size	8 × 8
Network area	400 (m) × 400 (m)
Application	CBR
Packet size	512 B
Transmission rate	0.2 packets/sec
Number of nodes	{100, 150, 200, 250, 300}
Number of sinks	1
Mobility	None

5.2. Simulation Results with No Failures

5.2.1. Packet Delivery Ratio

Figure 5 shows the packet delivery ratio (PDR) against the number of nodes in the network for MGEGR, GEGR, LAR, and DYMO protocols. Due to the nature of GEGR protocol as a unipath routing scheme, forwarded packets encounter higher number of holes, as shown in **Figure 6**, since they do not have other alternative paths to select from. This situation explains the low PDR for low node densities compared to MGEGR. However, MGEGR protocol is more stable and able to maintain high PDR performance against increasing number of nodes, due to its reliance on disjoint multi-paths from the source to the sink and its ability to avoid discovered paths with holes. Although LAR achieves better PDR performance than MGEGR, GEGR and DYMO for number of nodes equal to 100, it cannot sustain high packet delivery ratios when the number of nodes becomes greater than 150 in the network, due to collisions and loss of packets. In regard to DYMO, **Figure 5** shows that its performance is completely inappropriate to dense wireless sensor networks with periodic transmission of data to the sink. The continuous drop in DYMO PDR with the increasing number of nodes indicates the excessive effect of collisions on the transmitted packets, where the source nodes are forced to continuously apply the route discovery process.

5.2.2. End-to-End Delay

Figure 7 shows the relationship between the average end-to-end delay and the number of nodes for MGEGR, GEGR, LAR, and DYMO protocols. The end-to-end delay for the GEGR and MGEGR protocols are minimal compared to the other two protocols, due to the optimal or near-optimal grid-based routing. For the LAR protocol, the end-to-end delay is slightly higher than GEGR and MGEGR, and it shows adequate performance. However, the DYMO routing cannot function adequately in a dense wireless network with periodic transmission from nodes to the sink, since it is not able to maintain sustainable routes to the sink.

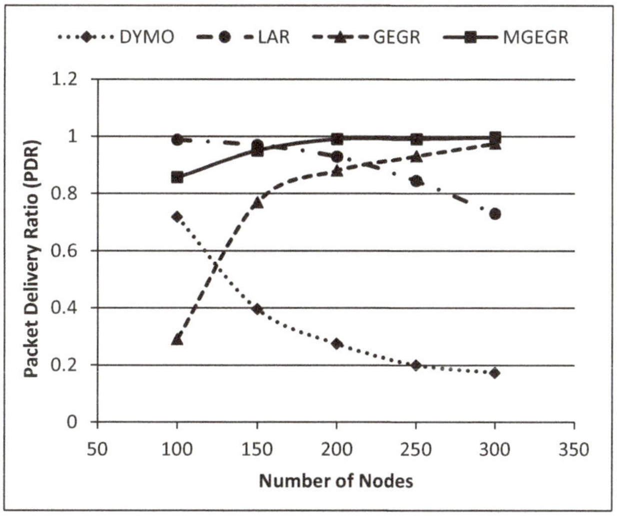

Figure 5. Packet delivery ratio vs. number of nodes for random deployment.

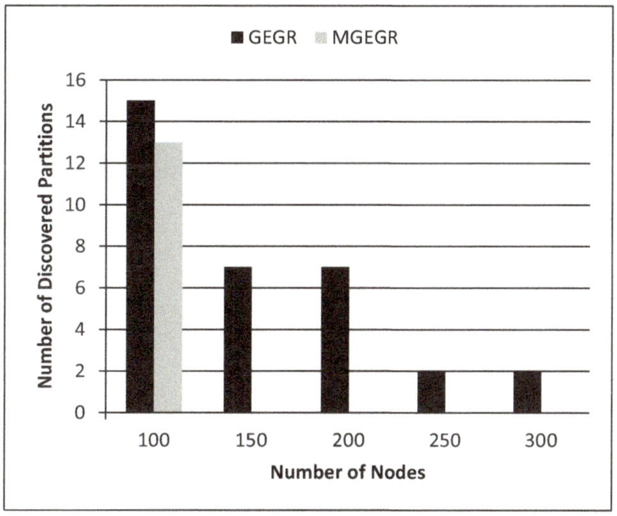

Figure 6. Number of discovered holes with respect to number of nodes.

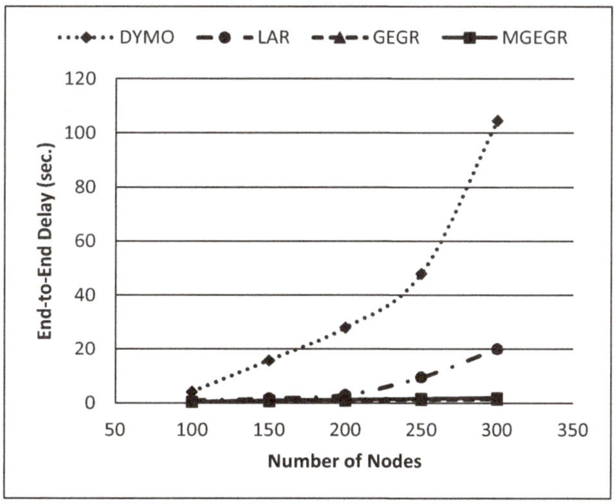

Figure 7. End-to-end delay vs. number of nodes for random deployment.

5.2.3. Energy Consumption

Figure 8 shows the relationship between the network energy consumption and the number of nodes in for MGEGR, GEGR, LAR, and DYMO protocols. The GEGR protocol consumes less energy than MGEGR and the other two protocols, mainly due to its optimal routing paths. Although the LAR protocol energy consumption is close to GEGR and MGEGR at number of nodes of 100 and 150, its energy consumption starts to increase at 200 nodes rapidly with the increase in the number of nodes. Definitely, LAR protocol cannot achieve the required scalability in the number of nodes for WSN, due to the increase of the number of control packets and the increase of packet retransmission, as supported by the continuous decline in its PDR values, and gradual increase of the end-to-end delay with the increase in the number of nodes. The performance of the DYMO protocol energy consumption is misleading, because it shows proper level of energy consumption, but that due to the small number of packets being transmitted and delivered to the sink.

5.3. Comparisons between GEGR and MGEGR Using Varying Failing Ratios

In this work, we have tested the PDR of GEGR and MGEGR schemes under varying failing ratios for number of nodes equal to 100, 200, and 300. We have applied uniform failing ratios of {0.0, 0.1, 0.15, 0.2, 0.25, and 0.3} at each originated packet transmission from each gateway. The first time a hole is discovered by the GEGR protocol is at 10 seconds of the simulation time, for all failing ratios and at all number of nodes considered. This situation indicates that the network is partitioned in the case of the GEGR protocol from the starting time of data forwarding. On the other hand, the first time a hole is discovered by the MGEGR protocol is at 15 seconds of the simulation time for all failing ratios at 100 nodes only. However, we can draw comparisons between the two schemes based on their PDRs vs. the failing ratios, as shown in **Figure 9** and **Figure 10**. Obviously, MGEGR protocol shows better PDR performance than GEGR for number of nodes equal to 100 and 200, where the nodes density is relatively low, due to its ability to avoid paths with existing holes or dead gateways. In general, both schemes can have dead gateways replaced by newly elected ones, when there is more than one gateway in the same cell. Therefore, the PDR performance of GEGR becomes close or the same to MGEGR at number of nodes equal to 300.

6. Conclusion

WSNs have high potential to be the solution to a wide range of application areas. However, there are many challenges that need to be overcome due to the special characteristics of WSNs. In this paper, we have proposed the MGEGR scheme as a multipath grid-based protocol for solving the routing problem in dense WSNs. In conclusion, the combined results of the proposed MGEGR scheme, in terms of stable high PDR values, low end-to-end delays, as well as low energy consumption, indicate its characteristics as an efficient and low cost

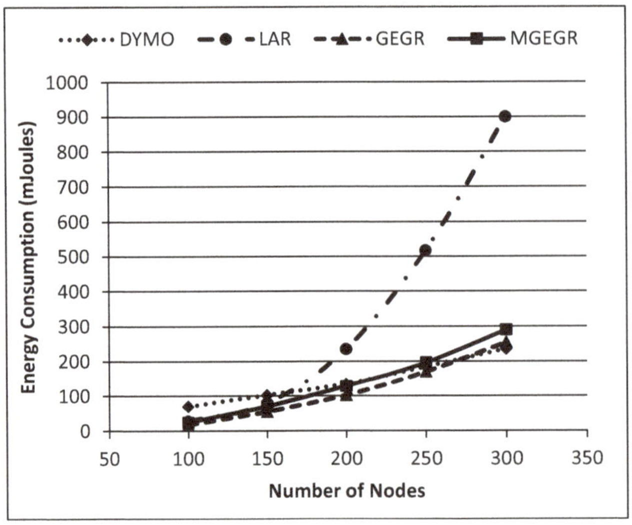

Figure 8. Energy consumption vs. number of nodes for random deployment.

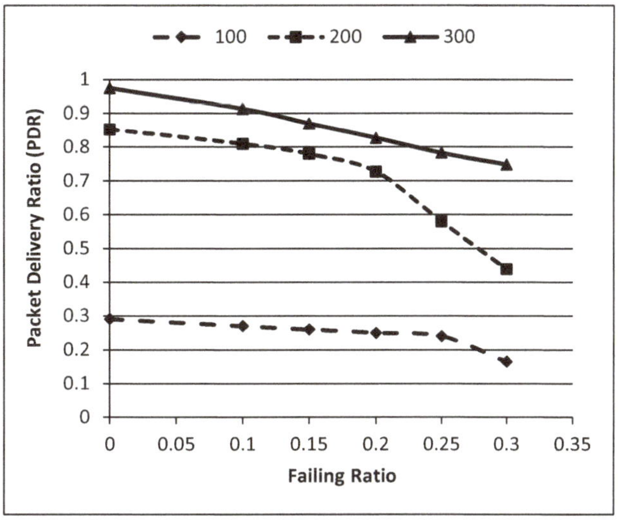

Figure 9. GEGR Packet delivery ratio vs. failing ratio for random deployment.

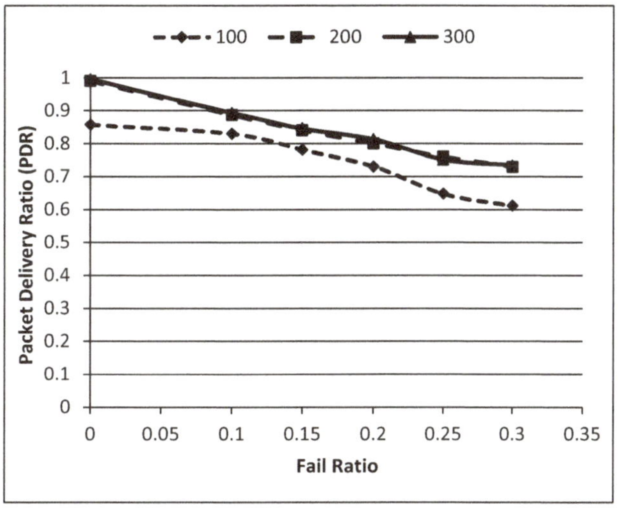

Figure 10. MGEGR Packet delivery ratio vs. failing ratio for random deployment.

geographic routing using grid-based approach. The proposed scheme achieves network size scalability based on clustering using a 2-D logical grid, and by limiting broadcasting transmissions. The MGEGR scheme guarantees optimal or near optimal path lengths, and can sustain high packet delivery ratios under relatively low node densities with varying failure ratios by avoiding discovered holes. Our simulation results show the effectiveness of the proposed scheme compared to a unipath approach of the protocol and common on-demand routing protocols in maintaining high PDR and low energy consumption at low and high nodes densities. For future work, we intend to address the problems of congestions and the existence of holes due to random deployment or sensor nodes failures.

References

[1] Yick, J., Mukherjee, B. and Ghosal, D. (2008) Wireless Sensor Network Survey. *Computer Networks*, **52**, 2292-2330. http://dx.doi.org/10.1016/j.comnet.2008.04.002

[2] Akkaya, K. and Younis, M. (2005) A Survey on Routing Protocols for Wireless Sensor Networks. *Ad Hoc Networks*, **3**, 325-349. http://dx.doi.org/10.1016/j.adhoc.2003.09.010

[3] Abbasi, A.A. and Younis, M. (2007) A Survey on Clustering Algorithms for Wireless Sensor Networks. *Computer Communications*, **30**, 2826-2841. http://dx.doi.org/10.1016/j.comcom.2007.05.024

[4] Al-Karaki, J. and Kamal, A. (2004) Routing Techniques in Wireless Sensor Networks: A Survey. *IEEE Wireless Communication*, **11**, 6-28. http://dx.doi.org/10.1109/MWC.2004.1368893

[5] Liao, W.-H., Tseng, Y.-C. and Sheu, J.-P. (2001) Grid: A Fully Location-Aware Routing Protocol for Mobile Ad Hoc Networks. *Telecommunication Systems*, **18**, 37-60. http://dx.doi.org/10.1023/A:1016735301732

[6] Yu, L., Wang, N., Zhang, W. and Zheng, C. (2006) GROUP: A Grid-Clustering Routing Protocol for Wireless Sensor Networks. *Proceedings of the International Conference on Wireless Communications, Networking and Mobile Computing*, (*WiCOM* 2006), 22-24 September 2006, 1-5. http://dx.doi.org/10.1109/WiCOM.2006.287

[7] Akl, R., Kadiyala, P. and Haidar, M. (2009) Non-Uniform Grid-Based Coordinated Routing in Wireless Sensor Networks. *Journal of Sensors*, Article ID: 491349. http://dx.doi.org/10.1155/2009/491349

[8] Shrestha, D.M., Kim, C. and Ko, U.-B. (2011) A Reliable Multi-Grid Protocol for Tactical MANETs. *Research in Applied Computation Symposium*, (*RACS*'11), Miami, 2-5 November 2011, 164-169.

[9] Sha, K., Gehlot, J. and Greve, R. (2013) Multipath Routing Techniques in Wireless Sensor Networks: A Survey. *Wireless Personal Communications*, **70**, 807-829. http://dx.doi.org/10.1007/s11277-012-0723-2

[10] Wu, Z., Song, H., Jiang, S. and Xu, X. (2007) Energy-Aware Grid Multipath Routing Protocol in MANET. *Proceedings of the First Asia International Conference on Modelling & Simulation*, (*AMS*'07), Phuket, 7-30 March 2007, 36-41. http://dx.doi.org/10.1109/AMS.2007.36

[11] Wu, Z., Song, H., Jiang, S. and Xu, X. (2007) A Grid-Based Stable Backup Routing Algorithm in MANETs. *International Conference on Multimedia and Ubiquitous Engineering*, (*MUE*'07), Seoul, 26-28 April 2007, 680-685.

[12] Banimelhem, O. and Khasawneh, S. (2012) GMCAR: Grid-Based Multipath with Congestion Avoidance Routing Protocolin Wireless Sensor Networks. *Ad Hoc Networks*, **10**, 1346-1361. http://dx.doi.org/10.1016/j.adhoc.2012.03.015

[13] Arafeh, B., Day, K., Touzene, A. and Alzeidi, N. (2013) GEGR: A Grid-Based Enabled Geographic Routing in Wireless Sensornetworks. *IEEE Malaysia International Conference on Communication*, (*MICC* 2013), Kuala Lumpur, 26-28 November 2013. http://dx.doi.org/10.1109/MICC.2013.6805856

[14] Chakeres, I.D. and Perkins, C.E. (2007) Dynamic MANET on Demand (DYMO) Routing Protocol. *Internet-Draft Version* 09, IETF.

[15] Ko, Y.-B. and Vaidya, N.H. (2000) Location-Aided Routing (LAR) in Mobile Ad Hoc Networks. *Wireless Networks*, **6**, 307-321. http://dx.doi.org/10.1023/A:1019106118419

[16] Perkins, C.E. (1999) Ad-Hoc on-Demand Distance Vector Routing. *Proceedings of the Second IEEE Workshop on Mobile Computing Systems and Applications*, (*WMCSA*'99), 25-26 February 1999, 90-100. http://dx.doi.org/10.1109/MCSA.1999.749281

19

Distributed Target Location in Wireless Sensors Network: An Approach Using FPGA and Artificial Neural Network

Mauro Rodrigo Larrat Frota e Silva*, Glaucio Haroldo Silva de Carvalho, Dionne Cavalcante Monteiro, Leomário Silva Machado

Exact and Natural Sciences Institute, Faculty of Computer Science, Federal University of Pará, Belém, Brazil
Email: *maurolarrat@gmail.com

Abstract

This paper analyzes the implementation of an algorithm into a FPGA embedded and distributed target location method using the Received Signal Strength Indicator (RSSI). The objective is to show a method in which an embedded feedforward Artificial Neural Network (ANN) can estimate target location in a distributed fashion against anchor failure. We discuss the lack of FPGA implementation of equivalent methods and the benefits of using a robust platform. We introduce the description of the implementation and we explain the operation of the proposed method, followed by the calculated errors due to inherent Elliott function approximation and the discretization of decimal values used as free parameters in ANN. Furthermore, we show some target location estimation points in function of different numbers of anchor failures. Our contribution is to show that an FPGA embedded ANN implementation, with a few layers, can rapidly estimate target location in a distributed fashion and in presence of failures of anchor nodes considering accuracy, precision and execution time.

Keywords

Wireless Sensors Network, RSSI, FPGA, Artificial Neural Network, Distributed Localization Methods

1. Introduction

Distributed target location algorithms provide lower power consumption and fault tolerance of sensors during

*Corresponding author.

the localization process when compared with centralized localization algorithms [1]. The main difference between these two types of localization algorithms is the local of occurrence of the processing target location information. In distributed localization, the target sends its location information to (at least) two or more base stations in the network (or anchor nodes), so that the execution of processing is distributed between these ones, in order that the locating procedure provides a reasonable estimative in case of failure of one of the base stations. The target location information can be obtained in various ways, as shown in the related works, by means of sensors which are adjacent to them (likelihood based) or by techniques based on measured distances (time of arrival, angle of arrival, received signal strength indicator, etc.) by the transmitted signals.

This work considers the characteristics of Artificial Neural Network, such as its ability of offering a reasonable approximation in presence of some absence of feeding information and the parallel processing of its units. The inputs of ANN are associated with the RSSI of each anchor nodes and the outputs are in form of 2-D coordinates in a plane surface. With our proposed method, at least two inputs are necessary to estimate, with some accuracy and precision, the location of a target. Considering the embedded level of our method, we just need to implement one neuron and replicate it to the desired architecture of the ANN in a modular fashion.

The embedded platform chosen to build in the ANN is the FPGA. With that we could achieve 6×10^{-7} s at each processing unit of the ANN. Considering parallel advantage architecture, the layer of the ANN is executed at same time. The FPGA is a reconfigurable platform that operates at 50 MHz and is very affordable nowadays. It is designed to operate process in parallel so it has equivalency to the ANN. There are many vendors and low costs FPGAs to fit any design problem. Our proposed method is implemented in the compact and tiny DE0-Nano's solution from Altera, to show that it does not need a lot of resources to be built. There are many works that discuss the advantages of ANN implementations on applied in target location, rarely using FPGA as an applied platform. We focus on ANN embedded locating methods implemented in related works, but we emphasize that the ANN based locating target is a wide area of research and the hardware implementation is a branch we have pointed to; [2] discourses a parallel and distributed ANN platform model for target location; [3] uses an ANN embedded method for target tracking; [4] proposes a non-embedded generalized regression ANN together with a Centroid simulation method in a RSSI target location problem. Other papers on ANN and RSSI based methods can be found in [5]-[7].

This paper contributes to showing a practical experiment implemented in a FPGA together with ANN applied in distributed target location in WSN and the advantages associated with this method, and how it fits to solve most known issues on distributed localization. We go beyond the limit of simulation to show new details and considerations of the applied method by analysing the characteristics intrinsic of this implementation. Section 2 shows some recent related works that use a set of complex algebraic techniques, common used to distributed localization, to estimate the target location, as well as some works that make use of Artificial Neural Network to estimate the target location. Section 3 details the methodology of our work. Section 4 shows our results and Session 5 presents our conclusions.

2. Related Works

In [8] a practical experiment is carried out in an 8×6.4 m indoor area with furniture. To locating targets, an ANN method is software embedded into an Arduino platform which makes use of Xbee as radio modules. The measured RSS signal operates in 256 bps and follows the IEEE 802.15.4 standard through Zigbee protocol. The RSSI is acquired through the proprietary Application Interface implemented into the Zigbee. The results showed compare the time spent considering six training methods, including the training method we used in our work (Resilient Backpropagation Algorithm—RBP), and the total average error which each training method did show. The error is in the form of average distance location error, comparing the known position with the estimated one. They show the Leven berg-Marquardt method (LM) as the best training algorithm based on the total average error, and the RBP being the third, with an error of 0.98. They cited that when the memory and time spent for training is not so relevant, the BP and variants could be used. In comparison, our results achieved a total error of 0.14 related to the RBP method. In fact, after the training method conclusion, the implemented ANN matrix is not affected in the estimated process. The lower training error is considered an advantage as we lead with the quantization error due to the hardware implementation and function approximation. Despite the authors cite the time constraint to be considered in a real-time solution they didn't comment about the time the method they proposed takes to process; just associate the size/topology of the proposed ANN as a function of the time execution.

In [9] a more powerful microcontroller BS2PX-IC is used to embedded an ANN to locate a mobile robot. The microcontroller can operate with 125 kbps serial communication through the Wavetrend 433 MHz radio. They considered one ANN for estimating each coordinate, with just one hidden layer, as the previous reference did. The indoor environment of this experiment is about 2 m^2. They suggest complement the locating process with additional frameworks, as sonar GPS and additional angle information, but such improvement reflects in aggregating cost to project, to not mention to the increase of power consumption. The results show the errors in function of tracking and angle of the mobile robot, being the tracking RMSE equal to 0.06 and the angle RMSE about 0.12. It is highlighted the difficulties in implementing ANN with more than one output (ordered pair) and more than one hidden layer (more stored information) due to the platform limitations, even if real-time response is required. We consider it as an embedded software implementation dependency problem in references.

In [10] an ANN is embedded in a PIC18f452 microcontroller to estimate the position of a mobile robot. Except they use infrared sensor, the objective is the same: imbuing the best ANN generated through the best training method into a hardware platform and approximate the position of a target given the reading of sensors as inputs. The LM training algorithm showed the lowest training error before others training methods. It is not clearly how the authors tested the proposed method in practice, unless the measures about 90 × 80 cm in the graphical results. The ANN modeled has one input, one hidden layer (with three neurons) and one output. The authors cite the limitation about processing time and memory associated to the chosen hardware platform. The ANN as a position estimator has stated itself an attractive choice, but the platform in which it is implemented has significant implications.

More references on implementing an ANN to estimate targets position can be found in, [11]-[14]. In the next section we explain our methodology and the main characteristics of the proposed method.

3. Methodology

The proposed method considers the real outdoor experiment and the data base acquired from [15]. The data base consists of a set of 6-tuples containing the inputs and outputs of ANN, respectively, the RSSI from anchor nodes and the ordered pair representing the position of the target in a plane. In resume, it consists of four anchor nodes spaced in an out door environment measuring 20 × 22 m^2. The anchors receive a broadcast message from one target node located in some region inside this environment. The anchor nodes take the RSSI from that message and apply it to the proposed method. All anchor nodes are in line-of-sight with the target node. The scenario does not offer any obstacle to the signal other than the temperature and humidity as noise parameters. We choose this scenario appropriately to show how the method works.

In the next, we show how the method was implemented and moreover how it works algorithmically.

3.1. Method Implementation

We begin with the data treatment and ANN training algorithm, followed by deciding the ANN topology to be implemented in the FPGA. We consider four anchor nodes that will sequentially operate in the method, transmitting their distributed processing in an ordered path from anchor one to anchor four. To simulate failures on third and fourth anchor nodes, the 21000 data base samples were replicated twice, resulting in a total of 63000 samples. Thereby, we could implicitly apply zeros in the RSSI of replicated samples, related exclusively to the input of fourth anchor node and related exclusively to the input of third and fourth anchor nodes. We assume that, if the third anchor node fails, the fourth anchor fails together, as they will not receive the ordered distributed processing from the third anchor node.

The data base was divided in 75% for training and 25% for testing samples. The training and testing were preceded through the R language through the neural net package, using the Resilient Propagation (Rprop+) algorithm for training. **Table 1** shows the parameters assumed in the training step.

We assumed the Universal Approximation Theorem [16] to decide the topology of the ANN, related to number of hidden layers and neurons contained in each one. Our ANN has four inputs, two hidden layers containing four neurons each of them and one output layer containing two neurons.

Each anchor node was implemented in a FPGA Altera™ DE0-nano [17] using VHDL. The chosen platform is recently developed and is affordable as an educational development board. It is considerable tiny in size in comparison with others bigger plat forms that include same resources. The 50 MHz processing capacity offered by DE0-nanogives us assuredness to concentrate in enforcing ANN features without worrying with timing or

Table 1. Parameters for ANN training.

Parameter	Value
Algorithm	Resilient propagation
Activation function	Log sigmoid
Learning rate	0.001
Stop error achieved	0.14
Iterations	1000000
Error function	RMS

processing issues.

To be implemented in synthesizable form, the free parameters generated for the ANN were converted to a fixed point format [18], in which the integer part is a 10 bits two's complement and the mantissa has 6 bits. The total 16 bits is sufficient to convert the bigger number (or lesser) of the free parameters.

The activation function of neurons is approximated using the Elliott activation function [19], as shown in Equation (1).

$$f(x)=\left[0.55x/1+abs(x)\right]+0.5 \quad (1)$$

This approximation is due to the Log Sigmoid function being not synthesizable in hardware, *i.e.*, it can't be implemented because of its logarithm component. In other hand, Elliott function is ease applicable in training ANN due to its simple first derivative, which is used in versions of Back propagation training method. We can visualize the related approximation between the Log Sigmoid and the Elliott activation functions in **Figure 1**.

To evaluate how close the Elliott can approximate to the Log Sigmoid we computed in each function the interval considered into activation function in the ANN [0; 1] at a step resolution of 0.001. The RMSE of the output of the Elliott function, at these conditions, is equal to ±0.13462, so we can addict this error in the total error of our proposed method.

The description of ANN inside the FPGA is divided in modules: a neuron module, a layer module and an ANN module. The FPGA implementation allows the neurons at same layer being executed in parallel, achieving performance.

The neuron was implemented to execute the Elliott function's fundamentals calculation in parallel, as the dependent results do not conflict. The products of summations occur in parallel, followed by the parallel calculation in the dividend and the divisor, then, the resultant summation with the constant, as stated in **Figure 2**.

In **Figure 2**, $x_n w_n$ are the products of input and free parameters that is summed and passed to the Elliott Function as the resultant **a**. The threshold T is a constant subtracted from the constant 0.5 summed in Elliott function, this subtraction occurs in parallel with the rest of operations. Each parallel operation was descripted in an individual flagged process in VHDL, with executes as soon as the dependent operations conclude.

The two hidden layers and the output layers were designed as VHDL components formed by the neuron units. So, the complete ANN was designed as a component formed by the layers. This design allowed the connections being implemented by easily binding the component parts.

The anchor nodes, containing the ANN, can communicate with each other through the IEEE 802.15.4 Xbee radio standard athwart the 115.200 kbps UART communication. The oversampling is implemented in VHDL to integrate the 1.8 KHz UART frequency with the 50 MHz from the DE0-Nano.

In a top level VHDL modular design, our propose method can be described as a controller, a transceiver and ANN module, all implemented in each anchor node. The controller is modeled as a Moore finite state machine that is responsible for managing data transmission and ANN control. The implemented system can be seen as shown in **Figure 3**.

In the next section we describe the proposed method as an algorithm.

3.2. Method Description

The method considers one testing target. The target sends a broadcast message which is received on the four anchors. All anchors calculate the RSSI of the broadcast message through the power signal received. For

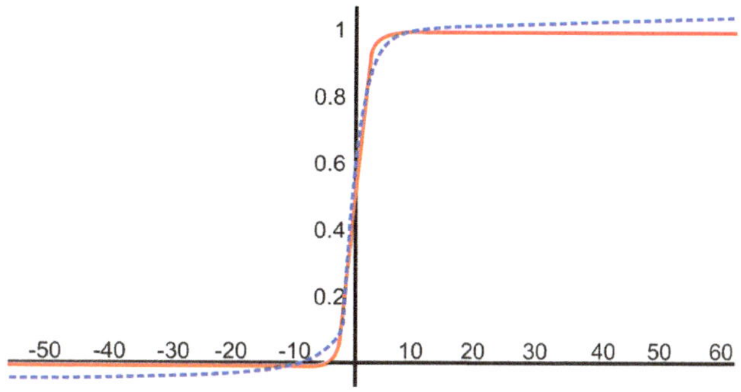

Figure 1. Relation between the continuous Log Sigmoid (in red) and the Elliott (in blue) functions.

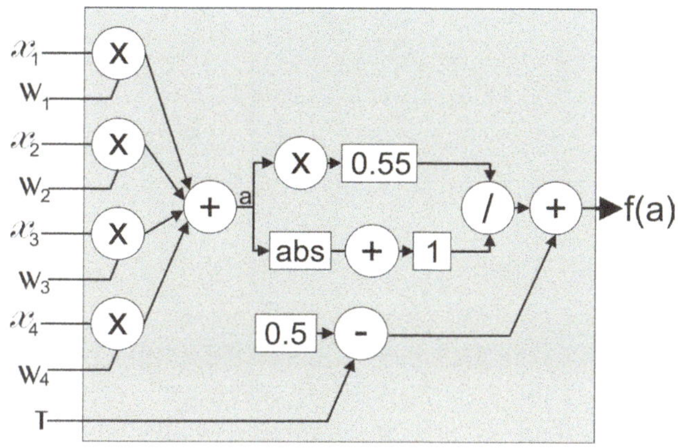

Figure 2. Neuron block diagram.

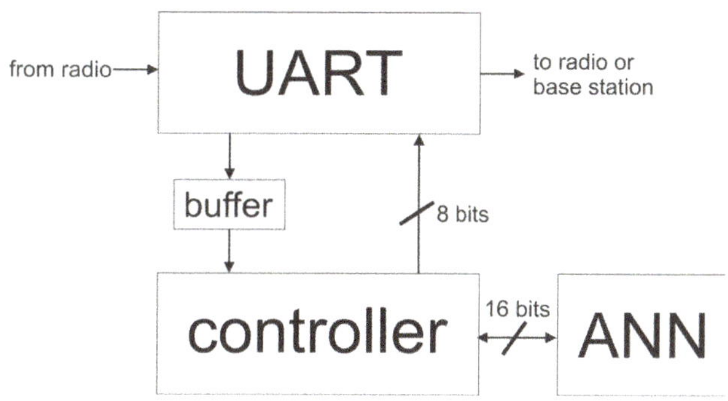

Figure 3. The block diagram of the whole system.

simplicity, we associate a token to identify the beginning anchor in the process. In an ideal case, the first anchor node sends its RSSI to the second anchor node, and the second anchor sends its RSSI and the RSSI of the first anchor to the third anchor. The third anchor sends the three RSSI to the fourth anchor.

As can be seen in **Figure 4**, the proposed method can be divided in seven mainly steps, vertically processed in each anchor node. The processing at same step is to be executed in parallel between anchors. If the fourth anchor node fails, the method can offer two estimated position (from anchor two and anchor three). If the third anchor fails, the method is still offering some reasonable response through second anchor. An optional but not implemented extension could be the last anchor (third or fourth) node returning the estimated position to the previous anchor. So the previous anchor can calculate some chosen media over the results, improving the accuracy and precision. The contribution of our proposed method is to explore the ANN to estimate with at least two inputs, different those methods based in triangulation, which needs at least three inputs.

In the next section we show our results and the errors associated to the anchor failures.

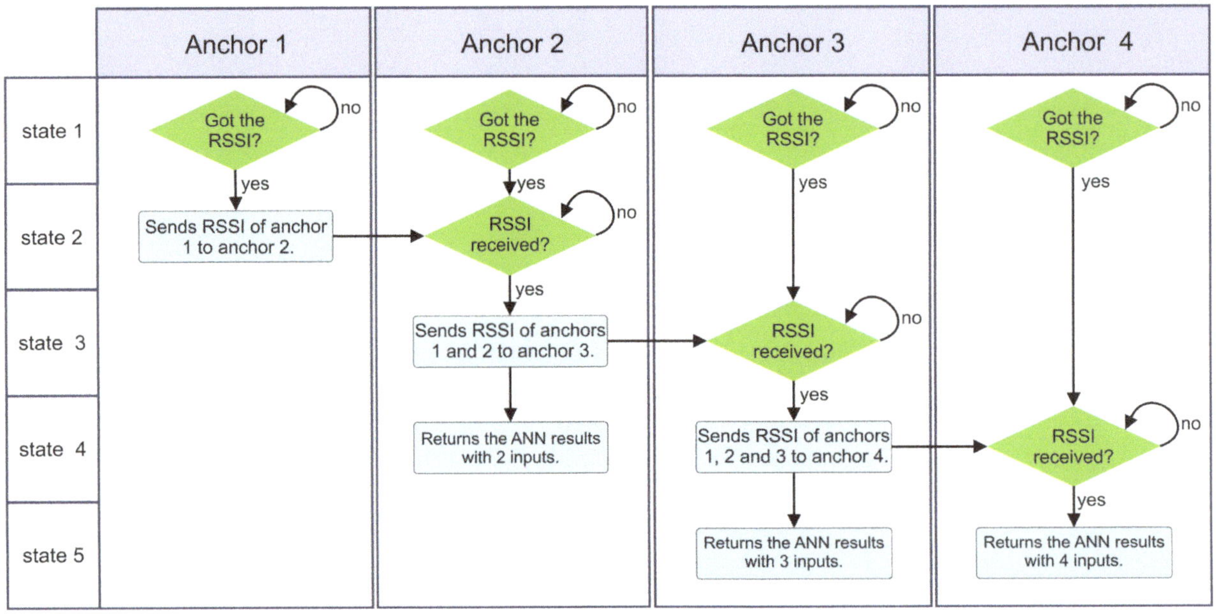

Figure 4. Flowchart of the proposed method.

4. Results

We took 25% of samples for testing the ANN and calculated the RMSE considering no failure and the failure of third or third and fourth anchors. **Table 2** shows the RMS calculated in meters.

We can calculate the total RMSE of the method as being the summation of the training RMSE associated with anchor failure and the RMSE of the discretized free parameters due to hardware implementation. This relation is expressed in Equation (2).

$$\text{RMSE}_{\text{total}} = f\left(\text{none}|\text{third}|\text{fourth}\right) \pm 1.25 \tag{2}$$

The function f in Equation (2) represents the RMSE considering anchor failures from **Table 2**. The constant 1.25 is the discretization RMSE generated in the free parameters due to binary representation. **Figure 3** shows some location points estimated in the region of 20 × 22 m, considering no failures and anchor's failure for the related cases.

We can visually observe in **Figure 5** the results for the three estimation cases. The estimation with no failures presents a high precision and the failure of just the fourth anchor is a good precision too, considering an approximation. The failure of the third (and consequently the fourth) anchor has a considerable imprecision.

Related with time execution, the neuron was simulated using the ModelSim™, from Altera™, and achieve a total time o execution of 1.75×10^{-7} seconds. This processing time is the same for each layer, as they execute all neurons in parallel. The total time execution of the ANN is 6.80×10^{-7}. In fact, the time execution of our proposed method is related to the distance the signal propagates from the target to anchor nodes and between the anchor nodes themselves. The time execution of our proposed method is faster compared with the time transmission. The time transmission is an inherent problem related to any method in location problems, and the reduction time during the location method can contribute to decrease this issue.

5. Conclusions

This work shows how the ANN can still be applied in WSN localization problem, with good approximations, and points the reconfigurable platform (FPGA) offering a way to achieve performance. The proposed method can be extended for multiple channels, if possible, allowing the multiple execution of the same algorithm in the FPGA, with a quite adjustment. The number of anchor nodes could be increased and grouped for estimating as cluster heads, for example. The media may be calculated for the three ANN estimations for better accuracy. The flexibility of FPGA and the ANN implementation allows this work to be extended to another new experiments to get better results.

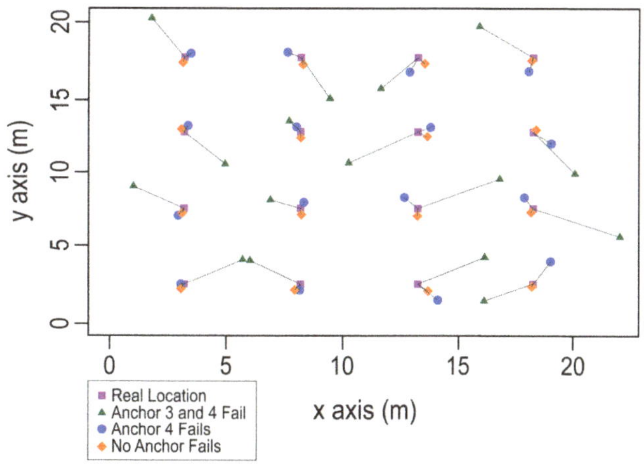

Figure 5. Visual testing results for 16 coordinates.

Table 2. The RMS error (m) considering no failures and failures of third and fourth anchors.

	No. anchor failure	Fourth anchor fails	Third and fourth anchor fail
Error on X	0.189077	0.199093	0.203451
Error on Y	0.177593	0.168486	0.194702

As future work we intend to perform location of multiple targets by applying this ANN to perform the task.

References

[1] Alhmiedat, T., Salem, A.O.A. and Taleb, A.A. (2013) An Improved Decentralized Approach for Tracking Multiple Mobile Targets through Zigbee WSNs. *International Journal of Wireless & Mobile Networks*, **5**, 61-76. http://dx.doi.org/10.5121/ijwmn.2013.5305

[2] Liu, L.Q. (2012) A Parallel and Distributed Computing Platform for Neural Networks Using Wireless Sensor Networks. Ph.D. Thesis, The University of Toledo, Toledo.

[3] Din, A., Bona, B., Morrissette, J., Hussain, M., Violante, M. and Naseem, M.F. (2012) Embedded Low Power Controller for Autonomous Landing of UAV Using Artificial Neural Network. *10th International Conference on Frontiers of Information Technology* (*FIT*), Islamabad, 17-19 December 2012, 196-203.

[4] Rahman, M.S., Park, Y. and Kim, K.-D. (2012) RSS-Based Indoor Localization Algorithm for Wireless Sensor Network Using Generalized Regression Neural Network. *Arabian Journal for Science and Engineering*, **37**, 1043-1053. http://dx.doi.org/10.1007/s13369-012-0218-1

[5] Kumar, S., Jeon, S.M. and Lee, S.R. (2014) Localization Estimation Using Artificial Intelligence Technique in Wireless Sensor Networks. *The Journal of Korea Information and Communications Society*, **39C**, 820-827. http://dx.doi.org/10.7840/kics.2014.39C.9.820

[6] Zheng, J. and Dehghani, A. (2012) Range-Free Localization in Wireless Sensor Networks with Neural Network Ensembles. *Journal of Sensor and Actuator Networks*, **1**, 254-271. http://dx.doi.org/10.3390/jsan1030254

[7] Bhardwaj, S. (2013) ANN for Node Localization in Wireless Sensor Network. *International Journal of Advanced Research in Electrical*, *Electronics and Instrumentation Engineering*, **2**, 1724-1731.

[8] Kumar, S. and Lee, S.-R. (2014) Localization with RSSI Values for Wireless Sensor Networks: An Artificial Neural Network Approach. *International Electronic Conference on Sensors and Applications*, 1 June 2014. http://dx.doi.org/10.3390/ecsa-1-d007

[9] Sabto, N.A. and Mutib, K.A. (2013) Autonomous Mobile Robot Localization Based on RSSI. *Journal of King Saud University—Computer and Information Sciences*, **25**, 137-143. http://dx.doi.org/10.1016/j.jksuci.2012.10.001

[10] Saad, A.-M.H.Y. and Alhady, S.S.N. (2014) Embedded Neural Network for Distance Recognition Using Distance Sensor. *1st International Conference of Recent Trends in Information and Communication Technologies*, Universiti Teknologi Malaysia, Johor, Malaysia, 182-191.

[11] Gogolák, L., Pletl, S. and Kukolj, D. (2011) Indoor Fingerprint Localization in WSN Environment Based on Neural

Network. *Proceedings of the IEEE 9th International Symposium on Intelligent Systems and Informatics* (*SISY*), Subotica, 8-10 September 2011, 293-296.

[12] Chen, C.-S. (2012) Artificial Neural Network for Location Estimation in Wireless Communication Systems. *Sensors*, **12**, 2798-2817. http://dx.doi.org/10.3390/s120302798

[13] Neves, A., Fonseca, H.C. and Ralha, C.G. (2013) Location Agent: A Study Using Different Wireless Protocols for Indoor Localization. *International Journal of Wireless Communications and Mobile Computing*, **1**, 1-6.

[14] Pajares, G. (2012) Sensors in Collaboration Increase Individual Potentialities. *Sensors*, **12**, 4892-4896. http://dx.doi.org/10.3390/s120404892

[15] Larrat, M., Machado, L. and Monteiro, D.C. (2013) Performance Evaluation of Lateration, KNN and Artificial Neural Networks Techniques Applied to Real Indoor and Outdoor Location in WSN. *Journal of Communication and Computer*, **12**, 72-81.

[16] Csáji, B.C. (2011) Approximation with Artificial Neural Networks. M.Sc. Thesis, Faculty of Sciences, Eötvös Loránd University, Budapest.

[17] Altera (2013) DE0 Nano User Manual v1.9.

[18] Bishop, D. (2010) Fixed Point Package User's Guide.

[19] Sibi, P., Allwyn Jones, S. and Siddarth, P. (2013) Analysis of Different Activation Functions Using Back Propagation Neural Networks. *Journal of Theoretical and Applied Information Technology*, **47**, 1344.

20

Ubiquitous Mobile Sensing for Water Quality Monitoring and Reporting within Lake Victoria Basin

Anthony Faustine, Aloys N. Mvuma

The University of Dodoma, Dodoma, Tanzania
Email: sambaiga@gmail.com, anmvuma@gmail.com

Abstract

As the human population growth and industry pressure in most developing countries continue to increase, effective water quality monitoring and evaluation has become critical for water resources management programs. This paper presents the ubiquitous mobile sensing system for water quality data collection and monitoring applications in developing countries. The system was designed based on the analysis of the existing solution. Open source hardware and software was used to develop the prototype of the system. Field testing of the system conducted in Nkokonjero, Uganda and Mwanza, Tanzania verified the functionalities of the system and its practical application in actual environment. Results show that proposed solution is able to collect and present data in a mobile environment.

Keywords

Arduino, Android, Ubiquitous Mobile Sensing, Water Quality

1. Introduction

The rapid development of society, economy and an increasing number of human activities in most developing countries speed-up contamination and damage to the water resources. As the result monitoring of water quality is essential in order to characterize water and identify changes or trends in water quality over time [1]. For decades, water quality monitoring and evaluation have depended on a low-resolution water quality monitoring. This approach is time-consuming and expensive and not appropriate for certain biological and chemical processes such as oxidation-reduction potential that need to be measured on-site to ensure accuracy [2]. Recently, Wireless Sensor Networks (WSN) have received considerable attention in water quality monitoring and control applications [3]-[7]. However, the practical application of WSN in water quality monitoring in developing countries is limited by following factors: 1) The initial investment cost of the WSN is considerably higher; 2) Physical security of the WSN nodes is also a challenge and 3) The sensor nodes are static over one particular location.

On the other hand, the rapid spread in mobile phone use in developing countries has increased the application of mobile data management [1]. Smart phones are becoming pervasive computing and communications platform [8]. The variety and number of mobile applications has also increased in recent years. This is because mobile phones are cheap, easy to use and can transmit multiple types of information including images and GPS points over long distances. The development and deployment of ubiquitous mobile sensing systems has become significant. However, the number of applications that leverage external sensing devices in water resources management is limited. The work presented by Aquaya [9] describes a water quality reporter system for data collection using featured phone. However, it doesn't provide the means to transfer the same data electronically to a central system and lacks sensing capability.

The use of ubiquitous sensor technologies and mobile data applications can improve the efficiency and accuracy of data reporting in the water sectors. This works extends the use of ubiquitous mobile sensing technologies by designing and developing mobile data capture and reporting applications that leverage external sensing devices in water resource management. It further discuses the role of this application in water quality management in rural and under-resourced areas in developing countries.

2. System Design

2.1. System Architecture

The proposed design for the Ubiquitous mobile Water quality Reporting system (u-WQR) for on-site sensing and reporting water quality data is presented. This system was developed as one component of the water governance system (WaGoSy) [10] designed to address water governance challenges around Lake Victoria Basin. Although more advanced techniques and technologies are already present in the world, u-WQR system is implemented using simpler and open source technology to reduce the system costs.

u-WQR system consists of four main components: 1) Ubiquitous Sensing Device (USD) which consists of Arduino microcontroller and sensor unit, 2) Communication module which connect the sensing device and the Android mobile phone, 3) The mobile application developed using the Android platform and 4) The WaGoSy web server. The Arduino microcontroller is the main controller that control and performs the necessary actions that need to be carried out by the system. The sensors and communication modules are directly interfaced to the main controller. The Arduino microcontroller communicate with mobile phone through a serial Bluetooth module. An overview of the proposed system architecture is shown in **Figure 1**.

2.2. Software Development of the u-WQR Mobile Application

Several platforms such as Windows Mobile, Symbian, iOS and Android can be used in the development of smart phone applications. The proposed u-WQR mobile application is developed using the Android platform. This is because Android platform is open source, supports background processes and rich inter-application communication and support most of the phones and handy devices on a market [8] [11]. Android Studio version 0.8.9 has been used for the development and implementation of the u-WQR mobile app. This is Android development environment based on IntelliJ IDEA [12].

The screenshots of the u-WQR mobile app developed is shown in **Figure 2** while the processing of the u-WQR mobile app is shown in **Figure 3**. The designed mobile app for the water quality reporting system provide, user authentication, serial Bluetooth connection to the USD, remote connection (via internet) to the WaGoSy web server and history of the data reported. To successfully connect and access the u-WQR the user has to enter the correct user name and password as shown in **Figure 3**. The u-WQR mobile app is locally developed using open source and freeware development tools.

2.3. The Proposed Ubiquitous Sensing Device

The ubiquitous sensing device (USD) consists of Arduino microcontroller, sensor unit and power supply. The sensor unit as shown in **Figure 4**, consists of several sensors used to detect water quality parameters. Four types of sensor pH sensor, electrical conductivity sensors (EC), oxidation reduction potential (ORP) and temperature

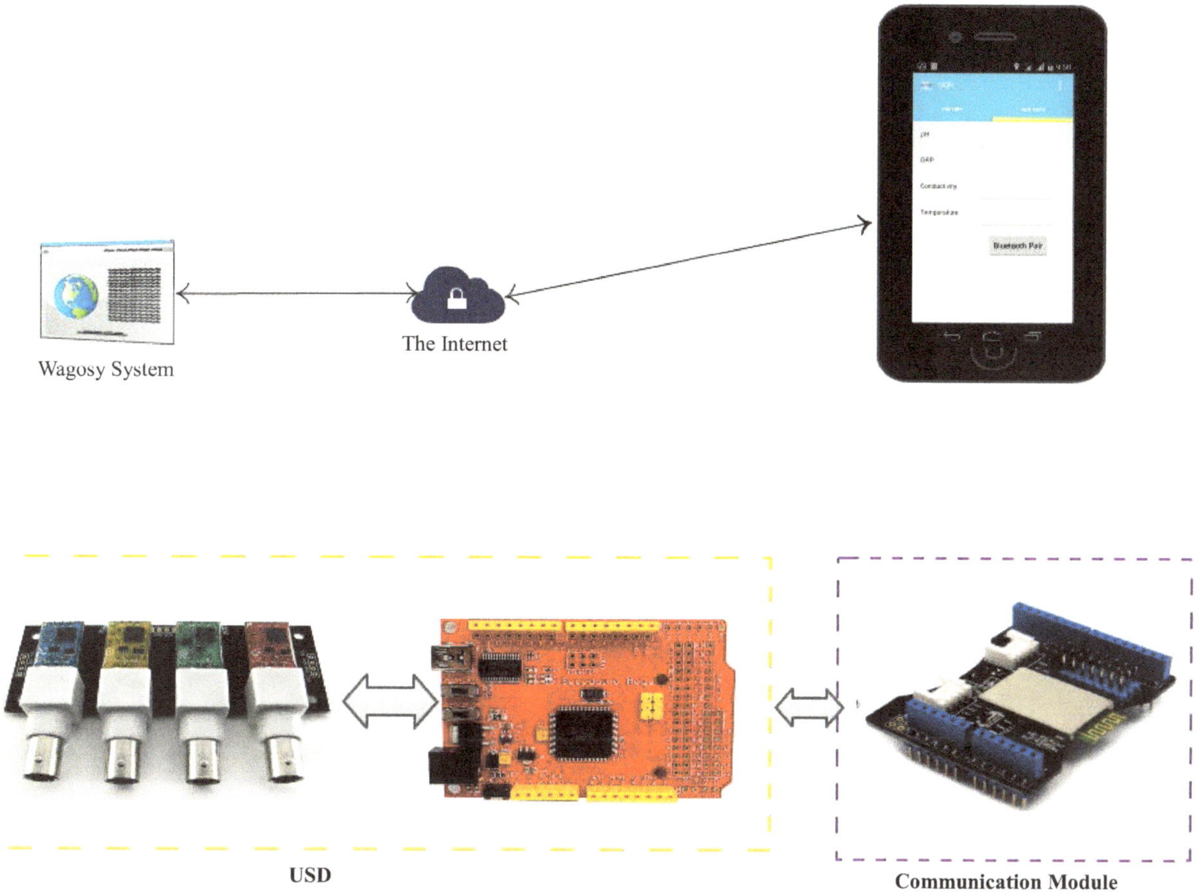

Figure 1. u-WQR system architecture.

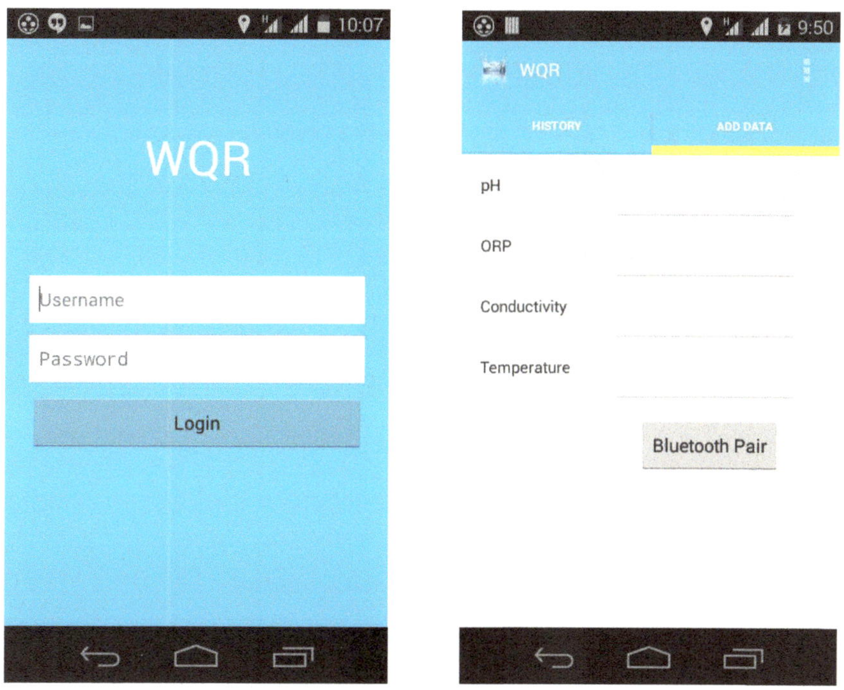

Figure 2. Screenshot of the proposedmobile app.

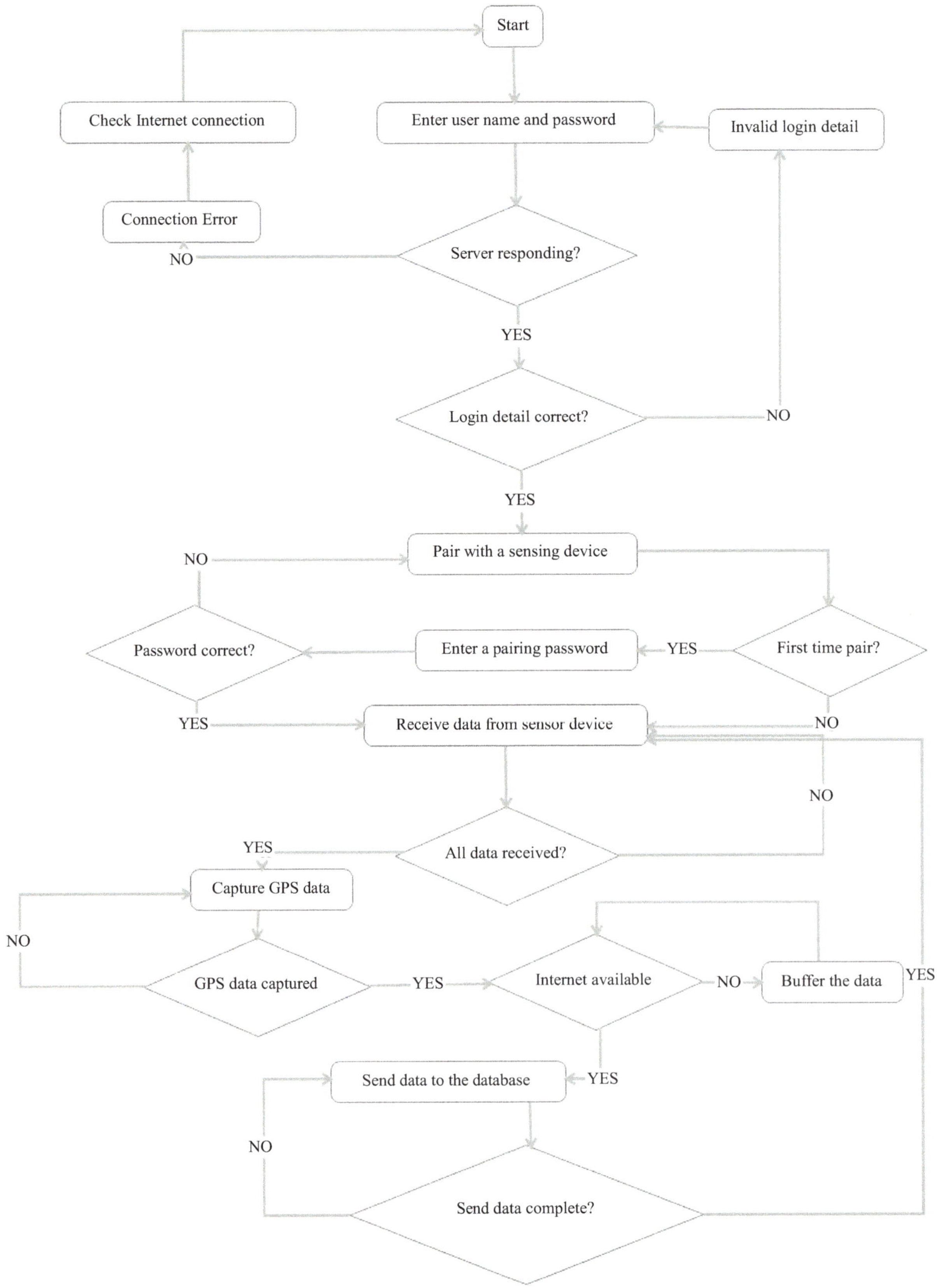

Figure 3. General processing of the proposed mobile app.

sensor are used to provide general characterization of water quality. Specification for these sensors is summarized in **Table 1**. The EC, pH and ORP sensors are interfaced to the microcontroller using their respective circuit from Atlas scientific with exception of the temperature sensor which is directly connected to the microcontroller. To connect all the sensors to the Arduino, multi circuit I2C carrier board Atlas scientific is used. The sensors were accurately calibrated to ensure that they operate correctly.

The process of acquiring data from sensors is shown in **Figure 5**. Seeeduino Mega [13] from Seeedstudio is used to acquire and process sensor data from the sensor unit. This microcontroller is derived from Arduino Mega and is based on the ATmega 256 having 70 digital input/output pins. In this work 3.7 V 6AH rechargeable polymer lithium ion battery is used to power this USD. The battery has longer life and excellent long-term and self-discharge rates. However it provide less than 5V required by most components on the USD. As the result the LiPo Rider Pro [14] is used to boost the output from a battery voltage to 5 V. LiPo Rider Pro, further allows to charge the lithium battery through USB.

2.4. The Proposed Communication Module

The recently announced Android Open Accessory Protocol (AAP) from Google allows external sensors to interact with an Android-powered device in a special accessory mode [8]. To use this approach to connect Arduino and the mobile phone you need to implement USB Host stack on the Arduino board. However, the device board is more complicated as it must provide power for both devices. The USB host API of Android can also be used to enable Arduino board to interact with an android device. With this solution the device board is very simple but works only with Android devices that have an USB Host port.

In addition, there are APIs to access the Bluetooth, Wi-Fi, and Near Field Communication (NFC) radios on Android devices. These communication APIs allow Android devices to connect to an even wider variety of external sensors. However, building applications that utilize external sensors is relatively complex from the application developers' perspective [8].

In this work, a serial Bluetooth module is used to provide communication between USD and the u-WQR mobile app. This is because Bluetooth is a popular method of communication between devices. Many smart phones today have the capability to communicate using Bluetooth and it is also easy to get cheap Bluetooth radio. The Bluetooth Shield from Seeedstudio is used to implement a communication. This shield integrates a serial

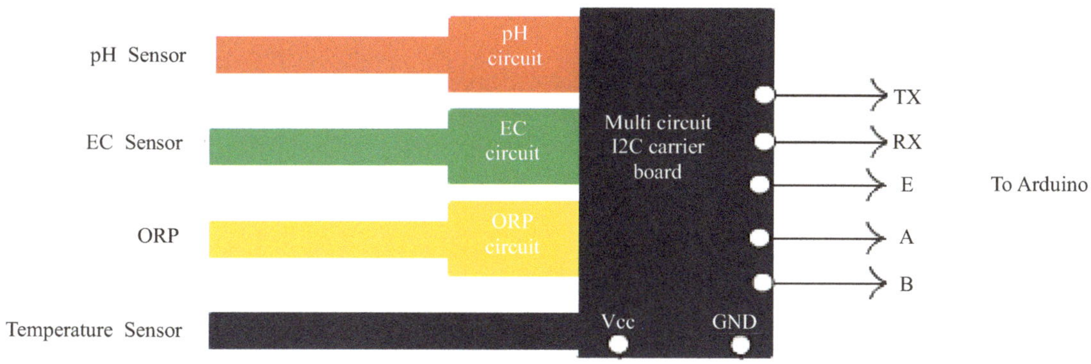

Figure 4. Sensor unit.

Table 1. Sensor specification.

Sensor	Manufacture	Model	Range
pH	Atlas Scientific	ENV-40-pH	0 - 14 (Na^+ error at >12.3 pH)
EC	Atlas Scientific	ENV-40-EC-K0.1	(K = 0.1) 0.1 - 0.5 μs/cm to 50,000 μs/cm
ORP	Atlas Scientific	ENV-40-ORP	+/− 2000 mV
Temperature	Maxim Integrated	DS18B20	−55˚C to +125˚C

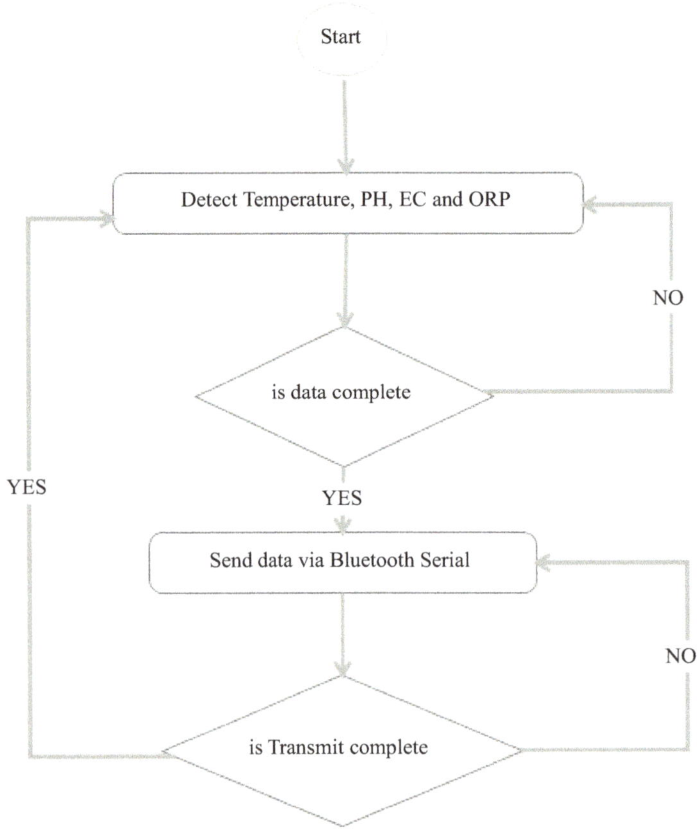

Figure 5. General processing of the proposed USD.

Bluetooth module which can be easily used with Arduino for transparent wireless serial communication. This module has two modes: master and slaver device and it is configured to operate in master mode in the proposed system.

3. Results and Discussion

The ubiquitous mobile sensing for water quality test and reporting system proposed in this paper was fully developed and tested to demonstrate its feasibility and effectiveness. **Figure 6** show the complete u-WQR system during field test. The proposed system has all the necessary feature for mobile data reporting in water quality management applications. It allows decentralized actors to submit water source monitoring data via cell phones to a central database. The system provides visualization module which enable administrator to visualize the datasets. Users can view different parameters of water quality data from different locations in graphical forms as shown in **Figure 7**. From the graphs, user can observe the behavior of each parameter and alert the public in case of abnormality. The system can be configured to send SMS and or email alerts to managers when data reveal problems. The system total cost including sensing device, mobile phones and the Bluetooth module stands at $750.

Field testing of the system conducted in Nkokonjero, Uganda and Mwanza, Tanzania verified the functionalities of the system and its practical application in actual environment. Through pilots in these two sites we have identified opportunities and challenges associated with the implementation of the u-WQR system in developing countries. For example, we noted that mobile data coverage is weak in many rural areas which limit instant data delivery. The system has a distinct advantage over traditional paper based reporting. Since it use electronic data collection which is relatively inexpensive, allow a well-coordinated and continuous monitoring of water quality from remote locations to centralized managers. Most stakeholders find the u-WQR system easy to use and require little training. On the other hand administrators and government authority managers were excited to see this system in place.

Figure 6. Field test of the u-WQR sytem.

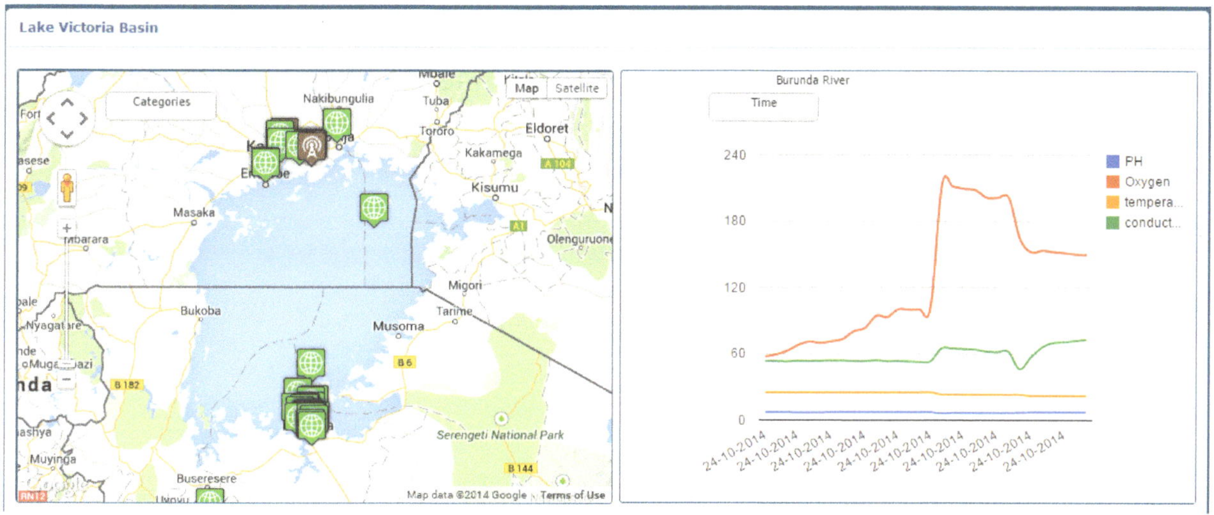

Figure 7. Visualizations of the u-WQR data.

4. Conclusion

The objective of the work presented in this paper was to realize the ubiquitous mobile sensing system for water quality monitoring and reporting. The system has been successfully designed and prototyped to capture and report the water quality parameters using an Android Bluetooth-enabled phone and Arduino based sensor device. The use of ubiquitous mobile sensing application in water quality monitoring and reporting has important implications for water resources management programs in developing countries. Prospective future works include incorporating water image processing and stabilizing the sensor device for field data collection.

Acknowledgements

Research team acknowledges Lake Victoria Research Initiative (VicRes), Inter-University Council for East Africa (IUCEA) and Swedish International Development Agency (SIDA) for supporting the research of which this paper is one of the outputs. The team also acknowledges the kind and material support received from stake-holders at different levels in the Lake Victoria Basin during the entire period of the research. The team wishes to thank the management and staff of Busitema University, Kigali Institute of Science and Technology and the University of Dodoma for their administrative support.

References

[1] DHI GRAS (2014) Water Quality Monitoring from Space, Baselines and Up-to-Date Information.

http://worldwide.dhigroup.com/

[2] Rao, A.S., Gubbi, J. and Palaniswami, M. (2013) Design of Low-Cost Autonomous Water Quality Monitoring System. *Advances in Computing, Communications and Informatics* (*ICACCI*).

[3] Xu, G., Shen, W. and Wang, X. (2014) Applications of Wireless Sensor Networks in Marine Environment Monitoring: A Survey. *Sensors* (*Basel*), **14**, 16932-16954. http://dx.doi.org/10.3390/s140916932

[4] Zia, H., Harris, N.R. and Merrett, G.V. (2014) Water Quality Monitoring Control and Management Framework Using Collaborative Wireless Sensor Networks. *Proceedings of the* 11*th International Conference on Hydro Informatics*, New York City.

[5] Kumar, R.K., *et al.* (2014) Solar Based Advanced Water Quality Monitoring System Using Wireless Sensor Network. *International Journal of Science, Engineering and Technology Research* (*IJSETR*), **3**, 385-389.

[6] Gokulram, K. and Dhakshinamoorthy, T. (2014) Intelligent Pollution Monitoring Using Wireless Sensor Networks. *International Journal of Research in Engineering and Technology*.

[7] Zennaro, M., *et al.* (2009) On the Design of a Water Quality Wireless Sensor Network (WQWSN): An Application to Water Quality Monitoring in Malawi. *International Conference on Parallel Processing Workshops*, 22-25 September 2009, Vienna, 330-336.

[8] Chaudhri, R., *et al.* (2012) Open Data Kit Sensors : Mobile Data Collection with Wired and Wireless Sensors.

[9] Aquaya (2013) Testing the Tests: Aquaya Evaluates Aquatest on a Global Scale. http://www.aquaya.org/testing-the-tests-aquaya-evaluates-aquatest-on-a-global-scale/

[10] Wagosy System. http://wagosy.com/

[11] Kumar, S. (2014) Ubiquitous Smart Homes System Using Android Application. *International Journal of Computer Networks & Communications* (*IJCNC*).

[12] Android Studio. https://developer.android.com/sdk/installing/studio.html

[13] Seeedstudio. Seeeduino Mega. http://www.seeedstudio.com/depot/Seeeduino-Mega-p-717.html

[14] LiPo Rider Pro. http://www.seeedstudio.com/wiki/index.php?title=Lipo_Rider_Pro

21

Energy Efficient WSN-SAT System for Surveillance of Satellites

Hafiz Bilal Khalil, Naeem Abas, Shoaib Rauf

Department of Electrical Engineering, University of Gujrat, Gujrat, Pakistan
Email: Engr.bilalkhalil@yahoo.com, naeemkalair@uog.edu.pk, engr.shoaib@uog.edu.pk

Abstract

Satellite communication plays a vital role in the arena of telecommunication. The main objective of this manuscript is to examine the threats and problems like space debris, space weather and space weapons that could damage or disable the satellite system. In consequence, we discuss the role of wireless sensor networks (WSN) and nodes in space based technologies. The scope of this paper is to minimize the threats and vulnerabilities to the satellite systems and based on that, design the next generation and state of the art satellite communication system. We proposed the solutions for the vulnerabilities and surveillance of satellite system by integrating the energy efficient and cost effective WSN to the satellite system.

Keywords

Energy Efficiency, Satellite Communication, Wireless Sensor Networks, Space Threats, Space Debris

1. Introduction

Satellite communication system enormously developed in the first decade of the 21st century. At the same time with the pace of technology augmentation, satellite communication has developed its applications for military operations, education, business, transportation tracking, navigation, television broadcasting, teleconferences, weather forecasting, health and many more. Most of the developed countries are now satellite dependent in order to conduct their routine work. The satellite system becomes more vulnerable to failure due to various factors, such as the space debris, space weather and anti-satellite missiles [1] [2]. In 2010 the International Telecommunication Union (ITU) in its recommendation S.1003-2 has declared that there are 1100 known spacecraft and rocket bodies near the Geosynchronous orbit (GSO) and one-third of them are currently active [3]. Due to the Earth-based observations, we have limited information about the geostationary orbit (GEO) environment. In low

Earth orbit (LEO) we can detect the smallest object having 30 cm dimensions while in best conditions the smallest object that can be detected in GEO is almost 1 m in dimensions [4]. This debris can produce more debris in collision with the inactive or lost satellite while it is also the risk for the active satellites. To reduce the collisions with debris we must have more knowledge about the geostationary environment. We can send a satellite for the observation, but after that mission it will also cause to create more fragmentations [5].

In addition, space weather is another problem faced by the satellite system as the space weather is unpredictable and more dynamic. Space weather is the byproduct of solar activity and has high energy electrons that affect the satellite. The surface of the Sun radiates the high energy electron during the solar activity. These electrons travel towards the Earth and also cause vulnerabilities in GEO system. However, these satellites are shielded for protection, but with the passage of time this shielding becomes weakened and degraded. When a satellite is damaged it is very difficult to get it back and analyze the reasons behind the failure. It becomes necessary to monitor the effects on the surface of the satellite at the end of its operational life. So, sending another satellite to monitor the effects is not economical and wise. We need an alternate system that should be low cost, energy efficient, flexible and small in size to sense and communicate the required information without creating more debris.

After extensive research, we found all these characteristics in the wireless sensor networks (WSN). We proposed integrated wireless sensor-satellite system (I-WSAT) system to approach the goals. Then WSN is widely used in monitoring, military operations, inventory control, health, tracking, sensing of vibration, intuiting heat and various atmospheric parameters not only on earth but also under the sea [6]. However, these wireless sensor nodes cannot be implemented in the spacecraft surface and orbit [7]. Special purpose, optimized and modified nodes will be required to communicate the sensed information in the space missions. Besides this, for WSN nodes a radio communication link should be designed to directly communicate with satellite [8].

The objective of this paper is to monitor the space threats without sending satellites into the space. We suggest the Wireless sensor networks to monitor the orbits and outer surface of the satellites. In the following sections, this paper will describe the threats to the satellite system and need of WSN. Furthermore, we explain the challenges and opportunities in the implementation of WSN node in space and orbits.

2. Wireless Sensor Networks (WSN)

Wireless sensor network is the collection of small nodes that sense the information and transmit data to the main node. These micro sensors consists sensing module, radio module, digital signal processing unit and an energy source. Wireless sensor network works cooperatively they sends their own data and also act as router to send the sensed information of other nodes. Each node sends the information of other nodes towards the main node according to predefined routing protocols. Applications of WSN are exponentially expanded in monitoring, surveillance, intelligence, protection and control of the environment or hard to go areas. **Figure 1** describes the role

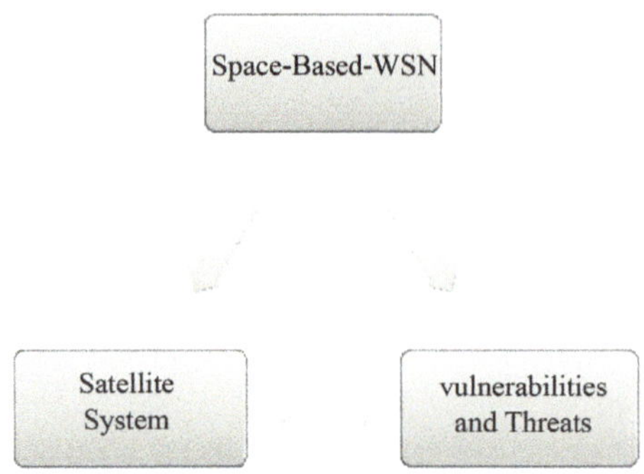

Space-Based-WSN	
Detection of Space Debris	Effects of Space Weather
Anti-satellite Missiles	Interception/Hiding of the satellite
Satellite-Node-Satellite	Emergency Communication

Figure 1. Role of WSN in satellite system.

of WSN for detection of various vulnerabilities like detection of debris, effects of space weather and emergency communication etc. in satellite system.

3. Vulnerabilities and Threats to Satellite System

3.1. Protection of GSO

There are some limitations to retrieve objects to the Earth from geostationary orbit (GSO) at the end of their operational life. In [3] ITU has also recommended that a region must be established above or below the GSO where the operational satellite may reside and maneuver. And at the end of operational life the satellites should be taken out of this region. Spacecraft operator should assure the availability of fuel to achieve the required maneuver at the end of satellite life. This dedicated region for active satellite should be free from debris and any other inactive satellite to reduce the risk of collision. To monitor the dedicated region, we cannot put additional satellite because it can cause some additional debris after its operational life. In this scenario WSN is the best options for the monitoring of debris of GSO.

3.2. Space Weather

Space weather has effects on the surface of the satellites and space weather is more dynamic than the predicted models. The space weather mostly affected the Geo stationary satellites [2]. So, Geo satellites are designed by taking the space weather into the account, but it is not enough. We need more information regarding this issue, so we need to study the inactive and failure satellites caused by the space weather. It is very difficult to bring back the satellite to study the weather effects on it. To study the space weather effects on the satellite WSN can play a vital role in continuous monitoring of the surface of satellite, and we can design more secure and space weather resistant satellites in the future. It can improve the life and efficiency of the GEO satellite and provide help in preparing the next generation communication satellites.

3.3. Space Weapon

Space weapons are the biggest threat to the satellite system. Anti-satellite weapons are capable to take the warhead into the space. Many countries such as Pakistan, China, Russia, Iran and India have made such ballistic missiles that can take the low-yield nuclear weapon into the space. Explosion of such type weapon in LEO can destroy the satellite [9] and shorten the life cycle of the satellite. China has already successfully conducted anti-satellite weapon (ASAT) experiment to destroy its weather satellite, which is alarming. The Low-altitude-short-duration-orbital ASAT and High-altitude-short-duration-orbital ASAT are interceptors that are launched from the ground into a temporary parking orbit from which they maneuvers to attack their respective targeted satellite. The on-orbit lifespan of temporarily parked interceptor is measured in hours; detection of this system is slightly more complex than direct ascent weapons. An operational satellite can be made hide from interceptors or radars using the WSN special nodes, which cover the original signal by transmitting the number of signals at the same frequency [10]. WSN can perform continues monitoring in their targeted region and can detect temporarily parked interceptor.WSN can be used to examine the nuclear effects in LEO and collect the information.

Space to space missile is another threat for the satellite security. Space to space missile system consists of a carrier satellite, missile and control system. A missile can be launched from space to hit the target in the space or another missile targeting the carrier satellite that is coming from the earth or other satellite [11]. WSN covers the larger area in space and gather the information about the other missile carrier satellites and give instructions for precautionary measures in emergency conditions.

Micro-satellite and Nano-satellite perform the satellite inspection and also use in imaging processes [11]. They also can be used as weapons to destroy the target satellite. It can be placed on an intercept course and programmed to hone in satellite or destroy the target satellite. Defense against such type attack would be difficult. But now WSN provides the solution, Nodes can detect any moving object along the satellite. The interceptor satellite can be detected before the explosion and warning can be submitted to the ground station with plenty of time. A precautionary measure can be taken to save that satellite from such type attacks.WSN can be used as parasitic communication, but according to the rules ITU-R S.524-8 and ITU-R SF.358-5 [3] interference should not take place among the other satellites. Nano and Pico satellite concept has already been introduced to interconnect the satellites and for the inspection of satellite [12]. WSN keeps monitoring the surroundings of satellite just like Pico-Satellite. WSN can be connected with active and inactive satellites to perform various activities. The synopsis of interconnected I-WSAT has been shown in **Figure 2**.

4. Challenges in Implementation of WSN in Space

4.1. Deployment

The deployment of WSN in space can be random or installed at premeditated chosen locations on the outer surface of the satellite or in the surroundings [13]. In physical environment the spreading of WSN nodes can be a continuous process, and nodes can be added time to time according to the requirement and nature of sensing. However, deployment of the nodes is a one-time activity and deployed on satellite or in space at predetermined locations and at the same time in the orbit. Because in case of satellite the replacing of nodes to improve the network and sensing efficiency is not an easy task. WSN nodes can be spread in kilometers around the satellite to detect the debris and cover or hide the satellite from the interceptors. At the initial stage of the space mission the WSN nodes can also be attached with the surface of satellite (as shown in **Figure 3**) to monitor the various parameters. The deployment cost, mass and classification of previous satellite mission is shown in **Table 1**.

4.2. Networking

In space there will be no infrastructure for networking. WSN nodes can make the ad-hoc network to communicate with each other and communicate with the ground station through the satellite. Each sensed information regarding to the space weather, change in the surface of the satellite, debris collision or any other valuable information should be communicated to the control system of satellite and also to the ground station through the satellite.

Ad hoc networking is the most suitable and preferred networking technique for the WSN nodes as the installation of network infrastructure is costly and not feasible. The combination of adhoc network and infrastructure based (satellite communication) [13] network is the way to communicate the sensed information between the sensor nodes and ground control station. Each sensor transmits the information to all the nodes, while one sink node or gateway sends the information to the satellite. Selection of the node that transmits the information to the satellite is important to increase the performance of the network; MNMU-RA [14] routing algorithm can be chosen to select the main node in WSN.

There are various technologies being used in WSN for networking such as ZigBee (IEEE 802.15.4 Standard), Wi-Fi (IEEE 802.11) and Bluetooth (IEEE 802.15.1) for low data rate and short distance communication. Some recent research has investigated the potential of ultra wideband (UWB) technology for WSN [15]. As we

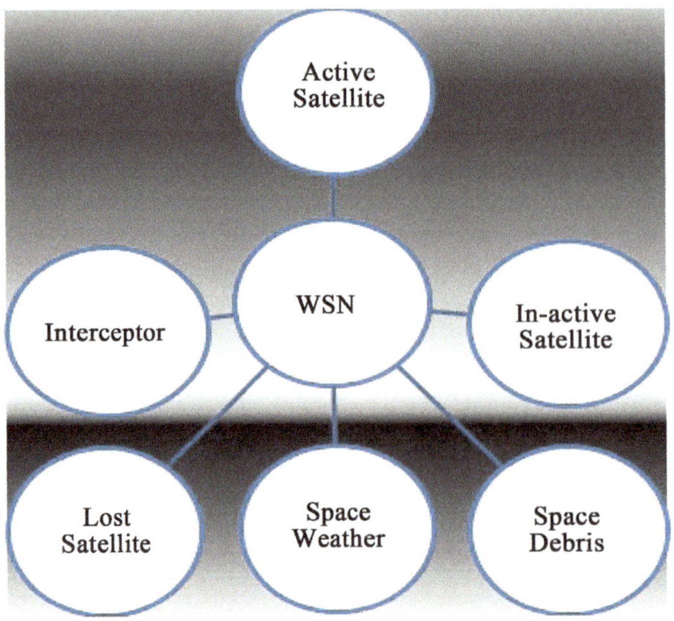

Figure 2. Synopsis of I-WSAT system.

Figure 3. Deployment of WSN on surface of satellite.

Table 1. Classification of satellites.

Names of satellite	Mass	Mission Cost	Classification
Inmar Sat-4	5945 Kg	\$0.1 - 2 B	Large
GIOVE-A	660 Kg	\$50 - 100 M	Medium
UK-DMC	166 Kg	\$10 - 50 M	Mini
PicoSat	67 Kg	\$2 - 10 M	Micro
SNAP	6.5 Kg	\$0.2 - 2 M	Nano
PalmSat	1 Kg	\$20 - 200 K	Pico
PCB Sat	250 g	\$3 - 20 K	Femto

concerned about the energy efficient WSN, the UWB has the characteristics of energy efficiency. We conduct the simulated experiment in a network and calculate the utilization of energy separately at various standards (Wi-Fi, Bluetooth and UWB) and other parameters are mentioned in **Table 2**. For simulation 20 nodes are randomly deployed ranging from 1 to 10 meters distance and moving with constant speed around an object (satellite). Same network and topology taken at each time and calculate the results. A comparison of energy consumption in different modes; sensing, receiving and transmission is shown in **Figure 5**. As the energy consumption is high during the transmission mode, but the results have shown the significant reduction in energy consumption during the transmission mode of WSN nodes.

We proposed satellite-node-satellite communication way to communicate (**Figure 4**). In this method two satellites can communicate with each other with the help of common connected WSN node. An active satellite can find or detect the inactive satellite with the help of nearby WSN nodes. An efficient link budget is required for the radio signal where WSN can directly communicate with the satellite [16]. While already satellite-satellite communication takes place in between the satellites. This technique will help to locate the lost or completely failed satellite in space with the help of near satellite and WSN nodes in that orbit.

4.3. Mobility

Space weather can also have an impact on WSN nodes and can cause in change of their location or even a failure of any single node. It is also a challenge for WSN in space, so the WSN nodes should have the capability to

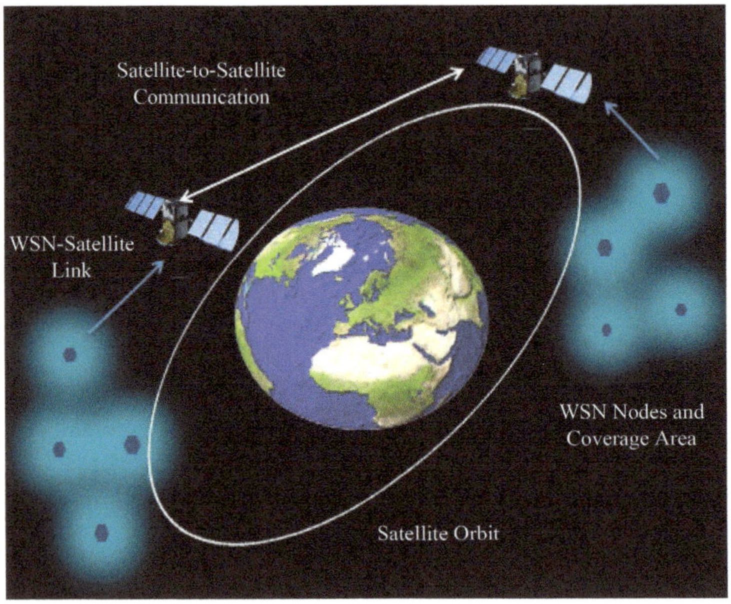

Figure 4. Satellite-WSN-Satellite communication.

Table 2. Parameters for simulation.

Features/operation	Unit/Detail
Sleep	**19 μA**
Receiving	**10 Ma**
Transmission	**17 Ma**
Radio Frequency	**2.4 GHz/900MHz/500MHz**
Bit rate	**420 Kbps**
Bandwidth	**25 MBPS/**
Data packet	**8 Bytes**
Number of Nodes	**20**
Minimum Distance	**~1 meter**
Maximum Distance	**10 meters**
Nodes to Main Node links	**4**
Main Nodes to satellite link	**1**

maintain the network and compute the new routing path in case of displacement or failure of any node. Networking protocol and algorithm should be designed to take into account that the orbital mobility can have a larger impact on the sensor nodes. For effective surveillance modified and efficient nodes are required, which may have a transmission range in kilometers and flexible to maintain the network and large coverage area. Space environment issues must be considered like shock, vibration and acceleration. Some other atmospheric such as corrosion and vacuum may also have effects on the network. The mobility of WSN nodes is uncontrolled and passive; the speed of movement of nodes should be compatible with the speed of under observation satellite.

4.4. Energy

The satellite system has used several integrated micro power systems for electrical energy such as solar cells, fuel cell, nuclear battery, chemical battery, induction and vibration. Energy is the tradeoff between size and cost/ network life [17]. Due to the smaller size, the limited energy source is also a big issue for WSN nodes. Energy is stored in small batteries but it can also be harvested from the environment for longer networking life. For space mission, we cannot depend only on batteries or fuel, WSN nodes should be embedded with the small solar panel just like the satellite system. Typical solar cells are fabricated with optimized silicon or gallium arsenide, for efficient power generation the integrated solar cell should be fabricated with CMOS but it is not yet practically implemented [18].

Energy conservation is directly proportional to the life of network. Most of the energy is consumed during the transmission of information, while sensing required minimum amount of energy during the idle mode. Routing algorithms play an important role in energy conservation, for the longer network life an energy efficient routing algorithm must be selected.

After the extensive research, we conclude that efficient utilization of energy, cost effectiveness and smaller size are the key elements for the future satellite missions. A comparison of energy utilization (at different stages) between UWB and our proposed solution has been shown in **Figure 5**. Simulated results have shown remarkable achievements in energy efficiency. Our proposed **Table 3** has shown the some other required features for the WSN-Satellite missions.

5. Conclusion

This manuscript describes both the wireless sensor networks and satellite systems. An integrated WSN-Satellite system was proposed with the aim to reduce the vulnerabilities and threats associated with the satellite based technologies. WSN can be a cost effective solution for many problems, but it is not an easy task to use WSN in the space. We describe the challenges in the integration of WSN with the satellite system and proposed various

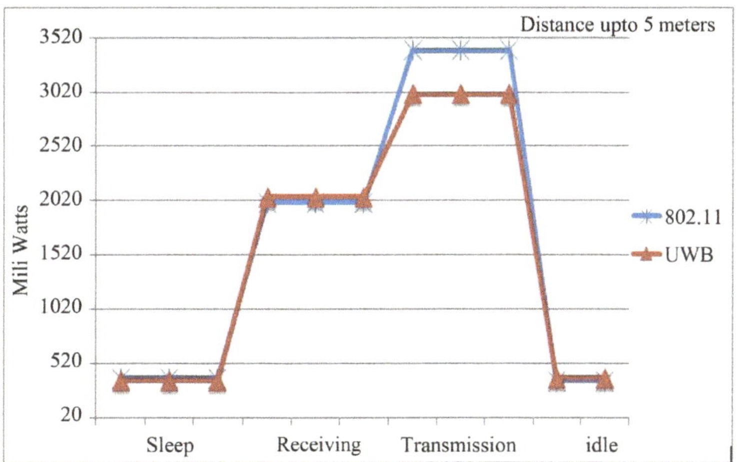

Figure 5. Comparison of power consumption at different operation.

Table 3. Requirement for space WSN nodes.

WSN Features	*Requirement for Space*
Size	As low as Possible to reduce debris
Mass	Minimum
Power	Maximum life cycle required
Communication	Node-to-node, Node to satellite, satellite-Node-satellite.
Number of Nodes	<100
Altitude	Appropriate for (LEO, GSO or mission)
Coverage	Satellite surface and surroundings
Routing	Co-operative,
Resilience	Vacuum, Space Weather,

modifications required in traditional WSN. An advantage of WSN over the parasitic and monitoring satellite is that typical satellite needs explosion or deterioration after the life spans, which increase the fragmentation debris and can be harmful for other satellites in the same orbit. WSN requires no such type explosion. ASAT weapons and interceptors are major threats for the developed and satellite dependent countries; WSN is an effective system to tackle such type of problems. In future, after the required modification and hardware optimization in the nodes WSN-Satellite system would be economical, effective and debris free system for the space missions and surveillance.

References

[1] Huntington, J. (2007) Improving Satellite Protection with Nanotechnology. Air WarColl Maxwell AFB AL Center for Strategy and Technology, United States Air Force (USAF) Proof. http://oai.dtic.mil/oai/oai?verb=getRecord&metadataPrefix=html&identifier=ADA474825

[2] Iucci, N., *et al.* (2005) Space Weather Conditions and Spacecraft Anomalies in Different Orbits. *Space Weather* 3.1.

[3] Recommendation ITU-R S.1003-2 (12/2010) Environmental Protection of Geo-Stationary-Satellite Orbit. http://www.itu.int/rec/R-REC-S.1003/_page.print

[4] Schildknecht, T., *et al.* (2004) Optical Observations of Space Debris in GEO and in Highly-Eccentric Orbits. *Advances in Space Research*, **34**, 901-911. http://dx.doi.org/10.1016/j.asr.2003.01.009

[5] Kitazawa, Y. (2012) Effective Detection of Low-Luminosity GEO Objects Using Population and Motion Predictions. Defense Technical Information Center, Fort Belvoir, VA.

[6] Celandroni, N., *et al.* (2013) A Survey of Architectures and Scenarios in Satellite-Based Wireless Sensor Networks: System Design Aspects. *International Journal of Satellite Communications and Networking*, **31**, 1-38. http://dx.doi.org/10.1002/sat.1019

[7] Akbulut, A., *et al.* (2011) Wireless Sensor Networks for Space and Solar-System Missions. 5*th International Conference on Recent Advances in Space Technologies* (*RAST*), Istanbul, 9-11 June 2011, 616-618. http://dx.doi.org/10.1109/RAST.2011.5966912

[8] Poulakis, M.I., Vassaki, S. and Panagopoulos, A.D. (2013) Satellite-Based Wireless Sensor Networks: Radio Communication Link Design. 7*th European Conference on Antennas and Propagation* (*EuCAP*), Gothenburg, 8-12 April 2013, 2620-2624.

[9] Hagt, E. (2007) China's ASAT Test: Strategic Response. *China Security*, **3**, 31-51.

[10] Kim, K., Lee, M.W. and Lim, J.S. (2012) Spreading Technique of Satellite Beacon to Avoid Jamming Attacks. 14*th International Conference on Advanced Communication Technology* (*ICACT*), PyeongChang, 19-22 February 2012, 778-781.

[11] Wilson, T. (2001) Threats to United States Space Capabilities. Prepared for the Commission to Assess United States National Security Space Management and Organization.

[12] Funase, R., *et al.* (2007) Technology Demonstration on University of Tokyo's Pico-Satellite "XI-V" and Its Effective Operation Result Using Ground Station Network. *Acta Astronautica*, **61**, 707-711. http://dx.doi.org/10.1016/j.actaastro.2006.12.032

[13] Romer, K. and Mattern, F. (2004) The Design Space of Wireless Sensor Networks. *IEEE Wireless Communications*, **11**, 54-61. http://dx.doi.org/10.1109/MWC.2004.1368897

[14] Khalil, H. and Zaidi, S. (2012) MNMU-RA: Most Nearest Most Used Routing Algorithm for Greening the Wireless Sensor Networks. *Wireless Sensor Network*, **4**, 162-166. http://dx.doi.org/10.4236/wsn.2012.46023

[15] Mahfuz, M.U. and Ahmed, K.M. (2005) A Review of Micro-Nano-Scale Wireless Sensor Networks for Environmental Protection: Prospects and Challenges. *Science and Technology of Advanced Materials*, **6**, 302-306. http://dx.doi.org/10.1016/j.stam.2005.02.008

[16] Akyildiz, I.F., Su, W., Sankarasubramaniam, Y. and Cayirci, E. (2002) Wireless Sensor Networks: A Survey. *Computer Networks*, **38**, 393-422. http://dx.doi.org/10.1016/S1389-1286(01)00302-4

[17] Barnhart, D.J. (2008) Very Small Satellite Design for Space Sensor Networks. Surrey Univ Guildford (United Kingdom) Surrey Space Centre, Guildford.

[18] Edmonson, W., Edmonson, W., Neogi, N. and Herencia-Zapana, H. (2014) Small Satellite Systems Design Methodology: A Formal and Agile Design Process. 8*th Annual IEEE Systems Conference* (*SysCon*), Ottawa, 31 March-3 April 2014, 518-524. http://dx.doi.org/10.1109/syscon.2014.6819305

22

Survey on Routing in Pocket Switched Network

Ratna R. Sarkar, Khadija Rasul, Amitabha Chakrabarty

Department of Computer Science and Engineering, BRAC University, Dhaka, Bangladesh
Email: ratnacse2013@gmail.com, amitabha@bracu.ac.bd, khadija.rsl@bracu.ac.bd

Abstract

Pocket switched Network (PSN) is an emerging research magnet that has been attracting researchers. This paper provides an introduction to PSN and covers some of its basic features and challenges. Continuous connectivity is difficult in infrastructure independent communication. In this paper, we will see how PSN provides an effective alternative. We also explain some routing protocols that can be incorporated for effective data forwarding. This paper also discusses possible applications and areas where PSN can be effectively used and some ideas that may foster future research in PSN routing.

Keywords

Pocket Switched networks, Connectivity, PSN Routing protocols, Forwarding

1. Introduction

Since the beginning of Internet, TCP/IP (Transmission Control Protocol/Internet Protocol) is the most used network protocol. For years, we know Internet has been successfully connecting communicating devices worldwide. TCP/IP plays most important role for achieving this effectively. The basic principle for end to end data transfer is based on TCP/IP. But when end to end connectivity is broken or intermittent, then TCP/IP may not work properly with reliability, in many cases, it can completely fail to transfer data from source to destination. This problem occurs mainly in remote areas, or villages that lack basic infrastructure to support internet. In such circumstances, wireless communication networks which are independent of end to end connectivity between nodes can provide better result. One such solution is Pocket Switched Network (PSN). It is a fairly new research area which can work properly without specific infrastructure. PSN is independent of end to end connectivity between humans and it can provide better result in extreme environment which TCP/IP cannot handle. PSN falls into the category of another older network known as Intermittently-Connected Mobile Ad-hoc Networks (ICMANET), which is also popularly known as Delay Tolerant Network. ICMANET/DTN is one of the well-established areas in the field of wireless communication. This network architecture is proposed that can work properly with limited expectations of end to end connectivity and resources [1]. Networks under this class are potentially deployed in challenged environments using isolated mobile devices with limited resources. It is typically different from traditional mobile ad-hoc networks (MANETs). Because, in an ICMANET, paths between two nodes are intermittent and communication is established only by multi-hop paths between two nodes. That means there is no end-to-end path between nodes. DTN architecture can be applicable for interoperability between and among challenged networks (characterized by latency, bandwidth limitation, error probability, and node longevity or path stability). The concept of region and gateway are included in DTN architecture and region boundaries are used as interconnection points between dissimilar network protocols. DTN can be used to change the basic service model, system interface and poor performance presented in some networks.

As mentioned earlier, PSN falls under the category of DTN. PSN and DTN mainly differ in their information carriers. PSN employs human beings for data forwarding, while DTN utilizes any possible carrier, including human beings, to disseminate data. Nowadays, large number of mobile devices is used by people and data collected from them show that "small world" features [2] are strongly existent in them. Small world features are best described by social graph/model theories. Study shows human related network relations are less volatile and long-termed than network depending on simple node mobility. Thus for human-related extreme networking situation, there was need of more social-based DTN than any other types of DTN. From that, the need of PSN was introduced, as PSN is the only type of DTN that uses the knowledge of social characteristics to make better forwarding and routing decision.

PSN is an alternative approach to computer network architecture whose aim is to address the technical issues in heterogeneous network. PSN experience lack of continuous network connectivity and enable data transfer when mobile nodes are only intermittently connected. Since the connectivity is not expected to be consistent in PSN, it employs what is called store-carry-forward mechanism. In this, intermediate nodes carry data packet when they receive it and forward to next node as when contact is established. PSN takes the advantage of human mobility to distribute data from source to destination [3]-[5]. PSN does not require any assistance of infrastructure. That's why PSN is applicable to rural area and developing regions to realize low cost communications. Mobile devices [6] (like mobile phones, personal digital assistances (PDAs) and laptops etc.) sometime are carried by people in some physical space with high number of nodes and contacts density. Such situation may occur that to individuals at conferences, around office spaces and in case of social communications. Networks that exist in such type of environment are examples of PSNs [7] [8] in which both mobility and multihop forwarding can be supported for communication.

In this paper, we discussed about different features and challenges of PSN routing. We also discussed different possible applications of PSN and different categories of PSN routing protocols. Finally this paper focuses some ideas that may foster future research in PSN routing.

2. Features of PSN

PSN takes the advantages of both human mobility and also intra-networks or internetworks to transfer data. PSN has been created to provide network services for mobile users in such places where they are localized truly between Islands of connectivity.

There is huge amount of portable devices such as laptops, PDAs and mobile phones, active on localized wireless bandwidth (like Bluetooth), storage capacity and CPU power. The only resource needed to run this portable device is power. With first development of power engineering and advancement in battery technologies, once charged it is possible for portable device to last for a week, while remaining in constant network contact. We expect that this innovation will continue, allowing devices to participate in wireless networks while minimizing power consumption. In case of increasing the rate of successful forwarding in PSN, data may travel through via multiple nodes. With development of technology memory has become cheap and accessible for all. However, in PSN nodes do not only carry their own message but store-carry-forward other nodes' messages as well. So, efficient memory management is a necessity here.

In this type of network, nodes may be mobile or fixed devices within a certain location [9] like a computer with blue tooth, a laptop with a Wi-Fi network, a mobile phone etc. Nodes usually refer to any network enabled component that has the property of receiving and forwarding messages. The main objective of PSN is to make the use of every communication opportunity, and the physical mobility of the devices for transporting data from the source to destinations in situation where there might not be any base station or central hub. Some features of

PSN problem space are described as follows.

2.1. Store-Carry-Forward Approach

In PSN, network links may be disrupted for a long time, therefore end to end paths are assumed not to exist. That's why store-carry-forward mechanism is the main mechanism for data delivery and main-objective of routing to achieve maximum delivery ratio.

This mechanism's is a real life analogy is postal service. In postal service letters are passed through several post offices, then they are processed and finally forward to their desire destination. This store-carry-forward method works similar to the concept of postal services. In store-carry-forward mechanism message or a chunk of message is forwarded and store in the nodes until it successfully reached to its desire destination.

Figure 1 shows a graphical representation of store-carry-forward approach and also shows how message or a chunk of message is propagated through a network. In this figure, the circle represents a node and the box represents its storage capability. So in **Figure 1**, each node is relaying its messages to its desire destination and also has storage to store message or a chunk of message, until the message/messages has reached its destination or the node has meet another suitable node.

2.2. User Mobility

Human mobility plays a key role in PSN. Mobility gives rise to new localized opportunities. User mobility also increases the use of network bandwidth as large amount of data needs to be carried around the network. The probability of node mobility is measured by using different mobility model. Some of the models are briefly discussed below:

1) Random waypoint model [10] is one of the most popular models of this category. In Random waypoint model, a node randomly chooses destination (waypoint) and moves towards the destination. In this model nodes are distributed over the simulation area randomly and they stop for a constant time. After the completion of waiting period each node chooses a waypoint (destination). Random waypoint model is used to simulate of wireless communication networks for humans and also used to represent cellular structured network. Other different approaches like the random-direction model [11], random-border model [12], and the modified-random-direction model [11] are various type of fully random movement with different node density distributions.

2) There are also different categories of mobility models [13]. Like, Manhattan-grid model [14] where initially nodes are randomly distributed on the streets (total simulation area is divided into several square blocks named as streets). Each node chooses a direction and a velocity. If a node reaches a corner, then the node changes direction with a certain probability. The velocity of node is also changed over time.

3) Obstacle mobility model [15] [16] is another mobility model that use Voronoi-diagrams to determine optimal path. The movement using this model is more realistic, but there is still no movement on optimal paths.

4) Reference-point-group-mobility model (RPGM) [17] models the movement of groups of nodes according to an arbitrary mobility model. According to this model, the actual position of a node is a random movement vector added to the position of node's reference point that is assigned to it. According to the arbitrary mobility model, the actual position of the node's reference point may change but the relative positions of the reference point inside a group do not change.

5) Clustered-mobility [18] is a mobility model that follows random based movement and uses similar approach

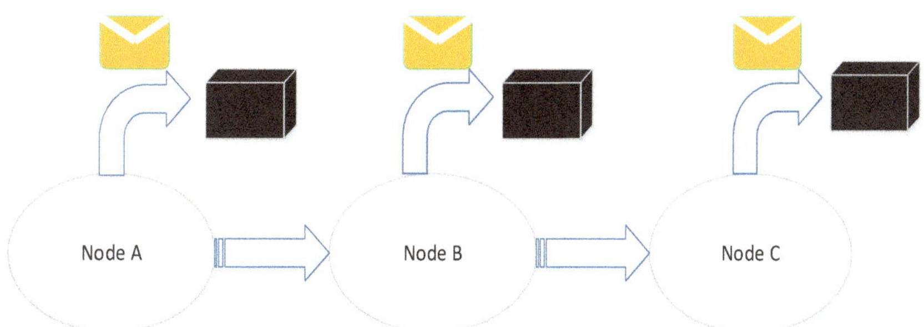

Figure 1. Store-carry-forward approach.

as random waypoint model. The difference is that the attraction of a point depends on the amount of nodes nearby. This model is applicable in different tactical scenarios.

6) Gauss-Markov model [19] is another example of mobility model. According to this model node's velocity and direction of the future depend on the current value. The new values are chosen based on a first order autoregressive process.

7) Smooth-random model [20] [21] is a mobility model where nodes are classified concerning their maximum velocity, preferred velocity, maximum acceleration, and deceleration. Velocity and direction may also be chosen by correlation among each other. By doing so, more realistic movements may be realized.

8) The area-graph-based mobility model [22] realizes sub-areas with higher node density. This model also realizes paths between higher node density and lower node density. The sub-areas are considered as vertices of the area graph and the paths are considered as edges. A weight or probability is assigned to each edge. A node moves inside the sub-area for a randomly chosen time according to the random-waypoint model. After this time, node chooses one path according to probabilities at the edges. Next, the node moves on the path to the next area.

9) There are some mobility models that can be applicable on social networks. The social-network-founded mobility model [23] is one of them. This model is based on interaction indicators among all pairs of nodes. If a node has larger interaction indicator, then the probability of social relationship of the node becomes high. Firstly, according to the nodes interaction indicator, they are grouped in clouds. Nodes are moved inside the clouds according to a random waypoint model where the waypoints are chosen according to the interaction indicators. Community based mobility model [24] is used for the classification of the nodes into groups and their movement inside the clouds.

Due to mobility of nodes, forwarding paths becomes unpredictable and probability of mobile reach ability becomes highly dynamic, thus mobility introduces new challenge to PSN.

2.3. Opportunistic Networking

Opportunistic networks are a very promising networking field. In opportunistic networks nodes build a self-organizing ad hoc network without requiring any pre-existing infrastructure [25] [26]. In opportunistic networking, nodes opportunistically exploit these contacts with other nodes to exchange messages to the appropriate destination. In opportunistic network nodes are able to make communication with each other even if there is no route exists to connect them and also collaborate them to exchange data from the source to destination [27].

Figure 2 shows the basic structure of an opportunistic network. As movements in PSN are not pre-defined thus a neighbor is met is only by chance and collaboration can happen only as opportunity arises. In this type of network, at first a node has to find a neighbor node for starting collaboration. Then they start to exchange data. Nodes pass data to its neighbor node that it has discovered. Data is distributed among all nodes through information sprinkler. Information sprinkler [9] is a dedicated node, which is less likely to be mobile and works like any other opportunistic network nodes. It uses data sharing protocol. The information sprinkler can collect information from other nodes whenever these neighboring ,nodes belonging to sprinkler's community or any other community, approaches the sprinkler and then the sprinkler distribute the information to other information sprinkler within a short time period. That means in this opportunistic network messages are distributed hop by hop towards the destination.

In case of data forwarding in PSNs, internetwork links are crucial. Both the intranetworking and internetworking opportunities allows PSNs to provide highly robust networking for users, to transfer data. That means

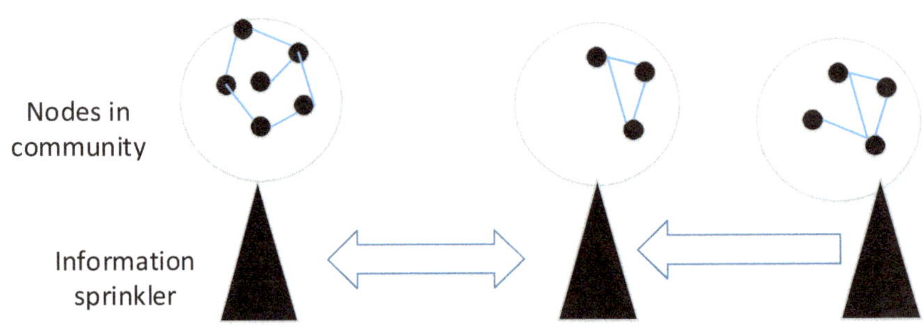

Figure 2. Opportunistic network.

opportunistic networking allow users to make communications with each other in such unavoidable circumstances and also use physical movement of the user to transfer important messages.

2.4. Social Properties

As already mentioned earlier, small world features or social properties are strongly evident in PSN [2]. This is because, PSN comprises of human and humans happen to be social being, one such social nature is the ability to form social graph from the data obtained through PSN. A social graph is a global map which shows how nodes are connected. It is also gives insight to many other social metrics like friendship, community, centrality and modularity. Most of these terms can be found in the work on graph theory by Girvan and Newman *et al.* [28]. Some definition of common social properties found in PSN is described briefly below:

1) For PSN social graphs, each vertex represent a person carrying device and each edge represents frequency of interaction between these devices.

2) Centrality is the measure of how important a node is or how a node can helps connect other nodes in the graph. A node with high centrality value represents a strongly interacting node, which can be a good candidate for being data carrier to other nodes. For example betweenness centrality of a node measures the number of shortest path passing through it. Higher the betweenness centrality value, better bridging node it is for data exchanging purpose.

3) Similarity value can also be found through social graph. Similarity is the measure of number of common neighbor between individual nodes.

4) Friendship is another concept from sociology which refers to personal relationship between individual nodes.

5) Modularity which is a function that calculates the quality of newly formed communities within a network. In that process community formation, modularity can be used to decide whether to further divide a network into newer communities or to keep it as it is.

3. Challenges of PSN Routing

PSN can be defined as a communication paradigm that can take the advantages of human mobility and intra-network or internetwork connectivity. As continuous end to end connectivity is absent so successful delivery of data becomes difficult. Here in this section we will discuss about some challenges which we may face in successfully deployment PSNs. These challenges are also beacons for researcher, willing to improve PSN, to work on.

3.1. Mobility

PSNs are formed of human communities. So it must include mobility. That means node mobility is an important issue in PSN. In PSN, it is difficult to build end to end paths for delivering message from source to destination, because user's undefined mobility is an issue here. Due to the node mobility, it becomes unpredictable to meet a node with other nodes in the network. That's why continuous end to end connectivity is hardly possible. That means, routing in such network means finding temporal path between sources to destination [29], but finding an efficient path between sources to destination, *i.e.* successful routing of data without proper knowledge about the dynamic network topology, is a difficult task [30].

3.2. Selfishness

As PSN is formed by human beings and they are selfish in nature. Human operated devices forming a PSN may not be willing to forward data for others, except their own desired destinations. PSN's store-carry-forward nature requires nodes to be willing to carry other's data but the presence of selfish nodes causing a great challenge in successful data transmission. So, necessary mechanisms should be employed to encourage the node to forward data and get rid of their selfish behaviors.

3.3. Data Forwarding

In PSN data forwarding are occurred based on some policy, as it supports both of intranetworking and internetworking. In the case of local connectivity, nodes forward messages according to the knowledge of their local environment. But to acquire accurate knowledge about the entire local environment is a key challenge because PSN is independent of any predefined infrastructure. As PSN has some limitation on resources, so availability of storage and energy are also the key challenges here.

In PSN, data are forwarded to the carrier nodes that are close to destination. But the main challenge is in the development of a method that will appropriately determine a carrier node, which will provide good forwarding opportunities to the destination, for a given message.

When global connectivity is available, a node forward messages directly to other suitable nodes that are globally connected. Finding which nodes are globally more strongly connected is also a challenge in volatile dynamic infrastructure-less network environment.

3.4. Security

In PSN routing take place by the collaboration, dependency and cooperation among nodes. Nodes have to collect and exchange messages among themselves. In that case, nodes may face different security issues. As messages are exchanged among number of nodes, so it may be altered, affected or modified by a malicious relay [31]. On the other hand, some other security problem may occur such as redirection, eavesdropping, denial of service, fabrication, poisoning etc. So incentive mechanism, like those found in [32] for Wireless Sensor Network, should also be developed for secure and safe delivery of messages and for preserving user's privacy in PSN.

3.5. Scalability & Clustering

Most of the routing protocols in PSN are flat. This flat approach is suitable for small network but is not scalable. In this case, clustering may provide a better result that can make groups among the mobile nodes with similar mobility pattern into same cluster. So, nodes belonging to same cluster can help each other to reduce overhead, make routing scalable and also share of resources. Clustering approaches can be useful to unfold large networks into different communities [33]. In order to uncover overlapping nodes in complex networks, clustering approach based on particle walking and competing with each other by using random deterministic movement [34] can be used. Another type of clustering technique is discussed in the paper LABEL [35]. According to LABEL, each node within a community is assumed to have a label that informs other nodes about its label. It provides proper knowledge about the communities to forward messages from source to destination and significantly improve the forwarding efficiencies. Overall clustering is a very important field that has lot of scopes to research on. It is undoubtedly an important open issue for future work in PSN.

3.6. Energy and Storage Management

Mobile devices that are carried by human beings have very limited energy and storage capacity. For rapid and reliable data transmission it becomes unreasonable to utilize large amount of storage and energy. Existing PSN routing protocols do not consider low energy consumption and low storage space in their design objectives. These energy and storage management can be an effective metric to evaluate routing in PSN.

4. Applications of PSN

The main objective of PSN is to survive complex and challenging network environment. This includes surviving hardware failure as well as software or protocol failures. As PSN do not require the assistance of any infrastructure, they are applicable in rural and developing regions to allow low cost communication. PSNs can provide efficient communication in place where internet connectivity has disrupted due to infrastructure fails. Some applications of PSN are given below.

4.1. Remote Communication

There are many rural communication projects in remote villages to provide the access to Internet. Some of which is try to reduce the cost of communications using the way of asynchronous information transmission. For example, Wizzy digital courier service[1] provides Internet access for some village schools in South Africa. DAKNET [36] project focused on providing low cost Internet services to the rural areas of India. It is suggested

[1]Wizzy digital courier. http://agln.aspeninstitute.org/projects/wizzy-digital-courier

the use of physical means for delivering messages to areas which are not connected via traditional networks.

4.2. Disaster Managements

There are some models that are used for security and disaster communication, search and rescue communication. For example, post disaster mobility model [37] is used for this purpose. This model includes the impact of disasters on the transportation network and models for human and relief vehicle movement.

4.3 Social Network Analysis

PSN can be used for analyzing social network. Many research areas, starting from anthropology to E-commerce to engineering, need social network analysis [38]. As PSN deals with people, it can be used to collect data among social entities and then use them to further understand implications, relationships and patterns among peoples, which then can be used to develop new applications for different usages.

4.4. Disease Detection

These days using PSN to understand how epidemic diseases spread is a common tool in the field of epidemiology. PSN helps to track people to collect data for understanding disease, as simple as Flo to deadly as HIV, is spread. One such work is done by Hashemian *et al.* [39].

4.5. Community Detection

Community detection is a very common use of PSN. From the data collected using PSN, communities having higher modularity are often grouped together. Works that have done this are refereed in [40] and [41]. Another application of community detection through data tracking using PSN could be used to detect terrorist movement in inhabited jungles or deserts. Another use of Community detection through PSN could be to provide secure access control in social network by finding groups of people who help to spread spams in social network. Such work is done by Grier *et al.* [42].

5. Routing Protocols of PSN

Different routing protocols are already proposed for PSN in order to, successful delivery of data. Some popular PSN routing protocols are categorized and discussed briefly below.

5.1. Flooding Based Protocol

Flooding based protocol is one of the categories of routing protocol used in DTNs, passed down to PSN. Epidemic routing [43] falls in the category of flooding based protocol. This protocol mainly work based on the replication of messages. Messages are distributed to every node present in the network, with an aim for messages eventually reached their destination. Epidemic routing protocol guaranteed successful transmission of messages. The main benefit of this routing protocol is that it has low latency in case of message delivery. This protocol increases the overhead drastically in terms of traffic congestion and energy. Different versions of the epidemic routing protocol [44] [45] have also been proposed in order to reduce message overhead by considering different constraints such as time limit, maximum hop count, forwarding probability, or applying different techniques to inform nodes about the successful delivery of the message. Another benefit of flooding based protocol is that it does not require any prior local or global knowledge about the network. In this protocol messages are continuously replicated to every node in the network that results high overhead, consumption of energy and traffic congestions.

Flooding based protocol can also be based on tree structure [46]. Here the decision of how to make copies of message and ensuring the number of copies of message is an important issue. Two hop relay [46] is another category of flooding based protocol. According to this protocol, if there are n nodes around the source are directly connected to the source node, then n copies of message are generated to the source and transmitted to those nodes. Spray and Wait [47] routing protocol is an example of variation in this category. According to this protocol in spray phase a number of message copies are forwarded by the source node to the same number of other nodes (known as relays). In wait phase of this protocol, if destination is not found in the spraying phase each of the nodes carrying a message copy will forward the message only to its destination.

These protocols can provide better result with respect to average message delivery delay but at the cost of huge number of transmissions per message delivered. Flood based protocols are simple to implement in order to achieve good performance.

5.2. Direct Pass

Direct Delivery Routing [48] protocol falls in the category of direct pass. It is the simplest protocol to transmit message or a chunk to its desire destination.

In this protocol a node called as *source node* will deliver message only to *destination node* is encountered by it. So direct communication between source and destination node is necessary for successful delivery of message. No relay node is necessary here for successful delivery of message between source node and destination node. Here, in **Figure 3**, we can see a graphical representation of direct delivery routing approach. In this figure, *node A* has a message to deliver and according to the direct pass approach when *node A* encounter the desire destination then it will deliver the message.

This routing protocol always selects the direct path for message delivery between source and destination. Direct delivery routing protocol does not require any information about the network. Because of this simplicity, this protocol does not need to consume too many resources. However this direct delivery routing protocol only works, when source node and destination node comes into direct contact with each other and is accompanied by huge delay.

First Contact Routing [48] is another type of direct pass routing protocol. In this protocol the relay node that come to first contact to the source node work cooperatively to increase the probability of successful message delivery. First contact routing approach also increases the use of bandwidth and storage. It has same advantages and limitations of direct pass routing algorithm.

5.3. Probabilistic Model Based Protocol

Probabilistic model based protocol is another category of routing protocol that is effectively used in PSN. PROPHET [49] routing protocol is a popular example of this category. This protocol is mainly proposed in order to increase the delivery probability in a social network. According to PROPHET if a node visits the same place several times, then there is a probability to repeat this in future. In this protocol each node uses a metric called probabilistic metric in order to deliver message to a reliable node. When two nodes meet, they exchange their stored information which contains delivery probability and encounter related information. Here the probability of message delivery can be calculated by using transitive delivery probabilities. When *node i* and *node j* meet one another and their elapsed time unit is k then the delivery probability can represented by the Equation (1) below:

$$P_{i,j}(k+1) = \left(1 - P_{i,j}(k)\right) * P_0 + P_{i,j}(k) \quad (1)$$

In this protocol forwarding decision is made based on the delivery probability and a message is delivered to a node with high delivery probability to reach the destination. PROPHET has lower overhead, small use of storage space, reduce power consumption and also increase the probability to deliver message to its desire destination than routing protocols of other categories. But in PROPHET average delay is increased to deliver a message from the source to destination. In this protocol forwarding decision is made based on the delivery probability and a message is delivered to a node with high delivery probability to reach the destination.

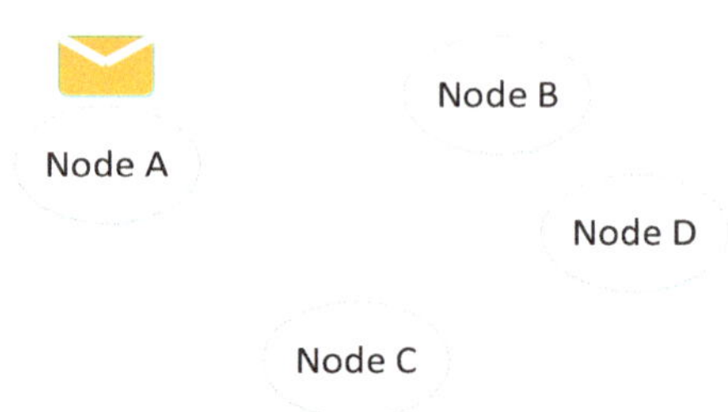

Figure 3. Direct delivery routing.

A similar probabilistic approach is proposed [50] to improve the performance of PROPHET routing protocol. This approach is based on predictability concept to remove message delivery delay and number of message drop. In this approach, an improvement factor is included in the probability calculation equation of PROPHET in order to improve the above mentioned factors.

MaxProp [51] routing protocol calculate the maximum probability of messages to be delivered. Messages in the buffer are prioritized. In this protocol the hop count value is lower and probability set is higher. According to this protocol the priority of packet is determined by calculating the probability of nodes meeting when hop count value exceeds the threshold value.

Plankton [52] routing algorithm also falls in this category. It is possible to improve routing performance in such type of network by controlling through message replication based on reliable contact prediction. Plankton considers two key ideas to predict contacts. Firstly, it classified the communication links into weak link or strong link. This classification is based on how successful a message is delivered to its desire destination using links. Secondly, strong links are identified based on the associative relationships that are observed in different time. In order to deliver a message it firstly assigns an initial replica quota and target delivery probability. Plankton controls message replication when message is duplicated to a node that has high contact probability or high delivery probability. Plankton reduces overhead by controlling replicas and also estimates contact probability in order to improve routing performance. It also dynamically adjust replication quota by estimating contact probabilities and delivery probabilities.

Context-Aware Adaptive Routing (CAR) [53] uses prediction to allow efficient routing. If a host want to send message to any other host it uses a Kalman Filter prediction [54] and multi-criteria decision theory [55] to choose the best next carrier for the message. Kalman Filter prediction techniques were originally developed based on automatic control systems theory and used in CAR to achieve more realistic prediction of the evolution of the context of a host and to optimize bandwidth usage. Multi-criteria decision theory is used to estimate overall delivery probability. According to CAR if both sender and receiver are in the same region of network then forwarding path is determined by synchronous routing protocol. If a message cannot be delivered synchronously then the best carrier for a message is chosen based on calculating the highest delivery probability that is synthesized locally context information. CAR reduces overhead in terms of the number of messages sent and can provide good performance. In dynamic environments, as the number of undetected nodes increases, the overall predictability level increases. This is a limitation of this protocol.

5.4. Social Network Based Protocol

Social Network based protocol is another category of routing protocol. Social network based protocol focuses mainly on social network features of humans for making routing decisions. Social structure helps to build forwarding paths, allowing two nodes to communicate over time using opportunistic contacts and intermediate nodes [56]. Human belongs to different communities' altogether make human society. This concept is used to build different type of social network based protocol. Routing of messages among different nodes (any network component that has the ability to receive and forward message) in PSN is occurred by detecting community structures throughout the network. Routing protocols that take this human community concept into consideration are fall in the category of social network based protocol.

Bubble-Rap [57] protocol falls in this category. It considers social network for making forwarding decision. In Bubble-Rap there are two important metrics through which nodes of the network is ranked. These two metrics are local ranking and global ranking. Local ranking of individual nodes means the popularity of the node within its own community. Global ranking indicates the popularity of an individual node within the whole network. Bubble-Rap protocol works based on two assumptions and these are: each node must belong to at least one community and each node must have a global ranking and also a local ranking. According to Bubble-Rap if the source node and destination node are in the same community, then firstly it checks whether the encountered node is also in the same community, if so the local ranking of the source node and encountered node is checked and if the local rank of encountered node is greater than the local rank of source node then the message is forwarded. If the source node and destination node are not in the same community, then this routing protocol is forwarded to the encountered node if the encountered node is in the same community of the destination node or if the global ranking of encountered node is higher.

Figure 4 shows an illustration of Bubble Rap forwarding from source (S) to destination (D). Here, source (S) and destination (D) nodes are specified by black color. On the other hand, black and green arrows show the Bubble Rap operations based on global centrality in global community and local centrality in D's community respectively.

Lobby Influence [58] is also fall in the category of social network based routing protocol. In this protocol message forwarding decision is made by considering the local and global popularity of the nodes and their lobby index. Lobby index is measured by the strength of the relationship with neighbor nodes. Lobby influence protocol allows more popular nodes to forward message to the less popular node that has high lobby index in the network. Lobby Influence protocol is an enhanced work of Bubble-Rap protocol. The main difference between these two protocols is in their forwarding decision. As we discussed earlier Bubble-Rap takes forwarding decision based on the node popularity. On the other hand, the forwarding decision of Lobby Influence depends not only on node popularity but also on Lobby index.

In this protocol there is consider three assumptions and they are:

1) Each node must have a label and must associate with at least one community;

2) Each node has one global rank and one local rank to identify global and local centrality. But a node may have multiple local ranks and multiple label;

3) Each node has its lobby index that indicates the strength of the relationship with its neighbors.

This protocol increase the probability of message delivery and at the same time decreases the overhead of more popular nodes. Another similar work that builds on [57] is discussed in [59] and this focuses on betweenness centrality value to enhance Bubble-rap's global routing decision.

Another social network based routing protocol is SimBet [60]. According to this protocol when two nodes meet one another, they are compared with one another by using SimBet utility. SimBet utility is a combination of similarity utility and betweenness utility. Similarity utility means number of common neighbors between a node and the destination. On the other hand, betweenness means number of indirectly connected nodes one node can link. That means these utilities can be estimated by the statistics of the direct links between itself and other nodes in the network. When two nodes meet they exchange their contact information with other and then determine SimBet utility to destination node. As a result the number of node and amount of exchanged data are proportionally increased. In SimBet routing, when node n, carrying a message for destination d, meets node m, n calculates betweenness and similarity value of node m. The functions used for calculating Similarity is shown in Equation (2) and Betweenness centrality are given in Equation (3) below:

$$SimUtil_n = \frac{Sim_n(d)}{Sim_n(d) + Sim_m(d)} \tag{2}$$

$$BetUtil_n = \frac{Bet_n}{Bet_n + Bet_m} \tag{3}$$

PRO [61] profile based routing protocol is another example of social network based protocol. PRO use community structure of social network, in order to cover maximum number of community to reach the destination.

In PRO, nodes that belong to the same community are called as local neighbors and nodes that are connected to other community is called as remote neighbors. PRO works through the regularity of inter-contact events between nodes. This regularity is measured in terms of meeting duration and number of meeting between nodes. In PRO each node maintain a local observation table which keep track of periodic inter contact events and updates

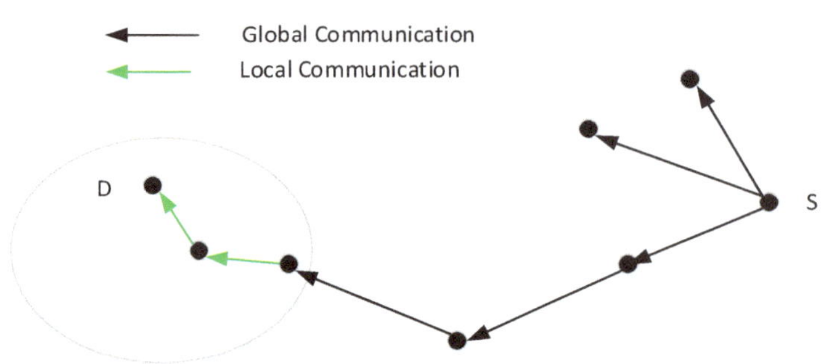

Figure 4. Illustration of bubble-rap routing.

the observation score. Observation score can be defined as a metric that indicate the probability of observing a node periodically. In case of forwarding PRO gives priority to the observation score. The nodes that observe the destination node regularly are more suitable candidates for forwarding directly to destination.PRO also checks the information dissemination score of nodes in the communication range. If the current nodes encounter another node, which is known as candidate node, having higher information dissemination score than the internal threshold score of the current node, then the message is forwarded to the candidate node. PRO also controls the number of copies for each message that can be forwarded by single node which is known as Forwarding-Quota.

PRO achieve fast and efficient routing in intermittently connected PSNs. That means PRO has low-delivery-latency and low-message-overhead. PRO is also self-learning, completely decentralized and local to nodes but sometime it can be costly.

CAS [62] Community-based Adaptive Spray routing protocol is also falls in this category. This protocol mainly based on two aspects, firstly, this protocol select the intermediate nodes that are closest to destination and secondly, it dynamically control the number of message copies according to taking into consideration the time-to-live of a message. According to the community graph firstly this protocol finds the shortest path to destination. In order to find gateway node gateway table is used here. The gateway table contains community ID, gateway node ID, the ID of the community that is connected by the gateway node, average inter contact time between gateway node and community and also a timestamp of last update of the table. Then optimal number of message copies needed to route is calculated. The gateway node is set as an intermediate target node and route specific number of messages. If the gateway node is the destination node then the process is ended. According to CAS, if a node is carried message of $L > 1$ (L is the number of message copies permitted to route the message to the intermediate target node) copies and encounters another node, then it checks whether this encountered node is inside its community. If they are in the same community, then it checks whether this encounter node is the intermediate target node. If so then the message is forwarded to the encountered node. Otherwise, $L/2$ copies of message are forwarded to the encountered node and the remaining copies of message is kept by the source node. If they are in different community then it checks whether the encountered node is the destination node, if so then copies of message is forwarded to the encountered node. The main goal of improving routing performance is to increase the delivery ratio and minimize resource consumption. In order to achieve this goal Community based adaptive spray (CAS) considers both of the mobility pattern and time-to-live of message. According to this time-to-live CAS dynamically control the number of message copies.

SANE [63] social-aware stateless forwarding is also an example of social network based protocol. In case of routing in PSN, SANE takes the advantage of both social aware and stateless approaches. SANE supports two communication services unicast and interest-cast. According to interest-cast a message should be forwarded to individuals whose interest profile closely resembles to destination. That means according to this service a user can communicate certain information to the maximum possible number of interested users, within a certain period of time. According to SANE a message has a header that contains a target interest profile, which is also known as message relevance profiles, this is an integer value that represents the number of replicas of the message that is allowed to a node to forward to other nodes and a time-to-live value. In human society individuals with similar interests often meet one another than individuals with reverse interest. This concept is used in SANE and then according to the above observation social-aware and stateless forwarding mechanism is designed.

Friendship based routing [64] is a social network based routing protocol in which forwarding decisions of messages are made based on temporally differentiated friendships. This protocol firstly identifies accurately the relations between nodes by considering a metric named social pressure metric (SPM) that motivate friends to meet to share their experiences. SPM considers three behavioral features of close friendship high frequency, longevity and regularity. Secondly, Local community is formatted by considering link qualities from nodes contact history. A node can define a community as a set of nodes with link quality itself larger than a threshold. This community will include only direct friends. But two nodes that are not directly close friends still can be close indirect friends. So finally, this protocol identifies the way of handling temporal differentiations of node relations by allowing different friendship communities in different periods. In case of forwarding, if a *node i* has a message for *noded* and meets with *node j*. Then *i* forwards the message to *j* if and only if *j* and *d's* current friendship community is same and *j* is a stronger friend of *d* than *i. i* will not forward the message , if *j* and *d* are not in the same friendship community but *j* has better link with *d.* This protocol can achieve better delivery ratio without maximizing the cost compared to other protocols. As this protocol allows different friendship communities in different period, it may require more space and that can increase cost.

In order to increase the performance of social network based protocols in large scale networks community aware framework (CAF) [65] can be used. CAF can be easily integrated with some of these social network based algorithms. This framework extension relies on the fact that social forwarding algorithms normally operates within the same sub-community. Here particular nodes are defined as MultiHomed (MH) nodes (nodes that belong to multiple sub communities) that circulate the message to other sub-communities. These MH nodes are ranked by the number of sub-communities they belong to (MHrank). CAF mainly relies on local social/contact information to predict future transfer opportunities. By integrating CAF with forwarding algorithm may increase delivery ratio, but sometime it can be costly.

6. Conclusions

In this article, we have discussed about basic features, challenges, routing protocols of different categories of PSN. We also focus on some open issues that may foster future research in PSN. This paper was written keeping in focus to introduce newcomer researchers to PSN and its routing protocols.

PSN becomes an important emerging facet of modern day networking. TCP/IP protocol is not suitable for modern day networking in extreme scenarios because it requires end to end connectivity. On the other hand, PSN takes the advantage of both human mobility and intranet/internet communication opportunities. PSN also does not require any end to end connectivity. PSN is mostly applicable in modern social networks where contacts are intermittent and challenged network conditions like disaster communication, search and rescue communication. PSN has also some challenges that are already discussed above. **Table 1** provides an overall summary of routing protocols that we have discussed in this work.

Our experience shows that security implementation remains one of the biggest challenges in PSN. As the data transfers are intermittent and data may end up with relay node for a longer period of time. So security of data should be maintained properly. While making this study, we have seen that merging multiple metrics of different protocols together can improve routing performance of PSN. However, care must be while merging as

Table 1. Comparison of different routing algorithms in PSN.

Type	Scheme	Delivery Ratio	Latency	Resource Usage
Flooding Based	Epidemic routing [43]	Maximum	High	Maximum
	Tree based Flooding [46]	Low	High	High
	Two hop relay [46]	Low	High	Low
	Spray and Wait [47]	High	Low	Low
Direct Pass		**Delivery Ratio**	**Latency**	**Resource Usage**
	Direct Delivery Routing [48]	Minimum	High	Low
	First Contact Routing [48]	Minimum	High	Low
Probabilistic model based		**Delivery probability**	**Power Consumption**	**Resource Usage**
	PROPHET [49]	High	Limited	Limited
	MaxProp [51]	High	Normal	Limited
	Plankton [52]	High	Limited	Limited
	CAR[53]	High	Limited	Limited
Social Network Based		**Community**	**Centrality**	**Similarity**
	Bubble-Rap [57]	Yes	Yes	No
	Lobby Influence [58]	Yes	Yes	No
	SimBet [60]	No	Yes	Yes
	PRO [61]	Yes	No	No
	CAS [62]	Yes	No	No
	SANE [63]	No	No	Yes
	Friendship based routing [64]	Yes	No	No

no general guaranteed rule is there and like any other ad-hoc network, PSN is also application-specific. Overall, it's an interesting research challenge to decide how to smartly incorporate ideas of different categories of PSN together to come up with a hybrid category.

Acknowledgements

We want to acknowledge BRAC University for allowing us to put this work together. The authors are grateful to the reviewers and the Editor of this journal for their time and valuable comments.

References

[1] Fall, K. (2003) A Delay-Tolerant Network Architecture for Challenged Internets. *SIGCOMM'*03, Karlsruhe, 25-29 August 2003, 27-34.

[2] Hossmann, T., Spyropoulos, T.T. and Legendre, F. (2011) Putting Contacts into Context: Mobility Modeling beyond Inter-Contact Times. *Proceedings of the Twelfth ACM International Symposium on Mobile Ad Hoc Networking and Computing*, Paris, 16-19 May 2011, Article No. 18. http://dx.doi.org/10.1145/2107502.2107526

[3] Wang, S., Liu, M., Cheng, X. and Song, M. (2012) Routing in Pocket Switched Networks. IEEE Wireless Communications.

[4] Chaintreau, A., Hui, P., Crowcroft, J., Diot, C., Gass, R. and Scott, J. (2006) Impact of Human Mobility on the Design of Opportunistic Forwarding Algorithms. *IEEE Transactions on Mobile Computing*, **6**, 606-620.

[5] Mtibaa, A., Chaintreau, A. and Diot, C. (2007) Popularity of Nodes in Pocket Switched Networks. *Proceedings of the ACM SIGCOMM*, Kyoto, 27-31 August 2007.

[6] Erramilli, V., Chaintreau, A., Crovella, M. and Diot, C. (2007) Diversity of Forwarding Paths in Pocket Switched Networks. *Proceedings of the 7th ACM SIGCOMM Conference on Internet Measurement*, San Diego, 24-26 October 2007, 161-174. http://dx.doi.org/10.1145/1298306.1298330

[7] Chaintreau, A., Hui, P., Crowcroft, J., Diot, C., Richard, G. and James, J. (2005) Pocket Switched Networks: Real-World Mobility and Its Consequences for Opportunistic Forwarding. Technical Reports, University of Cambridge, Cambridge.

[8] Hui, P., Chaintreau, A., Scott, J., Gass, R., Crowcroft, J. and Diot, C. (2005) Pocket Switched Networks and Human Mobility in Conference Environments. *Proceedings of the* 2005 *ACM SIGCOMM Workshop on Delay-Tolerant Networking*, Philadelphia, 22-26 August 2005, 244-251. http://dx.doi.org/10.1145/1080139.1080142

[9] Kaur, Er.U. and Kaur, Er.H. (2009) Routing Techniques for Opportunistic Networks and Security Issues. *National Conference on Computing, Communication and Control* (*CCC*-09), 2009, 155-161.

[10] Bettstetter, C., Hartenstein, H. and Pérez-Costa, X. (2004) Stochastic Properties of the Random Waypoint Mobility Model. *Wireless Networks*, **10**, 555-567. http://dx.doi.org/10.1023/B:WINE.0000036458.88990.e5

[11] Royer, E.M., Melliar-Smith, P.M. and Moser, L.E. (2001) An Analysis of the Optimum Node Density for Ad Hoc Mobile Networks. *Proceedings of the IEEE International Conference on Communication*, **3**, 857-861.

[12] Bettstetter, C. and Wagner, C. (2002) The Spatial Node Distribution of the Random Waypoint Mobility Model. *Proceedings of the* 1*st German Workshop on Mobile Ad-Hoc Networks*, Ulm, 25-26 March 2002, 41-58.

[13] Aschenbruck, N., Gerhards-Padilla, E. and Martini, P. (2008) A Survey on Mobility Models for Performance Analysis in Tactical Mobile Networks. *Journal of Telecommunications and Information Technology*, **2**, 54-61.

[14] Universal Mobile Telecommunications System (UMTS) (2010) Selection Procedures for the Choice of Radio Transmission Technologies of the UMTS (UMTS 30.03 Version 3.2.0). European Telecommunications Standards Institute (ETSI) TR 101 112 V3.2.0 (1998-04M).

[15] Jardosh, A., Belding-Royer, E.M., Almeroth, K.C. and Suri, S. (2003) Towards Realistic Mobility Models for Mobile Ad Hoc Networks. *Proceedings of the* 9*th Annual International Conference on Mobile Computing and Networking*, San Diego, 14-19 September 2003, 217-229. http://dx.doi.org/10.1145/938985.939008

[16] Jardosh, A.P., Belding-Royer, E.M., Almeroth, K.C. and Suri, S. (2005) Real world environment models for mobile network evaluation", *IEEE Journal on Selected Areas in Communications*, **23**, 622-632.

[17] Hong, X., Gerla, M., Pei, G. and Chiang, C.-C. (1999) A Group Mobility Model for Ad Hoc Wireless Networks. *Proceedings of the* 2*nd ACM International Workshop on Modeling, Analysis and Simulation of Wireless and Mobile Systems*, Seattle, 20 August 1999, 53-60.

[18] Lim, S., Yu, C. and Da, C.R. (2006) Clustered Mobility Model for Scale Free Wireless Networks. *Proceedings of the* 31*st IEEE Conference on Local Computer Networks* (*LCN* 2006), Tampa, 14-16 November 2006, 231-238.

[19] Liang, B. and Haas, Z.J. (2003) Predictive Distance-Based Mobility Management for Multidimensional PCS Networks.

IEEE/ACM Transactions on Networking, **11**, 718-732. http://dx.doi.org/10.1109/TNET.2003.815301

[20] Bettstetter, C. (2001) Mobility Modeling in Wireless Networks: Categorization, Smooth Movement, and Border Effects. *ACM SIGMOBILE Mobile Computing and Communications Review*, **5**, 55-66. http://dx.doi.org/10.1145/584051.584056

[21] Bettstetter, C. (2001) Smooth Is Better than Sharp: A Random Mobility Model for Simulation of Wireless Networks. In: *Proceedings of the 4th ACM International Workshop on Modeling, Analysis and Simulation of Wireless and Mobile Systems*, ACM Press, New York, 19-27. http://dx.doi.org/10.1145/381591.381600

[22] Bittner, S., Raffel, W.-U. and Scholz, M. (2005) The Area Graph-Based Mobility Model and Its Impact on Data Dissemination. *Proceedings of the Third IEEE International Conference on Pervasive Computing and Communications Workshops*, Kauai Island, 8-12 March 2005, 268-272. http://dx.doi.org/10.1109/percomw.2005.79

[23] Musolesi, M., Hailes, S. and Mascolo, C. (2004) An Ad Hoc Mobility Model Founded on Social Network Theory. In: *Proceedings of the 7th ACM International Symposium on Modeling, Analysis and Simulation of Wireless and Mobile Systems*, ACM Press, New York, 20-24. http://dx.doi.org/10.1145/1023663.1023669

[24] Musolesi, M. and Mascolo, C. (2006) A Community Based Mobility Model for Ad Hoc Network Research. In: *Proceedings of the Second International Workshop on Multi-Hop Ad Hoc Networks: From Theory to Reality*, ACM Press, New York, 31-38. http://dx.doi.org/10.1145/1132983.1132990

[25] Pelusi, L., Passarella, A. and Conti, M. (2006) Opportunistic Networking: Data Forwarding in Disconnected Mobile Ad Hoc Networks. *IEEE Communications Magazine*, **44**, 134-141. http://dx.doi.org/10.1109/MCOM.2006.248176

[26] Huang, C.M., Lan, K.C. and Tsai, C.Z. (2008) A Survey of Opportunistic Networks. *Proceedings of the 22nd International Conference on Advanced Information Networking and Applications-Workshops*, Okinawa, 25-28 March 2008, 1672-1677. http://dx.doi.org/10.1109/waina.2008.292

[27] Lilien, L., Gupta, A. and Yang, Z. (2007) Opportunistic Networks for Emergency Applications and Their Standard Implementation Framework. *Proceedings of the First International Workshop on Next Generation Networks for First Responders and Critical Infrastructure*, New Orleans, 11-13 April 2007, 588-593.

[28] Girvan, M. and Newman, M.E.J. (2002) Community Structure in Social and Biological Networks. *Proceedings of the National Academy of Sciences of the United States of America*, **99**, 7821-7826. http://dx.doi.org/10.1073/pnas.122653799

[29] Nguyen, A.-D., Senac, P. and Diaz, M. (2012) Understanding and Modeling the Small-World Phenomenon in Dynamic Networks. *Proceedings of the 15th ACM International Conference on Modeling, Analysis and Simulation of Wireless and Mobile Systems*, Paphos, 21-25 October 2012, 377. http://dx.doi.org/10.1145/2387238.2387301

[30] Nguyen, A.D., Senac, P. and Diaz, M. (2013) How Disorder Impacts Routing in Human-Centric Disruption Tolerant Networks. *Proceedings of the ACM SIGCOMM*, Hong Kong, 12-16 August 2013, 47-52.

[31] Hui, P., Chaintreau, A., Gass, R., Scott, J., Crowcroft, J. and Diot, C. (2005) Pocket Switched Networking: Challenges, Feasibility, and Implementation Issues. *Proceedings of the Second International IFIP Workshop*, *WAC* 2005, Athens, 2-5 October 2005, 1-12.

[32] Rasul, K., Nuerie, N. and Pathan, A.K. (2010) An Enhanced Tree-Based Key Management Scheme for Secure Communication in Wireless Sensor Network. *Proceedings of the 3rd IEEE International Workshop on Internet and Distributed Computing Systems*, Melbourne, 1-3 September 2010, 671-676.

[33] Blondel, V.D., Guillaume, J.L., Lambiotte, R. and Lefebvre, E. (2008) Fast Unfolding of Communities in Large Networks. *Journal of Statistical Mechanics: Theory and Experiment*, **2008**, Article ID: P10008.

[34] Breve, F., Zhao, L. and Quiles, M. (2009) Uncovering Overlap Community Structure in Complex Networks Using Particle Competition. In: Deng, H., *et al.*, Eds., *AICI* 2009, *LNAI* 5855, 619-628.

[35] Hui, P. and Crowcroft, J. (2007) How Small Labels Create Big Improvements. *Proceedings of the Fifth Annual IEEE International Conference on Pervasive Computing and Communications Workshops*, White Plains, 19-23 March 2007, 65-70. http://dx.doi.org/10.1109/percomw.2007.55

[36] Pentland, A., Fletcher, R. and Hasson, A. (2004) DakNet: Rethinking Connectivity in Developing Nations. *IEEE Computer*, **37**, 78-83. http://dx.doi.org/10.1109/MC.2004.1260729

[37] Uddin, M.Y.S., Nicol, D.M., Abdelzaher, T.F. and Kravets, R.H. (2009) A Post-Disaster Mobility Model for Delay Tolerant Networking. *Proceedings of the Winter Simulation Conference*, Austin, 13-16 December 2009, 2785-2796.

[38] Wasserman, S. and Faust, K. (1994) Social Network Analysis: Methods and Applications. Cambridge University Press, Cambridge. http://dx.doi.org/10.1017/CBO9780511815478

[39] Hashemian, M., Stanley, K. and Osgood, N. (2010) Flunet: Automated Tracking of Contacts during Flu Season. *Proceedings of the 6th International Workshop on Wireless Network Measurements*, Avignon, 31 May-4 June 2010, 348-353.

[40] Blondel, V.D., Guillaume, J.-L., Lambiotte, R. and Lefebvre, E. (2008) Fast Unfolding of Community Hierarchies in

Large Networks. *Journal of Statistical Mechanics: Theory and Experiment*, **2008**, Article ID: P10008. http://dx.doi.org/10.1088/1742-5468/2008/10/P10008

[41] Satuluri, V., Parthasarathy, S. and Ruan, Y. (2011) Local Graph Sparsification for Scalable Clustering. In: *Proceedings of the* 2011 *International Conference on Management of Data*, ACM Press, New York, 721-732. http://dx.doi.org/10.1145/1989323.1989399

[42] Grier, C., Thomas, K., Paxson, V. and Zhang, M. (2010) @Spam: The Underground on 140 Characters or Less. In: *Proceeding of the* 17*th ACM Conference on Computer and Communications Security*, ACM Press, New York, 27-37. http://dx.doi.org/10.1145/1866307.1866311

[43] Vahdat, A. and Becker, D. (2000) Epidemic Routing for Partially Connected Ad Hoc Networks. Technical Report, Duke University, Durham.

[44] Haas, Z.J. and Small, T. (2006) A New Networking Model for Biological Applications of Ad Hoc Sensor Networks. *IEEE/ACM Transactions on Networking*, **14**, 27-40. http://dx.doi.org/10.1109/TNET.2005.863461

[45] Zhang, X., Neglia, G., Kurose, J.F. and Towsley, D.F. (2007) Performance Modeling of Epidemic Routing. *Computer Networks*, **51**, 2867-2891. http://dx.doi.org/10.1016/j.comnet.2006.11.028

[46] Jones, E.P.C. and Ward, P.A.C. (2006) Routing Strategies for Delay-Tolerant Networks. ACM Computer Communication Review (CCR).

[47] Spyropoulos, T., Psounis, K. and Raghavendra, C.S. (2005) Spray and Wait: An Efficient Routing Scheme for Intermittently Connected Mobile Networks. In: *Proceedings of the* 2005 *ACM SIGCOMM Workshop on Delay-Tolerant Networking*, ACM Press, New York, 252-259.

[48] Mangrulkar, R.S. and Atique, M. (2011) Performance Evaluation of Flooding Based Delay Tolerant Routing Protocols. *Proceedings of the National Conference on Emerging Trends in Computer Science and Information Technology* (*ETCSIT*), Punjab, 25-26 February 2011, 35-40.

[49] Lindgren, A., Doria, A. and Schelen, O. (2004) Probabilistic Routing in Intermittently Connected Networks. *Proceedings of the Workshop on Service Assurance with Partial and Intermittent Resources*, Fortaleza, 1-6 August 2004, 239-254.

[50] Boudguig, M. and Abdali, A. (2013) New DTN Routing Algorithm. *IJCSI International Journal of Computer Science Issues*, **10**, 82-87.

[51] Burgess, J., Gallagher, B., Jensen, D. and Levine, B.N. (2006) MaxProp: Routing for Vehicle-Based Disruption-Tolerant Networks. *Proceedings of the* 25*th IEEE International Conference on Computer Communications*, Barcelona, 23-29 April 2006, 1-11.

[52] Guo, X.F. and Chan, M.C. (2013) Plankton: An Efficient DTN Routing Algorithm. *Proceedings of the* 10*th Annual IEEE International Conference on Sensing, Communications and Networking* (*SECON*), New Orleans, 24-27 June 2013, 550-558.

[53] Musolesi, M. and Mascolo, C. (2009) CAR: Context-Aware Adaptive Routing for Delay-Tolerant Mobile Networks. *IEEE Transactions on Mobile Computing*, **8**, 246-260. http://dx.doi.org/10.1109/TMC.2008.107

[54] Kalman, R.E. (1960) A New Approach to Linear Filtering and Prediction Problems. *Journal of Basic Engineering*, **82**, 34-45. http://dx.doi.org/10.1115/1.3662552

[55] Keeney, R.L. and Raiffa, H. (1976) Decisions with Multiple Objectives: Preference and Value Tradeoffs. John Wiley & Sons, New York.

[56] Mtibaa, A., Chaintreau, A. and Diot, C. (2008) Are You Moved by Your Social Network Application? In: *Proceedings of the First ACM SIGCOMM Workshop on Online Social Network*, ACM Press, New York, 67-72. http://dx.doi.org/10.1145/1397735.1397751

[57] Hui, P., Crowcroft, J. and Yoneki, E. (2008) Bubble Rap: Social-Based Forwarding in Delay Tolerant Networks. In: *Proceedings of the* 9*th ACM International Symposium on Mobile Ad Hoc Networking and Computing*, ACM Press, New York, 241-250. http://dx.doi.org/10.1145/1374618.1374652

[58] Khan, S.K.A., Mondragon, R.J. and Tokarchuk, L.N. (2012) Lobby Influence: Opportunistic Forwarding Algorithm Based on Human Social Relationship Patterns. *Proceedings of the* 2012 *IEEE International Conference on Pervasive Computing and Communications Workshops*, Lugano, 19-23 March 2012, 211-216. http://dx.doi.org/10.1109/percomw.2012.6197482

[59] Rasul, K., Chowdhury, S., Makaroff, D. and Stanley, K.G. (2014) Community-Based Forwarding for Low-Capacity Pocket Switched Networks. In: *Proceedings of the* 17*th ACM Conference on Modeling, Analysis and Simulation of Wireless and Mobile Systems*, ACM Press, New York, 249-257. http://dx.doi.org/10.1145/2641798.2641801

[60] Daly, E. and Haahr, M. (2007) Social Network Analysis for Routing in Disconnected Delay-Tolerant MANETs. In: *Proceedings of the* 8*th ACM International Symposium on Mobile Ad Hoc Networking and Computing*, ACM Press, New York, 32-40.

[61] Bayir, M.A. and Demirbas, M. (2010) PRO: A Profile-Based Routing Protocol for Pocket Switched Networks. *Proceedings of the Global Telecommunications Conference*, Miami, 6-10 December 2010, 1-5.

[62] Miao, J., Hasan, O., Mokhtar, S.B. and Brunie, L. (2012) A Self-Regulating Protocol for Efficient Routing in Mobile Delay Tolerant Networks. *Proceedings of the 6th IEEE International Conference on Digital Ecosystems Technologies (DEST)*, Campione d'Italia, 18-20 June 2012, 1-6.

[63] Mei, A., Morabito, G., Santi, P. and Stefa, J. (2011) Social-Aware Stateless Forwarding in Pocket Switched Networks. *Proceedings of the IEEE INFOCOM*, Shanghai, 10-15 April 2011, 251-255. http://dx.doi.org/10.1109/infcom.2011.5935076

[64] Bulut, E. and Szymanski, B.K. (2010) Friendship Based Routing in Delay Tolerant Mobile Social Networks. *Proceedings of the Global Telecommunications Conference*, Miami, 6-10 December 2010, 1-5. http://dx.doi.org/10.1109/glocom.2010.5683082

[65] Mtibaa, A. and Harras, K.A. (2011) Social Forwarding in Large Scale Network: Insights Based on Real Trace Analysis. *Proceedings of the 20th International Conference on Computer Communications and Networks (ICCCN)*, Maui, 31 July-4 August 2011, 1-8. http://dx.doi.org/10.1109/icccn.2011.6005907

Permissions

All chapters in this book were first published in WSN, by Scientific Research Publishing; hereby published with permission under the Creative Commons Attribution License or equivalent. Every chapter published in this book has been scrutinized by our experts. Their significance has been extensively debated. The topics covered herein carry significant findings which will fuel the growth of the discipline. They may even be implemented as practical applications or may be referred to as a beginning point for another development.

The contributors of this book come from diverse backgrounds, making this book a truly international effort. This book will bring forth new frontiers with its revolutionizing research information and detailed analysis of the nascent developments around the world.

We would like to thank all the contributing authors for lending their expertise to make the book truly unique. They have played a crucial role in the development of this book. Without their invaluable contributions this book wouldn't have been possible. They have made vital efforts to compile up to date information on the varied aspects of this subject to make this book a valuable addition to the collection of many professionals and students.

This book was conceptualized with the vision of imparting up-to-date information and advanced data in this field. To ensure the same, a matchless editorial board was set up. Every individual on the board went through rigorous rounds of assessment to prove their worth. After which they invested a large part of their time researching and compiling the most relevant data for our readers.

The editorial board has been involved in producing this book since its inception. They have spent rigorous hours researching and exploring the diverse topics which have resulted in the successful publishing of this book. They have passed on their knowledge of decades through this book. To expedite this challenging task, the publisher supported the team at every step. A small team of assistant editors was also appointed to further simplify the editing procedure and attain best results for the readers.

Apart from the editorial board, the designing team has also invested a significant amount of their time in understanding the subject and creating the most relevant covers. They scrutinized every image to scout for the most suitable representation of the subject and create an appropriate cover for the book.

The publishing team has been an ardent support to the editorial, designing and production team. Their endless efforts to recruit the best for this project, has resulted in the accomplishment of this book. They are a veteran in the field of academics and their pool of knowledge is as vast as their experience in printing. Their expertise and guidance has proved useful at every step. Their uncompromising quality standards have made this book an exceptional effort. Their encouragement from time to time has been an inspiration for everyone.

The publisher and the editorial board hope that this book will prove to be a valuable piece of knowledge for researchers, students, practitioners and scholars across the globe.

List of Contributors

Geetha V
Department of Information Technology, National Institute of Technology, Karnataka, Surathkal, Karnataka, India

K. Chandrasekaran
Department of Computer Science and Engineering, National Institute of Technology, Karnataka, Surathkal, Karnataka, India

Yu Hu, Yuanming Wu and Hongshuai Wang
School of Optoelectronic Information, University of Electronic Science and Technology, Chengdu, China

Kenya Sato, Naoya Sakamoto and Hideki Shimada
Department of Information Systems Design, Doshisha University, Kyoto, Japan

Juan Lu and Eric Campo
Univ de Toulouse, Toulouse, France
CNRS, LAAS, Toulouse, France

Adrien Van Den Bossche
Univ de Toulouse, Toulouse, France
CNRS, IRIT, Toulouse, France

Tsang-Ling Sheu, Yen-Hsi Kuo and Zi-Tsan Chou
Department of Electrical Engineering, National Sun Yat-Sen University, Kaohsiung Taiwan

Fawaz Alassery, Walid K. M. Ahmed, Mohsen Sarraf and Victor Lawrence
Electrical and Computer Engineering Department, Stevens Institute of Technology, Hoboken, NJ, USA

Shelei Li, Xueyong Ding and Tingting Yang
Polytechnical School, Sanya College, Sanya, China

Sanjeev Kumar Shah
Shri Venkateshwara University, Gajraula, India

Gyanendra Kumar Gupta
Graphic Era University, Dehradun, India

Kirankumar Y. Bendigeri
Department of Electronics, Jain University Bangalore, Karnataka, India

Jayashree D. Mallapur
Department of Electronics & Communication Engineering, Basaveshwar Engineering College, Bagalkot, Karnataka, India

Byeongwook Bae and Sunggu Lee
Department of Electrical Engineering, Pohang University of Science and Technology, Pohang, Korea

Junyoung Park
System LSI Division, Samsung Electronics Co., Ltd., Hwasung, Korea

Anthony Faustine, Aloys N. Mvuma, Hector J. Mongi, Maria C. Gabriel and Albino J. Tenge
The University of Dodoma, Dodoma, Tanzania

Samuel B. Kucel
Busitema University, Tororo, Uganda

Nour Brinis and Leila Azouz Saidane
National School of Computer Science, University of Manouba, Manouba, Tunisia

Mukhdeep Singh Manshahia and Satya Bir Singh
Department of Mathematics, Punjabi University, Patiala, India

Mayank Dave
Department of Computer Engineering, National Institute of Technology, Kurukshetra, India

Bei Liu and Yuanming Wu
School of Optoelectronic Information, University of Electronic Science and Technology, Chengdu, China

Qiang Ma, Abdullah Al-Dhelaan and Mznah Al-Rodhaan
Department of Computer Science, College of Computer & Information Sciences, King Saud University, Riyadh, Saudi Arabia

Marco Allegretti
Envisens Technologies s.r.l., Torino, Italy

Silvano Bertoldo
CINFAI (Consorzio Interuniversitario Nazionale per la Fisica delle Atmosfere e delle Idrosfere), c/o Dipartimento di Elettronica e Telecomunicazioni, Politecnico di Torino, Torino, Italy

Prachi Singh and Rajendra Kumar Gupta
Madhav Institute of Technology & Science Gwalior, Gwalior, India

Bassel Arafeh, Khaled Day, Abderezak Touzene and Nasser Alzeidi
Department of Computer Science, Sultan Qaboos University, Al Khoudh, Oman

Mauro Rodrigo Larrat Frota e Silva, Glaucio Haroldo Silva de Carvalho, Dionne Cavalcante Monteiro and Leomário Silva Machado
Exact and Natural Sciences Institute, Faculty of Computer Science, Federal University of Pará, Belém, Brazil

Anthony Faustine and Aloys N. Mvuma
The University of Dodoma, Dodoma, Tanzania

Hafiz Bilal Khalil, Naeem Abas and Shoaib Rauf
Department of Electrical Engineering, University of Gujrat, Gujrat, Pakistan

Ratna R. Sarkar, Khadija Rasul and Amitabha Chakrabarty
Department of Computer Science and Engineering, BRAC University, Dhaka, Bangladesh